Biochar Applications for Wastewater Treatment

Biochar Applications for Wastewater Treatment

Edited By

Daniel C.W. Tsang
State Key Laboratory of Clean Energy Utilization, Zhejiang University, China

Yuqing Sun
School of Agriculture at Sun Yat-Sen University, Guangzhou, Guangdong, China

This edition first published 2023
© 2023 John Wiley & Sons, Inc.

All rights reserved. No part of this publication may be reproduced, stored in a retrieval system, or transmitted, in any form or by any means, electronic, mechanical, photocopying, recording or otherwise, except as permitted by law. Advice on how to obtain permission to reuse material from this title is available at http://www.wiley.com/go/permissions.

The right of Daniel C.W. Tsang and Yuqing Sun to be identified as the author of this work has been asserted in accordance with law.

Registered Office
John Wiley & Sons, Inc., 111 River Street, Hoboken, NJ 07030, USA

For details of our global editorial offices, customer services, and more information about Wiley products visit us at www.wiley.com.

Wiley also publishes its books in a variety of electronic formats and by print-on-demand. Some content that appears in standard print versions of this book may not be available in other formats.

Trademarks: Wiley and the Wiley logo are trademarks or registered trademarks of John Wiley & Sons, Inc. and/or its affiliates in the United States and other countries and may not be used without written permission. All other trademarks are the property of their respective owners. John Wiley & Sons, Inc. is not associated with any product or vendor mentioned in this book.

Limit of Liability/Disclaimer of Warranty
In view of ongoing research, equipment modifications, changes in governmental regulations, and the constant flow of information relating to the use of experimental reagents, equipment, and devices, the reader is urged to review and evaluate the information provided in the package insert or instructions for each chemical, piece of equipment, reagent, or device for, among other things, any changes in the instructions or indication of usage and for added warnings and precautions. While the publisher and authors have used their best efforts in preparing this work, they make no representations or warranties with respect to the accuracy or completeness of the contents of this work and specifically disclaim all warranties, including without limitation any implied warranties of merchantability or fitness for a particular purpose. No warranty may be created or extended by sales representatives, written sales materials or promotional statements for this work. The fact that an organization, website, or product is referred to in this work as a citation and/or potential source of further information does not mean that the publisher and authors endorse the information or services the organization, website, or product may provide or recommendations it may make. This work is sold with the understanding that the publisher is not engaged in rendering professional services. The advice and strategies contained herein may not be suitable for your situation. You should consult with a specialist where appropriate. Further, readers should be aware that websites listed in this work may have changed or disappeared between when this work was written and when it is read. Neither the publisher nor authors shall be liable for any loss of profit or any other commercial damages, including but not limited to special, incidental, consequential, or other damages.

Library of Congress Cataloging-in-Publication Data
Names: Tsang, Daniel C. W., Editor. | Yuqing Sun, Editor. | John Wiley & Sons, publisher.
Title: Biochar applications for wastewater treatment / Daniel C. W. Tsang and Yuqing Sun.
Description: Hoboken, NJ : JW Wiley, 2023. | Includes bibliographical references and index.
Identifiers: LCCN 2023021267 (print) | LCCN 2023021268 (ebook) | ISBN 9781119764373 (hardback) |
   ISBN 9781119764380 (pdf) | ISBN 9781119764397 (epub) | ISBN 9781119764403 (ebook)
Subjects: LCSH: Biochar. | Sewage--Purification--Adsorption.
Classification: LCC TD753.5 .T73 2023 (print) | LCC TD753.5 (ebook) | DDC 631.8/6--dc23/eng/20230517
LC record available at https://lccn.loc.gov/2023021267
LC ebook record available at https://lccn.loc.gov/2023021268

Cover Design: Wiley
Cover Image: © roccomontoya/Getty Images

Set in 9.5/12.5pt STIXTwoText by Integra Software Services Pvt. Ltd, Pondicherry, India

Contents

Editors Biography *xi*
List of Contributors *xiii*
Preface *xv*

1 **Engineered Biochar** *1*
Yuqing Sun and Daniel C.W. Tsang
1.1 Overview of Biochar Production *2*
1.2 Biochar Properties and Characterization *4*
1.3 Pre- and Post-Modification of Biochar *9*
1.3.1 Physical Modification *10*
1.3.2 Chemical Modification *14*
1.3.3 Biochar Composites *16*
1.4 Sustainability Considerations *24*

2 **Adsorption of Nutrients** *29*
Yuqing Sun and Daniel C.W. Tsang
2.1 Nutrients in Wastewater *29*
2.2 Biochar Performance in Nutrients Removal from Wastewater *31*
2.2.1 Removal of Ammonium Using Modified and Pristine Biochars *31*
2.2.2 Removal of Nitrate Using Pristine and Modified Biochars *32*
2.2.3 Removal of Phosphate Using Pristine and Modified Biochars *33*
2.3 Biochar Mechanisms of Nutrients Removal from Wastewater *34*
2.3.1 Specific Surface Area *34*
2.3.2 Ion Exchange *34*
2.3.3 Surface Functional Groups *34*
2.3.4 Precipitation *35*
2.4 Factors Influencing Biochar Performance in Nutrients Removal *35*
2.4.1 Pyrolysis Temperature *35*
2.4.2 Metallic Oxides on Biochar *36*
2.4.3 Solution pH *36*
2.4.4 Contact Time *36*
2.4.5 Ambient Temperature *37*
2.4.6 Coexisting Ions *37*
2.5 Nutrients Desorption from Biochar *38*
2.5.1 Ammonium Desorption *38*
2.5.2 Nitrate Desorption *38*

2.5.3	Phosphorous Desorption *39*	
2.6	Nutrient-loaded Biochar as Potential Nutrient Suppliers *39*	
3	**Adsorption of Metals/Metalloids** *41*	
	Yuqing Sun and Daniel C.W. Tsang	
3.1	Metals/Metalloids in Wastewater *42*	
3.2	Mechanisms of Biochar for Adsorption of Metals/Metalloids *43*	
3.2.1	Physical Adsorption *43*	
3.2.2	Electrostatic Interaction *44*	
3.2.3	Ion Exchange *45*	
3.2.4	Surface Complexation *45*	
3.2.5	Precipitation *45*	
3.2.6	Reduction *46*	
3.3	Modified Biochar for Adsorption of Metals/Metalloids *46*	
3.3.1	Biochar/Layered Double Hydroxide Composites *46*	
3.3.2	Magnetic Biochar Composites *47*	
3.3.3	Biochar-Supported nZVI Composites *48*	
3.3.4	Comparison of Different Modification Methods for Metals/Metalloids *49*	
3.4	Biochar Recycling after Adsorption of Metals/Metalloids *51*	
4	**Adsorption of PPCPs** *53*	
	Yuqing Sun and Daniel C.W. Tsang	
4.1	PPCPs in Wastewater *54*	
4.2	Biochar Mechanisms for PPCPs Adsorption *55*	
4.2.1	π-π Interaction *55*	
4.2.2	Hydrogen Bonding *56*	
4.2.3	Electrostatic Interaction *56*	
4.2.4	Other Mechanisms *56*	
4.3	Factors Affecting PPCPs Adsorption by Biochar *57*	
4.3.1	Pyrolysis Temperature *57*	
4.3.2	Biochar Surface Modification *57*	
4.3.3	Properties of PPCPs *58*	
4.3.4	Environmental pH *59*	
4.3.5	Wastewater Composition *59*	
5	**Stormwater Biofiltration Media** *61*	
	Jingyi Gao, Yuqing Sun, and Daniel C.W. Tsang	
5.1	Introduction *62*	
5.2	Common Pollutants in Stormwater *64*	
5.3	Biochar for Biofiltration Media *66*	
5.3.1	Production of Biochar *66*	
5.3.2	Physicochemical Properties of Biochar *67*	
5.4	Removal of Pollutants in Biochar-Based Biofiltration Systems *67*	

5.4.1	Metals/Metalloids	67
5.4.2	Nutrient	70
5.4.3	Organic Chemicals	72
5.5	Microplastic in Urban Runoff	75
5.6	Challenge and Perspective	76
5.7	Conclusion	78

6 Biochar Solution for Anaerobic Digestion 89
Yanfei Tang, Wenjing Tian, and Daniel C.W. Tsang

6.1	Introduction	89
6.2	Application of BC as an Additive in Anaerobic Digestion	90
6.2.1	pH Buffering	90
6.2.2	Adsorption of Inhibitors	91
6.2.3	Effects on Microbial Growth and Activities	92
6.3	Effects of BC on Digestate Quality	99
6.4	Conclusions and Perspectives	100

7 Biochar-Assisted Anaerobic Ammonium Oxidation 105
Wenjing Tian, Yanfei Tang, Dongdong Ge, and Daniel C.W. Tsang

7.1	Overview of Anaerobic Ammonium Oxidation	105
7.1.1	Introduction	105
7.1.2	Constraints	107
7.2	Roles of Biochar in Promoting Anammox	108
7.2.1	pH and Inhibitor Buffer	111
7.2.2	Electron Transfer Promotion	112
7.2.3	Microbial Immobilization	113
7.3	Future Perspectives	114

8 Application of Biochar for Sludge Dewatering 121
Dongdong Ge, Nanwen Zhu, Mingjing He, and Daniel C.W. Tsang

8.1	Introduction	121
8.2	Preparation of Biochar-Based Sludge Conditioner	123
8.3	Efficacy of Biochar Conditioning on Enhanced Sludge Dewaterability	126
8.4	Variations of Sludge Physicochemical Characteristics via Biochar Conditioning	127
8.5	Technical Mechanism and Implementation Prospects	128

9 Effects of Biochar on Sludge Composting 137
Dong Li, Dongdong Ge, Yuqing Sun, and Daniel C.W. Tsang

9.1	Introduction	138
9.2	Effects of Biochar Addition on Sludge Composting	141
9.2.1	Effects on Compost Parameters Effect on C/N	141
9.2.2	Effects on Heavy Metals	142

9.2.3	Effects on Organic Matters *142*
9.2.4	Effects on Gaseous Emissions *143*
9.2.5	Effects on Microbial Community and Activities *145*
9.2.6	Effects on Quality of Sludge Compost *145*
9.3	Future Perspectives *146*
9.4	Summary *147*

10 Sludge Utilization as Biochar for Nutrient Recovery   *155*
Deng Pan, Dongdong Ge, and Daniel C.W. Tsang

10.1	Sewage Sludge (SS) Management *155*
10.2	Importance of Sludge as a Feedstock for Biochar *156*
10.3	Factors Affecting the Properties of SDBC *156*
10.3.1	Raw Material *159*
10.3.2	Temperature *159*
10.3.3	Heating Rates *159*
10.3.4	Retention Time *160*
10.4	Nutrients in SDBC *160*
10.4.1	Nitrogen (N) *160*
10.4.2	Phosphorus (P) *161*
10.4.3	Potassium (K) *161*
10.5	SDBC for Soil Amendment and Nutrient Utilization *161*
10.6	Current Challenges for SDBC *163*
10.7	Conclusions *164*

11 Biochar for Electrochemical Treatment of Wastewater   *171*
Dong Li, Yang Zheng, Yuqing Sun, and Daniel C.W. Tsang

11.1	Introduction *172*
11.2	Different Electrochemical Behavior of Biochar *173*
11.2.1	Electron Exchange *173*
11.2.2	Electron Donor or Acceptor *174*
11.2.3	Electrosorption Capacity *174*
11.3	Preparation of Biochar Electrode Materials *177*
11.3.1	Carbonization *177*
11.3.2	Activation *178*
11.3.3	Template *179*
11.3.4	Composite Materials *180*
11.4	Application in Electrochemical Wastewater Treatment *181*
11.4.1	Electrochemical Oxidation *181*
11.4.2	Electrochemical Deposition *182*
11.4.3	Electro-adsorption *182*
11.4.4	Electrochemical Disinfection *183*
11.5	Future Perspectives *183*
11.6	Summary *184*

12	**Peroxide-Based Biochar-Assisted Advanced Oxidation** *193*	
	Yang Cao, Qiaozhi Zhang, Yuqing Sun, and Daniel C.W. Tsang	
12.1	Introduction *193*	
12.2	Biochar-Based Catalysts *195*	
12.2.1	Pristine Biochar *196*	
12.2.2	Redox Metal-Loaded Biochar *197*	
12.2.3	Heteroatom-Doped Biochar *198*	
12.3	Peroxide-Based Advanced Oxidation *199*	
12.3.1	Fenton-Like System *199*	
12.3.2	Persulfate Activation System *201*	
12.3.3	Photocatalytic System *203*	
12.4	Conclusion and Future Perspectives *204*	
13	**Persulfate-Based Biochar-Assisted Advanced Oxidation** *213*	
	Mengdi Zhao, Zibo Xu, and Daniel C.W. Tsang	
13.1	Introduction *213*	
13.2	Activation Pathway and Reaction Mechanism of Persulfate by Biochar *214*	
13.2.1	Distinction between Different Pathways *214*	
13.2.2	Properties Necessitating the Generation of Radicals with PS *215*	
13.2.3	Nonradical Degradation with Biochar *215*	
13.2.4	Modifying Biochar for Enhanced Properties Related to the Degradation Process *216*	
13.3	Metal-Biochar Composites in Persulfate Activation System *217*	
13.3.1	Iron-Biochar *218*	
13.3.2	Copper-biochar *219*	
13.3.3	Cobalt Biochar *219*	
13.3.4	Biochar of Other Metal and Mixed Metal *220*	
13.4	Heteroatom-Doped Biochar for PS Activation *220*	
13.4.1	Nitrogen-doped Biochar *221*	
13.4.2	Sulfur-Doped Biochar *222*	
13.5	Conclusion and Perspectives *222*	
14	**Biochar-Enhanced Ozonation for Sewage Treatment** *229*	
	Dongdong Ge, Nanwen Zhu, Mingjing He, and Daniel C.W. Tsang	
14.1	Introduction *229*	
14.2	Preparation of Biochar-Based Catalyst for Ozonation *230*	
14.3	Efficacy of Biochar-Catalytic Ozonation on Sewage Treatment *232*	
14.4	Effects of Process Conditions on Biochar-Enhanced Ozonation Sewage Treatment *233*	
14.5	Technical Mechanism and Implementation Prospects *235*	

15 Biochar-Supported Odor Control 243
Jingyi Gao, Zibo Xu, and Daniel C.W. Tsang

- 15.1 Causes and Treatment of Odor 244
- 15.2 Odor Pollutants 245
- 15.3 Properties of Biochar for the Removal of Odor Pollutants 247
- 15.3.1 Surface Area and Total Pore Volume 249
- 15.3.2 Pore Size Distribution 250
- 15.3.3 Chemical Functional Group 252
- 15.3.4 Noncarbonized Organic Matter 253
- 15.3.5 Mineral constituents 253
- 15.4 Application of Biochar in Odor Control 254
- 15.4.1 Biochar as Adsorbent 254
- 15.4.2 Biochar as Additives 256
- 15.5 Conclusion and Perspective 260

16 Fate, Transport, and Impact of Biochar in the Environment 273
Deng Pan, Yuqing Sun, and Daniel C.W. Tsang

- 16.1 Transport Mechanism of Biochar in the Environment 274
- 16.2 Stability of Biochar 275
- 16.2.1 Physical Degradation of Biochar 275
- 16.2.2 Chemical Decomposition of Biochar 275
- 16.2.3 Microbial Decomposition of Biochar 276
- 16.3 Contaminants in Biochar and the Environmental Impact 277
- 16.3.1 Polycyclic Aromatic Hydrocarbons (PAHs) 278
- 16.3.2 Heavy Metals (HMs) 279
- 16.3.3 Persistent Free Radicals (PFRs) 280
- 16.3.4 Dioxins 281
- 16.3.5 Metal Cyanide (MCN) 281
- 16.3.6 Volatile Organic Compounds (VOCs) 282

17 Environmental and Economic Evaluation of Biochar Application in Wastewater and Sludge Treatment 289
Claudia Labianca, Sabino De Gisi, Michele Notarnicola, Xiaohong Zhu, and Daniel C.W. Tsang

- 17.1 Introduction 289
- 17.2 Environmental Evaluation 291
- 17.2.1 LCA Insights into Biochar Production and Applications 291
- 17.2.2 Main LCA Literature Studies of Biochar Applications in Wastewater and Sludge Treatments 295
- 17.3 Technical, Economic, and Sustainability Considerations 299
- 17.4 Future Trends 301
- 17.5 Conclusions 302

Index 309

Editors Biography

Daniel C.W. Tsang is Pao Yue-Kong Chair Professor in State Key Laboratory of Clean Energy Utilization at Zhejiang University. Dan strives to develop low-carbon engineering solutions to ensure sustainable urban development and attain carbon neutrality. Dan has published over 500+ articles in top 10% journal (h-index 107, Scopus), and serves as associate editor of *Science of the Total Environment, Journal of Environmental Management, Critical Reviews in Environmental Science and Technology*, and more. Dan was selected as Highly Cited Researchers 2022 in the academic fields of engineering as well as environment and ecology.

Yuqing Sun is an assistant professor at the School of Agriculture at Sun Yat-Sen University. Her research covers the customized design and application of engineered biochar in green and sustainable wastewater treatment. Dr. Sun has published 50+ publications in top 10% journals, including 16 Highly Cited Papers (Web of Science) with 4,600+ citations and h-index of 38 (Scopus). Dr. Sun serves as young editorial board member of *Critical Reviews in Environmental Science and Technology*. Dr. Sun was selected as Highly Cited Researchers 2022 in the academic field of cross field.

List of Contributors

Yang Cao
Department of Environmental Science and Engineering
Fudan University
Shanghai, China

Jingyi Gao
EIT Institute for Advanced Study
Ningbo, Zhejiang, China

Dongdong Ge
School of Environmental Science & Engineering
Shanghai Jiao Tong University
Shanghai, China

Sabino de Gisi
Department of Civil, Environmental, Land, Building Engineering and Chemistry (DICATECh)
Polytechnic University of Bari
Bari, Italy

Mingjing He
Department of Civil and Environmental Engineering
The Hong Kong Polytechnic University
Hung Hom, Kowloon
Hong Kong, China

Claudia Labianca
Department of Civil, Environmental, Land, Building Engineering and Chemistry (DICATECh)
Polytechnic University of Bari
Bari, Italy

Dong Li
Teleader Solid Waste Disposal (Shandong) Co., Ltd.
Jinan, China

Michele Notarnicola
Department of Civil, Environmental, Land, Building Engineering and Chemistry (DICATECh)
Polytechnic University of Bari
Bari, Italy

Deng Pan
EIT Institute for Advanced Study
Ningbo, China

Yuqing Sun
School of Agriculture, Sun Yat-Sen University
Guangzhou
Guangdong, China

Yanfei Tang
College of Environmental Science and Engineering
Tongji University
Shanghai, China

Wenjing Tian
Institute of Environment and Ecology
Chongqing University
Chongqing, China

Daniel C.W. Tsang
State Key Laboratory of Clean Energy Utilization
Zhejiang University
China

Zibo Xu
Department of Civil and Environmental Engineering
The Hong Kong Polytechnic University
Hong Kong, China

Qiaozhi Zhang
Department of Civil and Environmental Engineering
The Hong Kong Polytechnic University
Hong Kong, China

Mengdi Zhao
EIT Institute for Advanced Study
Ningbo, Zhejiang, China

Yang Zheng
School of Materials Science and Engineering
Ocean University of China
Qingdao, China

Nanwen Zhu
School of Environmental Science & Engineering
Shanghai Jiao Tong University
Shanghai, China

Xiaohong Zhu
Department of Civil and Environmental Engineering
The Hong Kong Polytechnic University
Hung Hom, Kowloon,
Hong Kong, China

Preface

The widely recognized terminology of biochar was first introduced in 2006, and has become increasingly important as a green and carbon-negative solution to some global problems, such as climate change and environmental pollution. The potential capacity of biochar as an effective, low-cost, and environment-friendly adsorbent and catalyst to remove various pollutants, which is related to its relatively large surface area and abundant surface functional groups, was quickly unveiled afterward. Seventeen years on, numerous scientists have worked on biochar technology, and it is proven that biochar with a science-informed and fit-for-purpose design can serve a promising agent for wastewater treatment. The booming biochar market in current years renders biochar a ready-to-implement technology for smart and sustainable wastewater treatment.

In this book we summarize recent research development on biochar production and emerging applications with a focus on the value-added utilization of biochar technology in wastewater treatment, succinctly summarizing different technologies for biochar production and characterization with an emphasis on feedstock selection and pre/post-treatment. The text discusses the mechanisms of biochar's various roles in different functions of wastewater treatment (i.e., adsorption, biofiltration, anaerobic degradation, sludge dewatering, sludge composting, nutrient recovery, advanced oxidation process, odor control, removal of pharmaceuticals and personal care products, removal of emerging contaminants, fate and transport in the environment, and life cycle assessment). It includes the latest research advances in manufacturing optimization and improvements to update the carbonaceous materials with desirable environmental functionalities.

Discussion and case studies are incorporated in treating municipal wastewater, industrial wastewater, agricultural wastewater, and stormwater to illustrate and emphasize the promising prospects of biochar technology in the treatment of various wastewater in actual utilization. Perspectives and future research directions of the emerging biochar technology in wastewater treatment are presented to provide insights for readers and researchers in biochar application for wastewater treatment.

State-of-the-art knowledge of biochar technology is crucial to sustainable wastewater treatment. Given our global targets of carbon neutrality, sustainable blueprints, human well-being, and one health for the planet, we hope this book will inspire interdisciplinary stakeholders to join hands and transfer knowledge to new generations for the sake of our sustainable future.

1

Engineered Biochar

Yuqing Sun[1] and Daniel C.W. Tsang[2]

[1] *School of Agriculture, Sun Yat-Sen University, Guangzhou, Guangdong, China*
[2] *State Key Laboratory of Clean Energy Utilization, Zhejiang University, China*

As a product of the thermochemical decomposition of biomass, biochar is produced under the condition of a limited-oxygen or oxygen-deprived environment. Many organic raw materials can be employed for biochar production, including urban organic solid wastes, manures, algae, sewage sludge, agricultural and forestry byproducts, and wood chips. The techniques for the thermochemical decomposition of the organic materials are various, including gasification, torrefaction, microwave heating, hydrothermal carbonization, and pyrolysis, which differ in production duration and thermal condition. The reasons that biochar has drawn much attention are twofold. First, biochar production can mitigate climate change by storing carbon in a relatively stable form, thereby reducing the emission of greenhouse gases into the Earth's atmosphere. Second, due to its porous texture, biochar has plentiful surface functional groups (SFG) and a large surface area, which makes it an environmentally friendly, low-cost, and effective adsorbent/catalyst. Many studies have reported that the production temperature and duration, decomposition techniques, and feedstock type could considerably affect the final product's physical and chemical properties. In addition, pre-treatment of the feedstock and post-treatments can also influence biochar's properties.

Biochar Applications for Wastewater Treatment, First Edition. Edited by Daniel C.W. Tsang and Yuqing Sun.
© 2023 John Wiley & Sons, Inc. Published 2023 by John Wiley & Sons, Inc.

1.1 Overview of Biochar Production

Both living and recently living organic materials can be used to produce biochar. They include lignocellulosic materials, such as forestry agricultural byproducts, plants, and wood. Nonlignocellulosic materials are also suitable, including organic municipal solid waste (MSW) and livestock's manure. On the other hand, due to the benefits of quick reaction time and high conversion ratio, thermochemical methods are the most preferred amongst a variety of biomass treatment techniques.

The International Biochar Initiative (IBI) states that "Biochar is a solid material obtained from the thermochemical conversion of biomass in an oxygen-limited environment" [1]. In other words, biomass is processed using the dry carbonization method (e.g., pyrolysis), yielding several products including biochar. Although the term *hydrochar* is analogous to *biochar*, it cannot be considered as biochar. Hydrochar is made employing hydrothermal carbonization, through which the organic raw material is heated with water at a temperature range of 200–300 °C. The material undergoes solid and liquid phases (i.e., slurry). Because biochar and hydrochar are essentially distinct, the hydrochar produced using hydrothermal carbonization is not equivalent to biochar.

Among the various thermochemical methods, gasification and dry torrefaction are unsuitable for making biochar. During dry torrefaction, biomass is heated to 200–300 °C and within an inert gas from half an hour to several hours. After the process, the organic raw materials loses energy in gaseous forms, accounting for merely 10% of the total energy, hence the final product's energy density is increased. Because of this, dry torrefaction has been used to pretreat the biomass before combustion. By contrast, gasification is essentially a process of partially combusting biomass at a higher temperature range (600–1200 °C) than pyrolysis, and with a duration of 10–20 seconds. The main reason for using gasification is to transform carbon-based material into gas mixtures (CO_2, CO, and H_2), such as producer gas and synthesis gas (syngas). However, IBI stresses that the production of biochar must avoid accumulation of toxicants in the final product and emission of unsafe air-pollutant.

Before processing for biochar, the wet organic material must be dried. However, the drying process is usually energy intensive, reducing the production's economic efficiency. That's why some forestry and agricultural wastes that possess low water content (less than 30%) after harvest are preferred sources of biochar. In addition, organic waste, such as livestock manure, sewage sludge, agroforestry waste, food waste, and organic fraction of MSW, are also commonly used as feedstock. Utilizing a variety of organic wastes to produce biochar is undoubtedly cost-efficient because the organic raw materials used in the production do not require arable land that could have been better used for growing energy and food crops. The consensus is that utilizing organic wastes for biochar production is eco-friendly and an ideal way to treat numerous wastes efficiently.

The main technology used to produce biochar is biomass pyrolysis. During the process, biomass is decomposed at a relatively high temperature and in an inert gas atmosphere (i.e., the inert gas N_2 with limited O_2). The organic constituents in the feedstock undergo thermochemical decomposition, releasing the gas phase and leaving solid phase residual (i.e., biochar). Next, the high-molecular-weight and polar compounds of the gas phase are cooled, producing a liquid phase (i.e., bio-oil), whereas the compounds that have lower molecular weight and primarily consist of noncondensable gases, such as H_2, CH_4, C_2H_2, CO, and CO_2, are left in the gas phase [2]. Slow pyrolysis can offer the highest yield of solid product (35 to 50 wt.%) and is most commonly adopted for producing biochar. The production takes place at temperatures ranging from 300 °C to 800 °C with heating rates of 5–10 °C min^{-1}, and the gas residence time varies from several minutes to a few hours.

During the pyrolysis process, cellulose starts to decompose between 240 °C and 350 °C, while the decomposition of hemicellulose starts between 200 °C and 260 °C, and lignin decomposes between 280 °C and 500 °C. In cellulose pyrolysis, cellulose first depolymerizes into oligosaccharides. The process of the cleavage of glycosidic bonds to form D-glucopyranose is as follows. It starts from the D-glucopyranose undergoing an intramolecular reorganization, producing levoglucosan. Levoglucosenone is then formed with dehydration of levoglucosan and subsequently proceeds through intermolecular condensation, dehydration, decarboxylation, and aromatization to create a complex solid porous structure (i.e., biochar). The pyrolyzing process of hemicellulose starts with the depolymerizing of hemicellulose into oligosaccharides, succeeded by the reorganization of oligosaccharides and cleavage of glycosidic linkages in xylan chains, forming 1, 4-anhydro-D-xylopyranose. This intermediate step is required for the formation of biochar, which undergoes intramolecular condensation, dehydration, decarboxylation, and aromatization. It is comparable to the pyrolyzing process of cellulose. In the pyrolyzing process of lignin, free radical reaction, among others, is the most crucial pathway. The β–O–4 bonds are cleaved in lignin to generate free radicals, which is the first stage of the free radical reaction. The generated free radicals capture protons of substances with weak O–H or C–H bonds (such as phenyl groups), yielding decomposition products, including vanillin and cresol [2]. They are passed on to other substances, causing chain propagation [2]. When the reaction progresses further and the two radicals collide, stable compounds are formed, ending the chain reactions of free radicals [2]. However, it is difficult to observe the free radicals being generated during pyrolysis [2]. The exact mechanism of lignin pyrolysis is still unclear, and fully understanding it remains a daunting task [2].

During the first stage (200–300 °C), the organic raw material starts to decompose, and during the second stage (300–600 °C), a large number of aliphatic functional groups originating from hemicellulose and cellulose start to form [3]. It is worth noting that, at 350 °C, hemicellulose and cellulose begin to degrade and create O–H bonding of phenol (1375 cm^{-1}), OH bending of alcohol (3300–3400 cm^{-1}), C–O stretching of conjugated ketones (1600 cm^{-1}), CH_3 and CH_2 alkanes bend (1375 and 1465 cm^{-1}), C–H stretching (855 cm^{-1}), aromatic C–C stretching (1475–1600 cm^{-1}), and symmetric C–O stretching (1110 cm^{-1}) [3].

During the final stage (600–900 °C), polycyclic structures and benzene derivatives (700 cm^{-1}) start to form by transformation [3]. As the pyrolysis temperature increases, the oxygen-containing functional groups gradually decrease. However, vast majority of the prominent spectral features that could be observed at a lower temperature disappear when the temperature reaches above 600–800 °C, and the FTIR spectrum is analogous to an ordered graphitic structure. Aliphatic structures, including carboxyl, C–O, ester, and ketone, are completely degraded at such a high temperature. As a result, biochar produced at a low temperature is predominantly hydrophilic because of a broad variety of oxygen-containing functional groups [3]. In contrast, biochars produced at a higher temperature were more likely to contain aromatic structures and skeletons. The aromatic rings of biochar lead to hydrophobic functional groups [3]. Therefore, higher temperatures mainly generate aromatic structures, while lower temperatures prominently generate oxygen-containing functional groups.

1.2 Biochar Properties and Characterization

Surface properties, such as elemental distribution, porous structure, charge density, and surface area, are crucial features because of their ability to affect the fundamental functions of microbial activity, nutrients, and water retention. Operating parameters, such as reaction vessel pressure, reaction residence time, pressure (catalyst, size, stirring method, and direction, etc.), maximum processing temperature (HTT), the heating rate of the process, pretreatment (chemical activation, crushing, and drying, etc.), and flow rate of the auxiliary input all affects biochar's final physical properties, regardless of the types of organic material used.

A higher fixed carbon content usually leads to a higher biochar yield. The biochar production decreased exponentially with increasing pyrolysis temperature for each of the studied organic material groups. The exponential decay curves fit the experimental data well for the wooden raw material group ($R^2 = 0.936$), livestock manure ($R^2 = 0.814$), and herbaceous raw material ($R^2 = 0.867$), but not for the biosolids group ($R^2 = 0.508$) and all raw materials group ($R^2 = 0.677$). Compared with the other three groups of raw materials, the physicochemical properties of biosolids usually vary greatly, depending on the sludge treatment method, wastewater treatment process, and wastewater source.

The N, H, and C contents in biochar can be determined with an elemental analysis, where the content of O is recorded as 100%, excluding the percentage of N, H, C, and ash (O (%) = 100% − N (%) − H (%) − C (%) − ash). In addition, the ratios of O/C and H/C, as well as N-related ratios, can be calculated. Biochars made from livestock manure are likely to have lower H, O, and C contents content because the higher volatile OC compounds in the feedstock are lost during the drying and carbonizing stages of pyrolysis. By contrast, lignocellulose biochar has higher O and H contents because the plant-based feedstock possesses the −OH groups connected with the lignin, hemicellulose, and cellulose components of the plant structure [4]. Biochars made with lignocellulosic material are likely to have a higher fixed carbon content that is not easily mineralized, representing the carbon materials that are not easily lost during pyrolysis.

The ratios of O/C and H/C indicate the aromaticity and polarity of biochars [3]. The carbonization of the biosorbents is assessed using the molar ratio of H/C. When the molar ratio of H/C decreases, it indicates the carbonization is at a higher degree, and a lower amount of plant residues is left (e.g., hemicellulose and cellulose) [3]. In addition, the hydrophilic quality of biochar is determined using the O/C molar ratio. As the pyrolysis temperature increases, the molar ratios of O/C and H/C increase because of the higher carbonization degree [3]. When the pyrolysis temperature of woody and herbaceous raw materials was higher than 600 °C and that of livestock manure and biosolids were higher than 700 °C, the ratio of O/C became less than 0.2, indicating a good carbon sequestration potential [3]. In general, the C/N ratios of the plant-based and herbaceous biochar were relatively higher, while those made from livestock wastes or biosolids generally had lower C/N ratios. In each group of feedstocks, the ratio of C/N increased with rising pyrolysis temperature.

In general, biochar made from livestock manure has more inorganic components than plant-based biochar, which is made from grass and wood chips. Compared to the biochar made from livestock manure, biochar produced from lignocellulosic materials contains fewer nutrients and has lower $CaCO_{3-eq}$ due to the lower content of ash. Mixing feedstock before the pyrolysis process provides flexibility in designing biochar to increase the concentration of specific nutrients in the soil or to decrease the concentration of certain compounds to balance plant nutrient requirements. As the temperature of pyrolysis increases, so does the inorganic elements' relative weight % (except those volatile elements such as N) because of the decreased hydrogen and oxygen content in the biomass and evaporation, yet their weight content remained almost constant.

The pH value of biochar is generally around 5–12, but it differs with the pyrolysis temperature and raw materials. Although some biochars made from herbaceous feedstocks, biosolids, and wooden materials exhibit neutral to slightly acidic pHs, especially when the pyrolysis temperature is below 400 °C, several studies suggest that most biochars tend to have an alkaline pH. Biochars made from livestock manure have higher pH and more ash contents. When different feedstocks were used, the pH of biochars all increased linearly with the pyrolysis temperature. The change in pH may be caused by the decline of acidic functional groups (–COOH) as the temperature climbs. In addition, the increased biochar content of carbonate and total base alkali cations also play a part in increasing biochar pH with a rising temperature. It is not difficult to notice that the sensitivity of the different feedstock groups to the pyrolysis temperature varies. The estimated rate of increase in pH was 0.6/100 °C ($R^2 = 0.657$) for the woody group, 0.5/100 °C ($R^2 = 0.504$) for the animal waste group, 1.4/100 °C ($R^2 = 0.785$) for the biosolid group, and 1.0/100°C ($R^2 = 0.803$) for the herbaceous group [5].

Some studies reported that, as temperature increases, biochar pH, carbon content, ash content, stability, and surface area rise, while O/C and H/C decline with rising pyrolysis temperature [3]. Evidence also showed that moderate positive correlations exist between temperature and the surface area of hardwood-, softwood-, and grass-, but not manure-derived biochar [3]. Comparably, biochars made with other materials usually have mild negative correlations with HTT, while MBC shows little correlation with O/C and HTT.

The surface charge of a particle is related to the point of zero charge (pH_{pzc}) and zeta potential, which can be used as indicators. If the solution pH is lower than the pH_{pzc} of biochar, the biochar surface will be positively charged. If the pH of the solution is higher than pH_{pzc} of biochar, then the biochar's surface will be negatively charged [6]. The charge difference on the surface of biochar is essential in deciding which of electrostatic attraction or surface complexation is the prominent adsorption mechanism.

The ability of certain materials to absorb and store positively charged ions, such as clays, soils, and biochars, can be commonly assessed using cation exchange capacity (CEC). The measure can indicate the negative charge on the biochar surface and thus reveals the biochar's ability to retain cations [7]. The biochars' CEC value increases with the degree of oxidation of carbon in the aromatic structure and the creation of carboxyl groups [7]. Biochars produced with biosolids exhibited the highest CECs, followed by biochars made with plant-based raw materials. Earlier research stated that the higher CEC values of biochars were a result of the facilitated formation of O-associated surface functional groups due to abundant minerals in the raw materials (e.g., K, Na, Ca, Mg, and P).

In order to understand the biochar's surface morphology, biochar samples were observed by reflected light optical microscopy, and scanning electron microscopy (SEM) was used to identify the microstructure. A backscattered electron image was observed from the Au-coated sample surface, representing average atomic abundance in a black and white image [8]. SEM and energy dispersive spectroscopy (EDS) with an acceleration voltage of 15 kV and beam current of 180 Pa was used to identify the mineral phases of the sample with an Au coating in a vacuum of 25 Pa [8]. SEM images collected at a scale of 200 μm revealed that the residues of structural material inside the wood-based biochar appeared to be fibers. A much closer image shows that the surface edge of the wood-based biochar has large (200 μm) and small porous structures and is covered with small particles. SEM images of biochar made with the manure of swine and poultry showed that the surface contained void spaces and pores, and was extensively covered with particles of irregular shapes. These particles might be salt accumulated during manure production and excreted by animals. Theoretically, particle accumulation is caused by electrostatic interactions happening in the electron-dense regions and led by ions in dissolved salts involved in redox reactions, in which they are attracted to sites. Transmission electron microscope (TEM) images (×800 K) reveal surface morphology of sugarcane bagasse biochar in greater detail [3]. In general, the raw organic materials' TEM image usually shows an amorphous structure. Conversely, the amorphous carbon starts to recombine at low pyrolysis temperatures (250–350 °C), which is caused by the hemicellulose and cellulose structures breakdown. At a higher pyrolyzing temperature (600 °C), the surface of biochar organizes into a regular crystal-like structure, most likely graphite [3].

N_2 sorption isotherms analysis can be used to investigate the surface area and pore characteristics. Biochars produced with agricultural wastes usually have larger surface areas than biochars made from animal manure or sludge. Pyrolysis processing of organic materials can enlarge the crystallites and make them well organized, which increases with HTT [8]. Increasing the temperature of pyrolysis from 250 °C to 500 °C can expand the BET surface area because of the increasing evolution of volatiles in the organic materials, leading to enhanced pore development in the biochars. The biochars' surface area usually increases with elevating HTT until reaching a temperature at which deformation occurs, causing a subsequent reduction in surface area to happen [8]. The pore sizes distributed within the micropores contribute the most to the total surface area. Micropores develop with higher temperatures and longer reaction residence times [8]. The elevated temperatures offer activation energy, and the longer retention time allows time for the reaction to complete, resulting in a higher order degree of the structure.

The Boehm titration method is based on the observing the reaction features of various bases and acids with distinct surface functional groups containing oxygen and quantitatively analyzing the functional groups on the substances' surface.

Semi-quantitative analysis of biochar functional groups was carried out using CPMAS ^{13}C NMR (solid-state ^{13}C cross-polarization magic-angle nuclear magnetic resonance). During NMR analysis, the biochar sample is exposed to radio frequency pulses and a strong magnetic field for a period of time. The ^{13}C nuclear active isotope among the ^{14}C atoms absorbs radio frequency in a narrow band and resonates at a frequency because of the shielding of nearby electrons, which offers a measurable magnetic spectrum, and differences in signal intensity are plotted along an *X*-axis [4]. These are chemical shifts in parts per million (ppm) [4]. Chemical shift regions are therefore assigned in the spectrum to core structural features, such as aromatic C (109–145 ppm), and in particular organic structures, such as C in unsubstituted aliphatic compounds (0–50 ppm), to perform the interpretation of ^{13}C NMR spectra [4]. Under each chemical shift region, the area is determined and reported as a distribution percentage of C [4]. The spectra showed that a higher proportion of aromatic C was found in the biochar with high pyrolyzed temperature (≥ 400 °C) [8]. The NMR spectral assignments of ^{13}C and C distribution percentage in the biochar are listed in Table 1.1 [4].

Fourier-transform infrared spectroscopy (FTIR) is another tool that is commonly used to analyze biochars because the method is nondestructive, the spectra can be quickly obtained, and the spectroscopic technique is easier to understand compared to NMR. As a vibrational spectroscopy method, FTIR offers a way to characterize the composition of organic and mineral samples. In FTIR spectroscopy, a number of ways can be used to present the samples to the instrument:

- *Transmission*, through which the IR beam penetrates the sample.
- *DRIFT or diffuse reflectance*, by which the sample absorbs the incident IR beam. The beam is then reflected re-diffusely and collected by the instrument.
- *Attenuated Total Reflection (ATR)*, where an optically dense crystal (such as diamond) is placed in contact with the sample. An infrared beam passes through the sample, and the absorbance information is reflected into the crystal and then transmitted by the detector [4].

Biochar samples are exposed to infrared radiation, and the bonds in the matrix absorb energy at the wavelengths of mid-infrared (4000 to 400 wavenumbers),

Table 1.1 The ^{13}C NMR spectral assignments to different C structures in biochar.

ppm	Chemical structures	Examples
0–50	Unsubstituted aliphatic	Alkanes, fatty acids
50–61	N-alkyl compounds	Amino acids, proteins
61–96	Carbohydrates	Cellulose
96–109	AnomericC	Monosaccharides
109–145	AromaticC	Benzene-like compounds
145–163	Phenolic C	Phenol, nitrophenol
163–190	Carboxylic C	Acetic acid
190–220	Ketonic C=O	Esters and amides

causing molecules to vibrate [4]. The molecules are excited to a higher state of energy, and the bonds will vibrate in one of several modes, depending on the mass of the bonded atoms and bond types [4]. The spectral wavelengths between 4000 and 550 wavenumbers (cm^{-1}) are usually used to scan the biochar samples [4]. Light absorption wavelengths are assigned to specific functional groups and core structures in mineral/impurities and biochar. In order to carry out FTIR analysis on carbonaceous solids, the pellet technique was utilized to prepare the samples, by which the dry biochars were uniformly mixed with the base substrate KBr powder. Biochars made at such low pyrolysis temperatures displayed a more diverse organic structural composition due to fewer volatile compounds loss. In real settings, as the pyrolysis temperature increases (400–600 °C), the number of aliphatic structure decrease manifested as stretching of the aliphatic C–H [7]. In the meantime, more aromatic structures are created with the enhancement or emergence of characteristic peaks in the C=O (1580–1700 cm^{-1}), C=C (1380–1450 cm^{-1}), C–C stretch, and C–H stretch (750–900 cm^{-1}) [7]. As the temperature of pyrolysis continues to climb (700–800 °C), the aromatic (3050–3000 cm^{-1}, 1600–1580 cm^{-1}) and hydroxyl groups (3200–3400 cm^{-1}) decrease slightly [7]. Also, the condensation and dehydration during biomass pyrolysis (400–600 °C) lead to the increase of aromatic structure and decrease of aliphatic structure [7].

In principle, aromatic C is deemed to have a stronger hydrophobic property than alkyl–C/O–alkyl–C, which means that pyrolyzing for biochar at a higher temperature (≥ 400 °C) would lead to a more hydrophobic product [8]. However, regardless of the biochar type, the hydrophilicity (contact angle ≥ 90°) of biochar pyrolyzed above 400 °C would increase. We hypothesize that HTT increases the structured regular spacing between planar results [8]. Interplanar distances also reduce with increased molecular order and organization, all of which lead to greater surface area per unit volume [8]. In addition, higher pyrolysis temperatures leading to a higher specific surface area and a bigger quantity of micropores might be the key factors for water retention capacity in biochar.

1.3 Pre- and Post-Modification of Biochar

The ability of raw biochar to remove pollutants is usually limited, especially at high concentrations. The main reasons for such limitation are threefold:

1) The functionality of the precursors after pyrolysis is low.
2) Biochars created at low temperatures have a small surface area and limited pore volume.
3) The size of the powdered biochar is so small that it is difficult to separate.

Biochar can be modified prior to, during, and/or after the process of pyrolysis. Currently, there are well-tested surface modification methods for carbon-based materials, such as impregnation methods, chemical modification, and physical modification. Among them, physical modifications, such as ball-milling, microwave-assisted pyrolysis, gas activation, and steam activation, could drastically affect the physiochemical properties of biochar. On the one hand, physical modification is environmentally friendly, low-cost, and simple because no toxic chemicals are used in the modification process. On the other hand, several studies have suggested that physical modification might be less effective than chemical modification in most scenarios. Therefore, chemical modification is more commonly used than physical modification, primarily including oxidants modification, alkali modification, and acid modification. Typical biochar composites include (1) microorganism–biochar composites, (2) carbonaceous engineering nanocomposites, (3) nonmetal heteroatom dopped biochar composites, (4) layered double hydroxide–biochar composites, (5) mineral–biochar composites, and (6) metal–biochar composites.

1.3.1 Physical Modification

Steam Activation

In the steam activation process, biochar obtained after pyrolysis of biomass in a limited-oxygen or oxygen-deprived atmosphere and at moderate temperatures (400–800 °C) was partially gasified with steam at 800–900 °C [9]. The steam triggers a series of reactions ($C + H_2O \rightarrow C(O) + H_2$; $2C + H_2 \rightarrow 2C(H)$; $CO + H_2O \rightarrow CO_2 + H_2$) [9], in which surface carbon first reacts with oxygen from the water to form surface oxide $C(O)$ and H_2 [10]. Furthermore, the generated $C(O)$ reacts with steam to form CO_2 and H_2 alleviating the surface activation and affecting the inhibition of biochar gasification [10]. Moreover, the generated H_2 reacts with surface site C to form $C(H)$ complexes [10]. Therefore, partial devolatilization occurs, and crystalline carbon forms on the biochar surface.

The cleavage of C=C bonds in the biochar during steam activation is comparable to the polymerization reactions to form C=C bonds in the biochar. The degree of self-aggregation increases significantly [11]. The activated biochar is composed of oxygenated functional groups such as ether, phenolic, carboxyl, and carbonyl, resulting in increased hydrophilicity. However, some studies suggested that steam activation has little or even negative effects on the properties of surface functional groups. The trapped products that are generated from incomplete pyrolysis are eliminated during steam activation and, in most cases, the carbon surface is oxidized because of the formation of syngas H_2 [12]. Thus, new pores are formed, and the diameter of the minor pores expands [12], leading to an increase in surface area. Thereafter, under the influence of process parameters – such as the activation time, water vapor flow, and activation temperature – reduced polarity, higher aromaticity (decreased value of H/C) and strength of delocalized π bonds, and increased specific surface area and pore volume are key properties of steam-activated biochar [13].

Compared with other approaches intended to increase the surface area, for example, increasing the pyrolysis temperature, the method that uses steam to assist pyrolysis has the advantages of simple operation and moderate energy-saving. Many studies have theorized that using steam to assist pyrolyzing process in biochar production could successfully halt organic aromatic pollutants such as polychlorinated biphenyls (PCBs) and polycyclic aromatic hydrocarbons (PAHs), because the cutback of oxygen-containing functional groups might lead to the above-described enhanced π-π interactions.

Gas Activation
Similar to steam activation, gas purging modification also consists of two procedures. The first is the pyrolysis of organic materials. The second is to modify biochar by removing carbon dioxide or ammonia. Compared with the standard reaction atmosphere, the presence of an active atmosphere drastically affects the functional groups, volatiles, and biochar yield [14]. The surface aromatic properties of sewage sludge biochar can be improved by using CO_2 as an atmosphere during pyrolysis [14]. When CO_2 is used as the reaction medium, the prepared biochar made from paper-mill sludge will result in a complex aggregate structure containing solid minerals (FeO, Fe_3O_4, and $CaCO_3$) and graphite carbon [14].

CO_2 has two crucial parts in the pyrolysis process: (1) it accelerates the thermal cracking of volatile organic compounds (VOCs) during pyrolysis and reacts with VOCs to increase the production of CO [14]; (2) it immediately leads to a substantial reduction (~40–60%) of condensable hydrocarbons (tars). The CO_2 medium helps the creation of both micropores and meso/macropores due to the intense carbon reformation at $\geq$ 710 °C through the Boudouard reaction (C + CO_2 → 2CO). CO_2 may also react with pore-clogging condensable hydrocarbons, for example, volatile organic compounds and tars, which can be converted to gaseous products to free the clogged pores and/or produce new ones. CO_2 medium induced more structural edge defects and fused aromatic rings in the biochar matrix during pyrolysis. This could be the result of enhanced dehydrogenation of organic matter with CO_2 purging at high temperatures (viz. Boudouard reaction: C + CO_2 → 2CO; biomass → biochar + tar + C_nH_m), which may create more vacancy and zigzag edges as structural dimensionality defects.

The basic nitrogen functional groups are introduced into the carbon matrix of biochar by ammonia gas purging [11]. During pyrolysis in an ammonia (NH_3) atmosphere, NH_3 may react with oxygen-containing species in the organic materials (such as furans, esters, aldehydes, and ketones) to create nitrogen (N)-containing heterocyclic compounds (such as indole, piperidine, pyridine, and pyrrole). At higher temperatures, ammonia decomposes to generate free radicals such as H, NH, and NH_2 [15], which react with the active sites allocated on the edge of the graphene layer in biochar to introduce nitrogen-containing functional groups [11].

Pyrolysis in a low-oxygen atmosphere is also a possible way to yield carbon with moderate porosity [14]. The oxygen content during the pyrolysis process could increase ash content and enlarge the specific surface area.

Microwave-Assisted Pyrolysis

Microwaves have a wavelength in the range of 0.001 to 1 m and are a form of electromagnetic radiation that can be used as a dielectric heating method at specific frequencies (typically 2.45 GHz) [16]. By converting electromagnetic energy into heat, microwave radiation can heat objects without direct physical contact, thus providing comparatively high heating rates during the process of pyrolysis. Its heating rates range from 0.1 °C/s to higher than 1000 °C/s [11]. With the merits of selective heating conditions, higher homogeneity, and low temperature, using microwaves in pyrolysis processes can produce higher-quality biochars. Using microwave as the heat source also has the benefits of reducing energy consumption and accelerating the rate of the chemical reaction and internal volume heating.

The microwave heating mechanism involves interfacial polarization, dipolar depolarization, and ionic conduction. In the pyrolysis process with microwave heating, it is central that the organic materials can efficiently absorb microwaves to induce polarization effects. It is believed that a high water content helps the absorption of microwaves because the water can rotate and align in dipoles, causing friction and collisions [17]. Microwave receptors, such as metal-based microwave receptors or activated carbon, are added during the pyrolyzing process to ensure an efficient absorption of microwaves and assist the reaction to attain the pyrolysis temperature and targeted heating rate.

Microwave radiation can change the morphology and chemical properties of the biochar being produced [14]. Compared with traditional pyrolysis methods, a higher degree of carbonization of biochar at a lower pyrolysis temperature can be realized by microwave-assisted pyrolysis. Utilizing microwave-assisted pyrolysis to produce biochar has been reported to offer a higher surface area and larger pore volume [11]. Deep and narrow pores are created with the help of microwaves. During the microwave-assisted pyrolysis process, tiny microplasma spots spread throughout the reaction medium. Microwave-assisted pyrolysis also helps generate gas, increase the local temperature, and in turn promote the formation of pores in the biochar. Using microwaves as the heating source, the hollowed carbon nanofibers are created during the pyrolyzing process of the palm kernel shells because the self-extrusion of volatile substances from the interior of the organic material and the resolidification of the resulting biochar surface are augmented by the formation of arc. Moreover, microwave irradiation can reduce the number of oxygen-containing functional groups [16].

Ball Milling

The mechanochemical technique seeks to find the right balance between producing defects through ball grinding and adjusting the size of a solid grain to nanoparticles (< 1000 nm). During the process, specific powder charges and a grinding medium are placed in a high-energy mill. The mechanisms of the ball-grinding process include ionic defects, electronic, structure, strain generation, and diffusion improvement, as well as interactions among the substrates [18]. The substrates and movement of grinding media create kinetic energy that can stretch or break chemical bonds of large molecules, resulting in the charge transfer, cleavage of glycosidic bonds, generation of electrochemical elements, and fragmentation of solid materials [18]. As the process has multiple phases and consists of several reactions, the mechanochemical technique improves the contact at the interfaces, thereby facilitating the reactions and removing products. Mechanochemical processes can lead to the sublimation and melting of reagents because thermal decomposition happens at the melting point of a particular compound [18]. The ball-grinding process also facilitates the creation of new surface structures, particle size reduction, and reagent contact. Therefore, the surface functional groups of biochars could be notably improved by ball grinding for broad implementation. The newly formed functional groups might be originated from broken chemical bonds of the materials because the energy generated during ball milling is associated with elevated temperature, higher defect density, and crystal deformation [18].

A variety of operational factors can change the properties of biochar during a ball grinding process. As a result, the optimization of ball-milling parameters is exceedingly crucial for tailoring the function and performance of ball-milled biochar [18]. The parameters include grinding temperature, reaction time, ball-size distribution, grinding speed, dry or wet grinding, and powder used to drive the grinding chamber [18]. The mass ratio of media to biochar and the solvent properties during a ball-grinding process is directly proportional to the surface area of the resulting biochar [18, 19].

Wet grinding and dry grinding are two distinct approaches to biochar modification techniques. The choice of grinding approach depends on the technical and economic analysis as well as the convenience of operation [18]. For materials that are difficult to filter, dry ball grinding usually work better [18]. In contrast, wet ball grinding is a labor-saving and green process that can be carried out at room temperature.

Wet ball milling of biochar using organic solvents such as heptane, hexane, and ethanol has been demonstrated to successfully expand the Brunauer-Emmett-Teller (BET) surface area of the resulting biochar [18]. The mass ratio of grinding media to biochar during a wet grinding process has proved to be the most critical parameter, through which the best result is obtained at 100:1 [18]. Although relatively ineffective, adding salt in dry ball grinding is proved to be a cost-effective manner that can expand the total surface area and micropore ratio of biochar [18]. As the biochar is ground in the spherical grinding media, the salt crystals are continuously broken down into finer grains, which provide friction and impact force to augment the biochar's surface properties and pore structure [18]. The size of biochar grains is also influenced by the ball mill's rotational speed and processing time [18].

1.3.2 Chemical Modification

Chemical modification processes include two-step modification and one-step modification processes. In the presence of activation chemicals, the activation and carbonization steps are achieved at the same time in a one-step chemical activation process. The two-step chemical activation entails the carbonization of the original raw material, succeeded by the activation of the carbonized product [9], in which chemical reagents are added, or pretreatment of the precursor is performed prior to the carbonization process. Chemical modification treatments are usually conducted by adding acids, bases, or inorganic salts (potassium carbonate, zinc chloride). Activation by acid/alkaline can be a fairly straightforward step of modification as it merely requires washing and mixing procedures in moderate conditions. Therefore, the step is a useful and applicable modifying technique in ecological remedying applications. Nevertheless, since a variety of chemical reagents are required, the impact to environment and cost efficiency need to be thoroughly assessed on each case individually. In addition, a relatively large amount of water is required to neutralize the pH after acid or alkali activation [2], otherwise excess acidity/alkalinity may be harmful to the environment when biochar is applied [17]. Furthermore, intentional oxidation employing ozone (O_3), ammonium persulfate [$(NH_4)_2S_2O_8$], potassium permanganate ($KMnO_4$), and hydrogen peroxide (H_2O_2) has been utilized to modify surface functional groups. The process of chemical activation is cost-efficient and is usually performed at relatively low temperatures and in a shorter time duration [11]. The chemical activation proves superior efficiency in the development of biochar microporosity, reduction of mineral matters, activation of carbonaceous materials, and increase of the surface functional groups [11].

Acid Activation

Washing with strong acids, such as hydrochloric–HCl acid, nitric–HNO_3, sulfuric–H_2SO_4, and phosphoric–H_3PO_4, has been investigated for aqueous oxidation, which can enhance surface acidity and change the porous structure of biochars [9]. Typically, acid activation of biochars is conducted by moving the biochars to an acid solution at a ratio of 1:10 and a temperature ranging from room temperature to 120 °C. The durations of reaction are generally varied from hours to days [11].

Phosphoric acid is one of the widely applied chemical modification activators and is more eco-friendly than other hazardous and corrosive reagents [9]. Phosphoric acid can break down aromatic, aliphatic, and lignocellulosic materials while forming phosphate and polyphosphate cross-bridges to prevent contraction or shrinkage during pore formation [9]. Because of its erosive character, oxidation treatment using HNO_3 has been proven to cause degradation of the micropore walls, causing a reduction in the total surface area to happen [9]. Likewise, H_2SO_4 treatment contributes to a reduction in the porosity of biochar from 10% to 40% and increased the size distribution of heterogeneous micropores [9]. Increasing the concentration of sulfuric acid used for activation from 5% to 30% has generally had a positive effect on increasing surface area [11]. Nevertheless, surface area reduction has been observed when the sulfuric acid concentration is excessively high. The dehydration of H_2SO_4 during pyrolysis might be detrimental to the surface area development because excess water vapor moves toward the surface structure [9]. Modification with citric acid and oxalic acid marginally reduced the surface area of the biochars. On the other hand, the combined modification of 30% oxalic acid and sulfuric acid drastically enlarged biochar surface area.

Generally, it is recognized that acidic functional groups such as amine and carboxyl groups can be introduced into carbonized surfaces by treatment with strong acids [20]. In light of the inherent oxidizing properties, H_3PO_4 or H_2SO_4 would also, to some extent, partially oxidize the C surface and augment carboxyl groups [9]. Surface carboxylation is mainly achieved by single-stage oxidation [9]. Citric acid and oxalic acid are weak acids that could similarly introduce the carboxyl group on the surface of biochar through esterification.

Acid activation, which in its essence is to introduce acid groups to the surface of the biochar, can lower the pH of the biochar and make alkaline materials less stable and more soluble [7]. However, acid activation might not be a feasible modifying option, if the organic materials to be modified contain heavy metals, as the heavy metals in the organic material could become more bioavailable. Heavy metals could be flushed out during acid activation and leaching tests are required to assess the possible changes in metal availability [17].

Alkaline Activation
Alkali activation of biochar employing sodium hydroxide (NaOH) and potassium hydroxide (KOH) can increase the surface basicity and oxygen content, at the same time, dissolving ash and concentrating organics such as lignin and cellulose to accelerate later activation. Sodium hydroxide treatment can lead to depolymerization. Potassium species (K_2O, K_2CO_3) may be formed during activation due to K^+ intercalation in the microcrystalline layer forming the condensed C

structure [9]. The potassium species might disperse into the internal structure of the biochar matrix, widening existing pores, and creating new pores for the product [9]. Nevertheless, it is worth noting that NaOH modification at lower temperatures (60–100 °C) is observed to cause a reduced surface area with only a few micropores [9]. The proportion of functional groups on the surface of the biochar changed slightly due to the dominance of C content in graphite [16]. The KOH activation process increases the silicon content available to plants due to the better solubility of silicon in alkaline environments, increasing the bioavailability of phytolith silicon. NaOH is considered more cost-efficient and less corrosive for carbon activation than KOH. On top of different types of alkali, the ratio between alkali and biochar could also considerably affect the properties of biochar. Nevertheless, KOH activation has both advantages and disadvantages. High concentrations (i.e., 5 M) would disrupt the rice husks' cellular structure under alkaline activation.

Oxidation
Conventionally, oxidation is carried out under reflux conditions in the presence of single or mixed inorganic acids, such as H_2SO_4 and HNO_3, and oxidizing agents, such as NaOCl, $KMnO_4$, and H_2O_2 [21]. The refluxing acids could occasionally be too detrimental or harsh to the physical aspects of carbons [21]. The modification with oxidizing agents could increase the content of oxygen-containing functional groups on the biochar. Hydrogen peroxide modification could reduce the pH and ash content of the resulting biochar and increase the oxygen-containing functional groups, especially carboxyl groups. It is analogous to hydrogen peroxide in that potassium permanganate modification would also increase the oxygen-containing functional groups. Moreover, the surface area of biochar can be increased with the modification of potassium permanganate.

1.3.3 Biochar Composites

Metal–Biochar Composites
Three synthesis methods are commonly used are applied to prepare metal–biochar composites:

1) Pretreatment of biomass/biochar with metal salts before pyrolysis
2) Chemical precipitation of metal-oxide particles onto biochar after pyrolysis
3) Pyrolyzing the organic materials rich in target metal element.

For pretreatment of feedstock/biochar with metal salts prior to pyrolysis, organic materials are initially impregnated with metal salts ($ZnCl_2$, $CaCl_2$, $MnCl_2$, $MgCl_2$, $AlCl_3$, $Fe(NO_3)_3$, and $FeCl_3$, etc.). Among them, organic based metal salt solutions and sulfur based metal salts are treated either under electric field or no electric field conditions. Metal ions in solution deposit on the surface or inside the raw material. Second, the pretreated feedstock/biochar is pyrolyzed at fairly low temperatures of 400 °C to 800 °C in a limited-oxygen or oxygen-deprived atmosphere after drying. During the process, the metal ions are converted into zero-valent metals or metal oxide nanoparticles (ZnO, CaO, MgO, Al_2O_3, MnO_2, and Fe_2O_3, etc.) on the biochar's surface. But, direct pyrolysis techniques have drawbacks such as intensive consumption of energy and unsuitable for raw materials with high water content. By thermally reducing granules of Fe-pretreated biochar powder in a reducing gas, thermal reduction can be used to produce magnetic metal powder [22]. Fe_3O_4 is produced at a thermal reduction temperature of 300–500 °C and Fe^0 is created at a thermal reduction temperature of 500–800 °C [22]. Because of its cost-efficiency, the thermal reduction of impregnating biochar in the gas phase at moderate temperatures is prefered, in so doing, resulting in preservation and minimal changes of the support material and high dispersion of Fe [22]. As a result, 600 °C is deemed as a suitable temperature for thermal reduction of Fe [22].

The second synthesis method of metal–biochar composites is to load metal oxide particles onto biochar. First, the organic materials are converted to boichar through the pyrolysis. Next, the biochar is immersed in the metal salts solutions, adsorbing metal ions to the biochar's pores and surface. In the end, the metal oxide particles are deposited on the biochar's surface to create a composite material using reduction method, pyrolysis method, and adjusting the pH the metal salt solution.

The third production method that is commonly used to prepare metal–biochar composites is to directly pyrolysis the organic materials rich in the target metal element. The metal elements in the organic materials are converted to metal-oxide particles through thermal pyrolysis.

The three approaches for preparing metal–biochar composites all have their own merits and downsides. Pretreatment of biomass with metal salts prior to pyrolysis is simple and inexpensive. The disadvantage is that, although this method can be used for production on a large scale, the type, composition, shape, and size of metal particles are difficult to control. Loading metal particles onto biochar can improve control over the type, composition, shape, and size of the particles, and multilayer, double-layer, and single-layer metal particles supported on biochar could be created. But this method is fairly complex and expensive. In contrast, directly pyrolyzing the organic materials rich in the target metal element is straightforward and cost-friendly; unfortunately, the sources of the organic raw materials rich in the target metal element are rare.

This novel approach can maximize the advantages of biochar and metal particles. First, the inherent shortcomings of biochar and metal particles can be overcome using the synthesis of metal–biochar composites. The loaded metal particles can be distributed and stabilized with metal–biochar composites, resulting in surface passivation, leaching, and reducing aggregation of metal particles. Biochar acts as a scaffold for anchoring and dispersing metal particles with minimum aggregation. Second, metal–biochar composites can also increase the active sites and oxygen-containing functional groups and change the properties of both, making metal–biochar composites display excellent functionality [18]. Third, metal–biochar composites can promote the reaction rate of pollutants by enhancing the catalytic/redox performance of the metal–biochar interface. Furthermore, metal–biochar composites could enhance biochar production and improve thermal stability. Therefore, these novel composite materials could be exceptionally useful in remedying environmental problems.

Mineral–Biochar Composites
Clay/mineral–biochar composites are made by combining a small amount of clays/clay minerals with biochar [23]. As a type of hydrous aluminosilicate mineral, clays consist of mixtures of fine-grained clay minerals and clay-sized crystals of other minerals, such as metal oxides, carbonates, and quartz. Clay minerals are a group of phyllosilicates with diameters less than 2 μm. They have a higher cation exchange capacity (CEC) due to the permanent negative charge brought about by isostructural substitution [23]. Phyllosilicate clay minerals also have variable charges at fracture edges due to unsatisfied bonding [23]. These variable charges are primarily responsible for the adsorption of anions by the clay minerals [23]. Clays and their minerals represent an indispensable, abundant, and low-cost group of geological materials and are popular adsorbents for pollutant adsorption studies [23]. Clay minerals are easy to mine and are nontoxic [23]. They have the advantages of intercalation, low cost, layered structure, and fairly high surface area.

In recent years, extensive research has been conducted on clay-biochar composites. Impregnation of biochar with clay minerals, such as kaolinite, bentonite, or montmorillonite, could change the composition and physical properties of biochar [23], thereby enhancing the adsorption capacity for polyatomic cations (e.g., NH_4^+) and oxyanions (e.g., PO_4^{3-}). In particular, clay–biochar composites combine the salient features of biochar and clay materials, exhibiting unique properties [24], such as compatibility, porous structure, and high carbon content, making them also suitable for reuse and regeneration.

There are essentially two preparation methods for making clay–biochar composites. The first one is to mix and pyrolyze raw materials with clay minerals, and the second one is to mix pyrolyzed biochar with clay minerals [25]. Specifically, the raw material is immersed in a stable clay suspension made by adding the powdered material to deionized water or other suitable solvent and heated in a muffle furnace at an appropriate temperature in a limited-oxygen atmosphere. On the other hand, biochar can also be straightforwardly steeped in a slurry made from acetic acid, deionized water, and clay minerals, stirred overnight, and then dried in an oven at 60 °C for 24 h. The most widely used weight ratios of clay for both procedures are as follows: biomass/biochar = 1:5, 1:4, 1:2, and 1:1.

The association of Al and Si could act as binders to facilitate the interactions between biochar and clay minerals [23]. Clay minerals attached to the biochar surface are introduced to the internal pores to form organomineral complexes [23]. Biochar would also react with multivalent cations existing in clay minerals to form organomineral complexes through various mechanisms such as ligand exchange and acid-base reaction [23]. These interactions depend on the functional groups of the biochar and clay mineral and the type of clay mineral and are similar to the interactions occurred between clay minerals and organic matter in the soil [23].

Layered Double Hydroxide–Biochar Composites
Layered double hydroxides (LDHs) exhibit unique physical and chemical properties that are similar to those of clay minerals. Among these properties, the valuable ion exchange capability, high mechanical and chemical stability, and the high specific surface area stand out, resulting in excellent adsorption properties.

The term *layered double hydroxides* refers to the two dimensionally organized structure with two metal cations in the lamella that can incorporate negative species into the interlamellar region, thereby, neutralizing the lamellae's positive charges. The general formula of LDHs is

$$\left[M^{2+}_{1-X} M^{3+}_X (OH)_2 \right]^{X+} A^{m-}_{X/m} \cdot nH_2O \tag{1}$$

where M^{2+} represents a divalent metallic cation, M^{3+} a trivalent metallic cation, A^{m-}, an anion interspersed with load m, x is the di- and tri-valent cations ratio, and n is the number of water moles. There are quite a few combinations of di- and tri-valent cations for LDHs' synthesis. The divalent cations generally utilized are: Ca^{2+}, Zn^{2+}, Cu^{2+}, Ni^{2+}, Co^{2+}, Fe^{2+}, Mn^{2+}, and Mg^{2+}; whereas the trivalent cations are: Ni^{3+}, Co^{3+}, Fe^{3+}, Mn^{3+}, Cr^{3+}, and Al^{3+}. To compensate for the positive charge of LDH lamella, several anionic species, either organic or inorganic anions, can be utilized, including halides (F^-, Cl^-, Br^-, and I^-); oxyanions (CO_3^{2-}, NO_3^-, SO_4^{2-}, and CrO_4^{2-}); complex anions ($[Fe(CN)_6]^{-4}$, $[NiCl_4]^{-2}$); polyoxometalates ($V_{10}O_{28}^{-6}$, $Mo_7O_{24}^{-6}$); and organic anions (such as alkylsulfates, carboxylates eporphyrins).

In most of the cases, biochar/LDH composites are prepared by the methodology based on a liquid phase deposition process and consists of preparing biochar starting from any method, then putting it in contact with the metal precursors in proportions for LDH formation, and finally applying a vigorous mixing process under basic pH. In this process, fine LDH particles will be deposited by precipitation over biochar surface. Then, the material should be subjected to the pH neutralization step through several washes with deionized water, and then the composite must be dried and sieved. In addition to the conventional methods just described, another form of composite preparation is the hydrothermal carbonization. The metallic salts are mixed in aqueous solution with the biochar under a constant agitation for 1 h. After that, the solution pH is adjusted using Na_2CO_3 and NaOH solution, assuring 2 h of contact. Then, the system is transferred to an autoclave at 100 °C for 1 h. The sludge obtained is dried at 80 °C for 12 h [26], bathed with deionized water and dried, again, at 70 °C for 12 h. In addition to post-precipitation, LDH/biochar can be synthesized by a process based on the co-precipitation followed by another heat treatment to improve the crystallinity. An alternative process developed was composite synthesis methodology called *electric field-assisted pyrolysis*. It consists of generating a type of electrochemical cell using the reagents by mixing the raw biochar and metal hydroxides under acid pH and constant agitation, passing a defined current density for 5 min. This process provided a material with a higher porous surface and a more homogeneous distribution of LDHs.

Nonmetal Heteroatom Dopped Biochar Composites
In recent years, the idea of metal-free heteroatom doping has become predominant in the carbonaceous filed. Earth-rich elements, including sulfur (S), boron (B), and nitrogen (N), were comprehensively applied to manufactured nanocarbons using doping techniques. The application of heteroatom doping technology, especially nitrogen doping (N-doping), has the best efficacy to enhance the dispersion of nanomaterials, improve the detection limit of sensors, and promote the catalysis of nanocarbons. Low-temperature processing (< 700 °C) of biochar produced with a low degree of graphitization is deemed unfitting for doping technique. Graphitized biochars with ordered crystalline domains produced at high temperatures allow heteroatom doping techniques to modify engineered biochars.

N-doping technologies can be divided into two categories according to distinctive nitrogen sources, namely internal nitrogen sources of organic materials and external nitrogen sources with purified ammonia or additives. According to the doping procedure in the certain preparation, the doping method could be broken into post-doping after carbonization and *in-situ* doping. The difference between them lies in the introduction time of the nitrogen source [27]. In particular, post-doping and *in-situ* doping refer to combining the nitrogen source with the organic materials and biochars individually, followed by heat treatments.

The critical point of N self-doping into biochar depends on the choice of the organic material. This designated organic material must contain a substantial proportion of the raw nitrogen content in its inherent macromolecular structure or the overall chemical composition [28]. Organic materials such as animal tissue, microbes, algae, and other nitrogen-rich biowaste could more plausibly ensure an overall high nitrogen level in the resulting biochars [28]. For the N-doping from exogenous nitrogen sources, the generally used nitrogen additives include organic substances (such as aniline, melamine, and urea) and inorganic substances (such as nitric acid, ammonium salts, and ammonia) [28, 27].

Contrary to endogenous N doping, adding or impregnating external N-rich materials is used to construct an artificial nitrogen-containing environment. Therefore, operating conditions may need to meet stricter and more detailed prerequisites. After activation using a variety of nitrogen sources with different chemical compositions and macromolecular structures, the configurations of nitrogen bonds introduced in the resulting biochars also varied according to operating settings. It is important to note that the profusion of oxygen functional groups in the organic materials is positively correlated with the N-doping level in the resulting biochars because the oxygen functional groups could immediately bind with N precursors or go through thermal decomposition to react with nitrogen precursors [27].

During thermal reforming of the organic material, ammonia purging could provide NH_3 molecules to react with the carbon skeleton, however, a higher temperature is required, and the reaction intensity is low. Ammonia activation usually occurs at the solid-gas interface and cause partial N-doping (i.e., aminated contents) [27]. The abundant –OH groups in agar powder played a key part by reacting with NH_3 molecules at high temperature to form amino groups (R–NH_2). The introduced amino groups further reacted with –OH to generate –C=N. NH_3 is also thought to react with carbonyl groups on biochar through a Maillard reaction with concomitant H_2 production. The carboxyl groups are also thought to create hydrogen bonds with NH_3 to facilitate N-doping, followed by conversion to graphitic nitrogen and pyridinic nitrogen. Therefore, ammonia purging favors the formation of microporous structures due to the consumption of oxygen functional groups.

Ammonium salt is another low-cost nitrogen source, which, compared with ammonia gas purging, could mitigate nitrogen waste caused by inefficient collisions of solid-gas molecules at the raw material interface and gas purging. During the thermal decomposition, the ammonium salt used appears to act as both a pore-creating structure modifier and a reducing N-doping agent, as it releases N_2O, N_2, or NH_3 at the same time. It can be hypothesized that the mechanism of N-dopant formation using ammonium salts is analogous to that of ammonia purge, where the NH_3 molecules first bind to defective sites terminated with oxygen functional groups and subsequently are converted to the corresponding N-dopants.

N-doping with organic additives is a convenient one-pot approach to acquiring N-doped biochar because it simply needs organic additives and homogeneous mixing of the organic material [27]. Urea is the most popular choice among all organic nitrogen additives because it has nitrogen and carbon contents and could be utilized to introduce carbon substrates and nitrogen atoms. N-doped biochar associated with urea activation usually exhibits high specific surface area owing to the decomposition of urea in the organic materials during pyrolysis [27]. This could be caused by the thermal instability of mixed urea opening pores and releasing NH_3 [27]. For operating conditions with organic additives, the temperature reaction of reaction should be sensibly regulated because nitrogen-rich sources at the specified peak temperatures, such as melamine and urea, incline to decompose into different intermediate products [27]. During the initial pyrolysis stage, carbon nitrides are formed at 300 °C, and various aminated moieties (i.e., C–N, N–H, and –NH_2) appeared on the biochar surface [27]. These N-based moieties can be coalesced and decomposed into the carbonized biomass lattice to produce a variety of nitrogen-bonding configurations (i.e., graphitic N, pyrolytic N, and pyridinic N) when the temperature is raised above 400 °C [27]. As the temperature rises above 700 °C, only a considerable portion of graphitic nitrogen remains. In addition, urea cannot disperse to the internal macrostructure of the organic materials, implying that initial modifications to the size of particles and/or the size of pores might promote the introduction of nitrogen content.

Sulfurization treatment is frequently utilized to introduce sulfur-containing functional groups (e.g., S=O, C=S, or C–S) to the carbon sorbents' surface [21]. The introduction of sulfur into carbon materials is achieved through reactions of carbon with sulfurizing agents such as dimethyl disulfide, H_2S, SO_2, K_2S, Na_2S, and elemental S [21]. Analogous to the nitrogenation treatment, sulfuration of carbon surfaces can also be achieved through different chemical reaction pathways that usually need acid or heat treatments [21]. A significant amount of sulfur (up to > 30% *wt.*) could bind to most carbon-sulfur surface groups in the form of thiocarbonyl groups [21]. On top of the advantage of introducing sulfur-containing functional groups, depending on the treatment conditions, the porous structure of carbon can be modified by sulfuration treatment with notable increases or decreases in pore volume and certain surface areas [21]. The reduction in pore volume and surface areas might be caused by sulfur-containing particles blocking the pores.

Carbonaceous Engineering Nanocomposites
Biochar can bind to carbonaceous materials with functional groups that are eligible to establish strong bonds with contaminants present in an aqueous media and on biochar surfaces. The most frequently used agents are carbon nanotubes (CNTs), carbonaceous nanomaterials, amines, and polysaccharides. This modification could be accomplished by mixing biochar with amino-rich polymers, such as chitosan or polyethyleneimine, or through simple chemical reactions. Among bulk carbon

materials, carbon nanotubes (CNTs) and graphene (G) are generally utilized in composite synthesis because of the strong binding affinity through chemical oxidation with –COOH and –OH groups. Because of the fairly high specific surface area and the porous structure of biochars [19], they can be employed as hosts to stabilize and distribute carbon nanomaterials, expanding the range of potential applications.

Microorganism–Biochar Composites
Biochar-immobilizing microorganisms (BIM) are precisely coordinated by impregnating pollutant-degrading microorganisms together with biochar, which enriches the degrading bacteria to form biofilms and improves the mass transfer of pollutants from the polluted environment to the degrading microbial community [13]. Biochar's porous structure provides space for the reproduction and growth of microorganisms. It could also offer a small amount of nutrients to protect the growth of microorganisms from the influence of external environmental factors. In particular, biochar could increase biosorption by changing environmental conditions and reducing pollutant concentrations to protect the growth of microorganisms from excessive pollutant concentrations. As a result, BIM is a potentially useful method to improve treatment of wastewater.

Worked like living cells, immobilized microorganism could physically or chemically prevent cells from their initial position in the external environment. Since BIM technique was originated from fixed enzyme technology, it employs similar immobilization approach. However, no general immobilization approach available at the moment, and the suitable immobilization approach needs to be decided according to the product characteristics and living habits of the objective microorganism. The fixation methods consist of cross-linking, covalent bonding, entrapment, and adsorption. Among them, embedding and adsorption methods account for most of the BIM preparation.

The adsorption method is based on the interaction of nonspecific forces (adhesion force and surface tension) between microbial surface functional groups and biochar (such as –C=O, –COOH, and –OH), to achieve the microbial colonization in biochar [13]. The adsorption method has the advantages of simple operation, low cost, and little effect on cell activity, and has been generally used in immobilization technology. The fixed substrate of the adsorption method, though, could be reused, the interactions between substrates and microorganisms are unstable and weak, causing the microorganisms to fall off easily [13]. Thus, this method is suitable for the fixation of living cell organisms [13].

Microorganisms are trapped in a polymeric carrier mesh. The grid structure could prevent the microorganisms from seeping out of the carrier, and small molecule products and substrates in the external environment could easily exit and enter the carrier [13]. Materials for embedding include synthetic polymers (eco-friendly decomposition ability and mechanical strength) and natural polymers (low microbial toxicity and high fixed density). The method has wide application, high particle strength, and low toxicity to microorganisms. However, it is not suitable for the degradation of macromolecular pollutants because only small molecular substrates can enter the interior of the particle. In addition, the diffusion of oxygen is affected as the mass transfer resistance increases, and the carriers cannot be reused. These are the reasons that most current studies only used polyvinyl alcohol and sodium alginate. Thus, this method has a wider scope of application than the adsorption method, and is more suitable for industrial applications that require stability [13].

The chemical covalent bonding, which works between the groups on the surface of the solid phase and the functional groups on the microbial cells [13], is utilized in the covalent bonding method to immobilize the cells. The technique also possesses the merits of good stability and tight binding. However, it gives a violent reaction when combining the groups. In addition, the operation is difficult to control and complicated, and only limited number of relevant cases have been studied so far.

1.4 Sustainability Considerations

In terms of biochar modification, eco-friendly methods without using toxic reagents are strongly encouraged. Magnetic modification, acid/base modification, and mineral modification are recommended because the reagents employed in these processes are either nontoxic or natural [17]. However, some new modification strategies involve the use of toxic substances, which may increase occupational safety risks. For example, while methanol can successfully introduce carbonyl groups to biochar surfaces, the compound can damage the optic nerve function (permanent blindness may be caused by ingesting 10 mL of pure methanol). In another example, a recent study investigated the possibility of employing biochar-supported nanoscale zero-valent iron (nZVI) to adsorb metals and catalyze the degradation of trichloroethylene. Toxic sodium borohydride ($NaBH_4$) was utilized in the reduction of Fe (II) to nZVI, but the potential risks were ignored. Though carbon nanotubes or grafting graphene onto the biochar's surface can successfully improve the affinity of biochar to organic pollutants with aromatic rings, its practical adoption to niche applications might be hindered by the complex procedure of nanocomposite creation. It has been argued that more modification methods and eco-friendly synthesis for producing engineered biochar should be suggested to reduce the environmental impact and cost.

It is worth noting that if the organic material itself contains pollutants (e.g., certain types of phytoremediation biomass, wood, and sewage sludge), the associated risks of pollutant release should not be ignored. Co-pyrolysis of uncontaminated biomass and contaminated organic materials could be a feasible method to lessen the possible risks of pollutant leaching. It is a trend that in co-pyrolysis processes, plastics have been utilized to promote hydrophobicity and facilitate electrostatic interactions between pollutants and biochar adsorbents in wastewater. But the latest studies have also suggested that, in some cases, co-pyrolysis of organic materials and plastics adds to the risk of microplastic release and migration, as the plastic material breaks down into smaller fragments during the pyrolysis process. Reducing the risks from traditional pollutants might lead to increased risks from new pollutants such as microplastics. Care should be taken to avoid such risks when choosing pyrolysis conditions and organic materials for production. Until recently, research regarding this topic has been very limited, but it will be necessary to create operational boundaries and safe practices.

References

1 Kambo, H.S. and Dutta, A., 2015. A comparative review of biochar and hydrochar in terms of production, physico-chemical properties and applications. *Renewable & Sustainable Energy Reviews*, 45, 359–378.
2 Lee, J., Sarmah, A.K., Kwon, E.E., 2018. Production and Formation of Biochar, in: Ok, Y.S., Tsang, D.C.W., Bolan, N., Novak, J.M., *Biochar from Biomass and Waste: Fundamentals and Applications*. Elsevier B.V, pp. 3–18.
3 Hassan, M., Liu, Y., Naidu, R., Parikh, S.J., Du, J., Qi, F., Willett, I.R., 2020. Influences of feedstock sources and pyrolysis temperature on the properties of biochar and functionality as adsorbents: A meta-analysis. *Science of the Total Environment*, 744, 140714.
4 Novak, J.M. and Johnson, M.G., 2018. Elemental and Spectroscopic Characterization of Low-temperature (350 °C) Lignocellulosic- and Manure-based Designer Biochars and Their Use as Soil Amendments, in: Ok, Y.S., Tsang, D.C.W., Bolan, N., Novak, J.M., *Biochar from Biomass and Waste: Fundamentals and Applications*. Elsevier B.V, pp. 37–58.
5 Li, S., Harris, S., Anandhi, A., Chen, G., 2019. Predicting biochar properties and functions based on feedstock and pyrolysis temperature: A review and data syntheses. *Journal of Cleaner Production*, 215, 890–902.
6 Tangsir, S., Hafshejani, L.D., Lhde, A., Maljanen, M., Hooshmand, A., Naseri, A., Moazed, H., Jokiniemi, J., Bhatnagar, A., 2016. Water defluoridation using Al_2O_3 nanoparticles synthesized by flame spray pyrolysis (FSP) method. *Chemical Engineering Journal*, 288, 198–206.
7 Xue, Q., Xie, S., Zhang, T., (2022) Biochar Production and Modification for Environmental Improvement, in: Tsang, D.C.W., Ok, Y.S., *Biochar in Agriculture for Achieving Sustainable Development Goals*. Elsevier B.V, pp. 181–191.

8 Jien, S.-H., (2018) Physical Characteristics of Biochars and Their Effects on Soil Physical Properties, in: Ok, Y.S., Tsang, D.C.W., Bolan, N., Novak, J.M., *Biochar from Biomass and Waste: Fundamentals and Applications*. Elsevier B.V., pp. 21–35.

9 Rajapaksha, A.U., Chen, S.S., Tsang, D.C.W., Zhang, M., Vithanage, M., Mandal, S., Gao, B., Bolan, N.S., Ok, Y.S., 2016. Engineered/designer biochar for contaminant removal/immobilization from soil and water: Potential and implication of biochar modification. *Chemosphere*, 148, 276–291.

10 Dissanayake, P.D., Palansooriya, K.M., Withana, P.A., Senadeera, S.S., Yuan, X., Ok, Y.S., Samaraweera, H., Wang, S., Mašek, O., Shang, J., 2022. Engineered Biochar as a Potential Adsorbent for Carbon Dioxide Capture, in: Tsang, D.C.W., Ok, Y.S., *Biochar in Agriculture for Achieving Sustainable Development Goals*. Elsevier B.V, pp. 345–359.

11 Rangabhashiyam, S. and Balasubramanian, P., 2019. The potential of lignocellulosic biomass precursors for biochar production: Performance, mechanism and wastewater application-A review. *Industrial Crops and Products*, 128, 405–423.

12 Tareq, R., Akter, N., Azam, M.S., 2018. Biochars and Biochar Composites: Low-cost Adsorbents for Environmental Remediation, in: Ok, Y.S., Tsang, D.C.W., Bolan, N., Novak, J.M., *Biochar from Biomass and Waste: Fundamentals and Applications*. Elsevier B.V, pp. 169–209.

13 Wu, C., Zhi, D., Yao, B., Zhou, Y., Yang, Y., Zhou, Y., 2022. Immobilization of microbes on biochar for water and soil remediation: A review. *Environmental Research*, 212, 113226.

14 Li, Y., Xing, B., Ding, Y., Han, X., Wang, S., 2020. A critical review of the production and advanced utilization of biochar via selective pyrolysis of lignocellulosic biomass. *Bioresource Technology*, 312, 123614.

15 Gao, Y., He, D., Wu, L., Wang, Z., Yao, Y.C., Huang, Z.H., Yang, H., Wang, M., 2020. Porous and ultrafine nitrogen-doped carbon nanofibers from bacterial cellulose with superior adsorption capacity for adsorption removal of low-concentration 4-chlorophenol. *Chemical Engineering Journal*, 420, 127411.

16 Jiao, Y., Li, D., Wang, M., Gong, T., Sun, M., Yang, T., 2021. A scientometric review of biochar preparation research from 2006 to 2019. *Biochar*, 3(3), 283–298.

17 Wang, L., Ok, Y.S., Tsang, D.C.W., Alessi, D.S., Rinklebe, J., Wang, H., Masek, O., Hou, R., O'Connor, D., Hou, D., 2020. New trends in biochar pyrolysis and modification strategies: Feedstock, pyrolysis conditions, sustainability concerns and implications for soil amendment. *Soil Use and Management*, 36(3), 358–386.

18 Kumar, M., Xiong, X., Wan, Z., Sun, Y., Tsang, D.C.W., Gupta, J., Gao, B., Cao, X., Tang, J., Ok, Y.S., 2020. Ball milling as a mechanochemical technology for fabrication of novel biochar nanomaterials. *Bioresource Technology*, 312, 123613.

19 Osman, A.I., Fawzy, S., Mohamed, S., Farghali, M., El-Azazy, M., Elgarahy, A.M., Fahim, R.A., Maksoud, M.I.A.A., Ajlan, A.A., Yousry, M., Saleem, Y., Rooney, D.W., 2022. Biochar for agronomy, animal farming, anaerobic digestion, composting, water treatment, soil remediation, construction, energy storage, and carbon sequestration: A review. *Environmental Chemistry Letters*, 20(4), 2385–2485.

20 Park, S.J. and Kim, B.J., 2004. Influence of plasma treatment on hydrogen chloride removal of activated carbon fibers. *Journal of Colloid and Interface Science*, 275(2), 590–595.

21 Yang, X., Wan, Y., Zheng, Y., He, F., Yu, Z., Huang, J., Wang, H., Ok, Y.S., Jiang, Y., Gao, B., 2019. Surface functional groups of carbon-based adsorbents and their roles in the removal of heavy metals from aqueous solutions: A critical review. *Chemical Engineering Journal*, 366, 608–621.

22 Lyu, H., Tang, J., Cui, M., Gao, B., Shen, B., 2020. Biochar/iron (BC/Fe) composites for soil and groundwater remediation: Synthesis, applications, and mechanisms. *Chemosphere*, 246, 125609.

23 Arif, M., Liu, G., Yousaf, B., Ahmed, R., Irshad, S., Ashraf, A., Zia-ur-Rehman, M., Rashid, M.S., 2021. Synthesis, characteristics and mechanistic insight into the clays and clay minerals-biochar surface interactions for contaminants removal-A review. *Journal of Cleaner Production*, 310, 127548.

24 Han, H., Rafiq, M.K., Zhou, T., Xu, R., Mašek, O., Li, X., 2019. A critical review of clay-based composites with enhanced adsorption performance for metal and organic pollutants. *Journal of Hazardous Materials*, 369, 780–796.

25 Cheng, N., Wang, B., Wu, P., Lee, X., Gao, B., 2021. Adsorption of emerging contaminants from water and wastewater by modified biochar: A review. *Environmental Pollution*, 273, 116448.

26 Porcari, A.R., Ptak, R.G., Borysko, K.Z., Breitenbach, J.M., Drach, J.C., Townsend, L.B., 2003. Synthesis and antiviral activity of 2-substituted analogs of triciribine. *Nucleosides Nucleotides & Nucleic Acids*, 22(12), 2171–2193.

27 Wan, Z., Sun, Y., Tsang, D.C.W., Khan, E., Yip, A.C.K., Ng, Y.H., Rinklebe, J., Ok, Y.S., 2020. Customised fabrication of nitrogen-doped biochar for environmental and energy applications. *Chemical Engineering Journal*, 401, 126136.

28 Kumar, A., Singh, E., Mishra, R., Lo, S.L., Kumar, S., 2022. A green approach towards sorption of CO_2 on waste derived biochar. *Environmental Research*, 214, 113954.

2

Adsorption of Nutrients

Yuqing Sun[1] and Daniel C.W. Tsang[2]

[1] School of Agriculture, Sun Yat-Sen University, Guangzhou, Guangdong, China
[2] State Key Laboratory of Clean Energy Utilization, Zhejiang University, China

Removing phosphorus (P) and nitrogen (N) by using different sorbents is generally deemed a cost-saving method to complement traditional ways. Although extensive studies have been done on biochars and their adsorption potentials for soluble P and N, the performance of a variety of biochar varies widely. This chapter discusses the adsorption capacity of biochars to remove P (PO_4–P) and N (NH_4–N and NO_3–N). The influencing factors and possible mechanisms are discussed. It is suggested that because of its slow-release property, biochar can be subsequently utilized as fertilizer through desorption.

2.1 Nutrients in Wastewater

As essential elements of living organisms, phosphorus (P) and nitrogen (N) are required in the functions and structures of the organic things. For example, P is necessary in energy transfer, cell membranes, and genetic material, and N in protein synthesis formation. A large number of P- and N-rich compounds are manufactured and utilized for domestic, agricultural, and industrial purposes. Because wastewater is high yield and high concentration, it is a source of nutrients. The most abundant wastewater sources are stormwater runoff, industrial, and household/urban (black or gray). The estimated total wastewater production globally is about 359.4×10^9 m^3 yr^{-1}, of which 52% is variously treated and 63% is collected. Most P and N are removed through wastewater treatment plants (WWTPs). The concentrations of P and N are reduced to levels following local guidelines and then finally discharged to bodies of water and natural waterways. However, they are still the major source of P and N in receiving natural waters. Furthermore, 48% of the wastewater generated globally is discharged into the natural environment without any treatment. Decomposition of organic matter in sediments, biological nitrogen fixation, and atmospheric precipitation, as well as anthropogenic sources, may also affect total P and N loading in water ecosystems. Anthropogenic nutrients of a disproportionate amount into water systems have resulted in the eutrophication of estuaries, reservoirs, lakes, and rivers worldwide, with subsequent algal blooms threatening the biodiversity of water ecosystems and drinking water security.

In the fluid phase, inorganic P and N are the leading causes of eutrophication. They primarily present as phosphate (PO_4^{3-}), nitrate (NO_3^-), and ammonium (NH_4^+). In order to eliminate excessed nutrients from wastewater, a variety of traditional approaches are employed, such as artificial wetlands, flocculation/coagulation, chemical precipitation, and biochemical degradation. These approaches typically eliminate useful nutrients from wastewater, but are costly to operate and maintain and generate large amounts of waste.

Biological (bioelectrochemical systems) or physical methods (stripping, adsorption, filtration, and membrane processes) are used to remove nitrogen from wastewater. As a widely used method to remove biological nitrogen, the modified Ludzak-Ettinger (MLE) process converts ammonia in wastewater to gaseous nitrogen and nitrate through denitrification and nitrification, respectively. A number of processes have been developed and applied on the basis of anaerobic ammonium oxidation (anammox), through which ammonia and nitrite are converted to gaseous nitrogen. High concentration of ammonia in sewage water impedes the activities of many microorganisms engaged in the conversion of nitrogen, which results in the coupled adoption of physicochemical approaches, including the airstrip process, magnesium ammonium phosphate (MAP) precipitation, and folding point chlorination [1].

Biological processes can also be used to remove phosphorus from sewage water [2], entailing phosphate-accumulating organisms (PAOs) in the aerobic/oxic (A/O) process. PAOs are capable of storing and releasing phosphate in reaction to periodic environmental settings. The Integrated Fixed-Film Activated Sludge Systems combined with struvite precipitation and Enhanced Biological Phosphorus removal (IFAS-EBPR) is one of the more effective technologies to remove phosphorus from sewage water. The costs of treating sewage water are mostly influenced by vacuum stripping 470 US \$ t^{-1}, IFAS-EBPR: 60 € P kg^{-1}, and energy prices [3]. Furthermore, since phosphate could be precipitated by a number of metal oxides and metals, chemical-precipitating procedures could be a complementary or alternative method to replace biological procedures.

As an effective and quick procedure to remove the pollutants from the liquid phase, adsorption efficiency depends on the conditions of the ambient electrolyte and the properties of the adsorbent. A number of sorbents have been assessed for their efficacy in removing P and N from natural waters, such as aluminum oxides, nanoparticles, polymeric ion exchangers, bentonites, and zeolites. Adsorption utilizing organic waste from agricultural production has become the most favored method because it is effective, user-friendly, and cost-effective for removing excessed nutrients in sewage water. Biochar is a product rich in carbon and made by pyrolyzing organic materials, such as urban sludge and organic agricultural waste, in a limited-oxygen or oxygen-deprived environment. The potentially low production cost and easy accessibility of a variety of biochars have made it a crucial substance for remedying wastewater pollutants in the last couple of decades.

Biochar has a series of functional properties, such as high cation exchange capacity (CEC), expansive surface area, porous texture, sound structure, and high carbon content. In addition, biochars have abundant surface functional groups, including carbonyl, alkyne, amide, alkene, ether, alkyl, amine, hydroxyl, and carboxyl. These functional groups are responsible for adsorbing nutrient ions in sewage water. However, because the functional groups are limited, unprocessed biochars do not have particular and high adsorption capacity for nutrient. Recently, with the goal of improving the adsorption of N and P, many attempts have been made to biochars through modifying surface and coating or impregnation with a variety of chemicals, such as organofunctional groups, Al, Ca, Mg, and Fe minerals. Many accounts have indicated that the efficiency of biochar to remove nutrient in water varies considerably with the kinds and characteristics of biochar used and the environment settings of the liquid phase.

2.2 Biochar Performance in Nutrients Removal from Wastewater

2.2.1 Removal of Ammonium Using Modified and Pristine Biochars

Extensive large-scale adsorption experiments were conducted to assess the efficiency of removing ammonium nitrogen (NH_4-N) using biochars made with different production temperatures and various organic raw materials. For instance, the potential applications of a variety of biochars in NH_4-N adsorption have been assessed, such as biochars made from reeds, cotton stalks, corncobs, peanut shells, wheat straw, and pine wood chips. The biochars' Q_{max} values for NH_4–N varied broadly throughout the reports, of which the mean value is 11.2 mg N/g [1]. In most cases, the described NH_4-N adsorbing capabilities of biochars were under 20.0 mg N/g [1]. But, some exceptional cases have been observed with NH_4–N adsorbing capability reaching 133 mg N/g.

A lot of efforts have been made to improve biochars' adsorbing capabilities for ammonium by altering biochar surface functional groups, enhancing precipitation of chemicals, or increasing CEC. For instance, the Q_{max} of ammonium can be drastically improved by adding materials with high CEC to the organic raw materials. These enhancing materials include magnesium, bentonite, and montmorillonite. It was also suggested that the adsorbing capability of phosphate-rich biochar for NH_4^+ could be significantly enhanced with the addition of Mg^{2+} by the precipitation of struvite ($MgNH_4PO_4$). Many modification techniques did not show promising results – for example, the result of using H_2O_2 to mildly oxidize biochars did not display a significant improvement in the adsorbing capability. The medium and average Q_{max} values of modified biochars for ammonium (13.9 mg N/g and 22.8 mg N/g, respectively) were considerably higher than those of the unmodified biochars (4.12 mg N/g and 11.2 mg N/g, respectively) [1], implying that modifications could be useful strategies to improve the efficiency of biochar for removing ammonium.

2.2.2 Removal of Nitrate Using Pristine and Modified Biochars

Although the nitrate-adsorbing capabilities of biochars have been tested, which were made with a broad range of organic materials and pyrolyzing conditions, few of them were verified to have the ability to effectively interact with nitrate. Most studies showed that unmodified biochar had little or no adsorbing capability for nitrate, while some types of biochars even incline to release rather than adsorb nitrate. Such low nitrate-removing capacities that are almost negligible are most likely caused by the electrostatic repulsion between the nitrate anion and the negatively charged biochar surface. A higher pyrolyzing temperature (> 600 °C) can enhance the adsorbing capabilities of biochars because it leads to physicochemical changes, such as lower oxygen-containing functional groups and higher surface area [4]. The increased area in the surface could also increase the number of adsorption sites of biochars, while the reduced oxygen-containing functional groups could decrease the electrostatic repulsion between nitrate and biochars [1]. Nevertheless, it is worth noting that even biochar nitrate absorption was detected, the efficiency was rather minimal with an average Q_{max} value of 1.78 mg N/g, suggesting that pristine biochars could not be a successful substance for removing nitrates in water bodies.

Extensive modification of biochars could help their adsorbing capability for nitrate, particularly if the modification hinders the electrostatic repulsion between biochars and nitrate ions. Common modification strategies include protonation of negatively charged functional groups (e.g., treating biochar with concentrated HCl) and metal/metal oxide (e.g., La and MgO) impregnation into biochar or biochar feedstocks [1]. The average Q_{max} of the modified biochar to nitrate was 7.42 mg N/g, about 3.8 times higher than that of the pristine biochars.

2.2.3 Removal of Phosphate Using Pristine and Modified Biochars

Consistent with the assessments for NO_3^-, the surface of biochars that are negatively charged generally repels phosphate anions. Therefore, the adsorbing capabilities of pristine biochars for inorganic phosphorus are usually low. Furthermore, PO_4-P was detected to be desorbed from the biochars rather than being adsorbed. The significantly low Ca/P and Mg/P ratios in the biochars were also theorized to be the reason for phosphorus desorption because the main binding mechanism of phosphorus on biochars might be the divalent cation bridging. The Q_{max} of biochars for phosphate ranged from 1.37 to 193 mg P/g with a mean of 28.9 mg P/g.

Similar to removing nitrate, the researchers tried to increase the adsorption of phosphate by adding metal oxides and metals to the organic raw materials. Among others, Mg is the most extensively examined because it could considerably improve the PO_4-P adsorbing capabilities of biochars caused by the strong divalent cation bridging between P and Mg. The intrinsic Mg found in the tissues of organic raw materials has also been demonstrated to have phosphate adsorption-enhancing effects. Nevertheless, Mg-modified biochars showed no considerable surge in PO_4-P adsorption in some studies. The difference in the effect of Mg-modified biochars on the adsorbing capabilities of PO_4-P might be caused by many factors, such as the pyrolyzing conditions, the concentration of added Mg, and differences in intrinsic biochar properties [1]. The Al doping process could successfully protect the negatively charged surface of biochars and form positive adsorption sites for PO_4-P, which serve as an "aluminum bridge" between PO_4-P and biochars [1], facilitating the adsorption/precipitation of phosphate. It is crucial to conduct further evaluations for the utilization of aluminum-doped biochars in treating water because the toxicity and solubility of aluminum-containing chemicals in water environments should also be considered, especially in natural water systems. Clearly, evidence has proved that modified biochars could considerably improve phosphate-removing efficiency. The average Q_{max} of modified biochar for phosphate was 143 mg P/g, about 4.95 times higher than that of the pristine biochars [3].

2.3 Biochar Mechanisms of Nutrients Removal from Wastewater

2.3.1 Specific Surface Area

As a critical parameter, the specific surface area (SA) is usually believed to have a regulating effect on the adsorbing capabilities of carbon-based substances. Since the density of adsorbing sites per unit mass is directly related to SA, a higher adsorbing capability is estimated due to the increase of SA. This phenomenon is recorded many times in various research reports. However, biochars with larger SA do not necessarily possess a higher adsorbing capability for NH_4^+, which suggests that SA is not the only determinant for the adsorbing capability for NH_4^+. It is important to note that the rise in SA was associated with the decrease in oxygen-containing functional groups, which might also act as a vital part in adsorbing NH_4^+. Although rising SA could promote NH_4^+ adsorption, this might not be the case for the adsorption of anions such as PO_4^{3-} and NO_3^- because biochar's surface is usually negatively charged.

2.3.2 Ion Exchange

The positively charged cations are in balance with the negatively charged surface of biochar in a water environment. It has been observed that biochar produced with low temperatures sometimes has a higher CEC, thus resulting in a higher NH_4^+ adsorbing capability. Such a phenomenon suggests that other cations are replaced with lower affinities for the surface sites of biochars, resulting in the adsorption of NH_4^+. Besides the pyrolyzing temperatures, the properties of the organic materials is directly related to the CEC of biochars. Modification by adding metals can improve the CEC of biochars, thereby enhancing the adsorbing capability for NH_4^+.

2.3.3 Surface Functional Groups

The main features of the surface chemistry of biochar are negatively charged surface functional groups at a lower pyrolyzing temperature and hydrophobicity at a higher pyrolyzing temperature, any of which often contain oxygen (e.g., COOH and −OH) [1]. Therefore, the resultant biochars usually have a certain affinity for NH_4^+. Nevertheless, the electrostatic attraction of PO_4^{3-} and NO_3^- to biochars is minimized by the lack of surface functional groups that carry a net positive charge [1]. The biochars produced with low-temperature-pyrolysis and higher mole ratios of O/C would possess much better NH_4^+ adsorbing capability because of electrostatic interactions or chemical bonding between oxygen-containing functional groups on NH_4^+ and the surface of biochars. The interaction between oxygen-containing functional groups and NH_4-N could be elucidated by investigating the property changes before and after adsorbing of NH_4^+ [1]. As far as it is learned, there are currently no strong indications from samples of biochars. The deprotonation and protonation of surface functional groups are affected by the pH of the solutions and can notably affect the ionic bonding between NH_4^+ and oxygen-containing functional groups (COO^- and $C=O$), thus affecting the final adsorbing capabilities of biochars for NH_4^+ [5].

2.3.4 Precipitation

The formation of calcium phosphate and magnesium phosphate on the surface of biochars is deemed to be the main mechanism for the removal of PO_4^{3-} from water. A positive correlation is also observed between the phosphate adsorption and biochars' Mg and Ca contents. Only a tiny portion of phosphorus loading is reversible by desorption, indicating the formation of stable phosphorus complexes on the surface of biochars. XPS and SEM images showed that nanoscale Mg–P precipitates were tightly attached to the surface of biochars.

2.4 Factors Influencing Biochar Performance in Nutrients Removal

2.4.1 Pyrolysis Temperature

Regardless of the pyrolyzing environments, pristine biochar generally has a very low capability for NO_3^- and PO_4^{3-}. Usually, an increase in the pyrolyzing temperature results in an increase in the surface area of the biochars. Nevertheless, this increase in the surface area does not always lead to higher N or P adsorbing capabilities. It is known that biochars produced at a lower range of temperatures could keep more acidic anionic functional groups, such as hydroxyl and carboxyl groups, and thus display greater complexation potential with NH_4^+ than biochars produced at a higher range of temperatures. The negative relationship between pyrolyzing temperatures and NH_4^+ Q_{max} is caused by the loss of oxygen-containing polar functional groups, H-, and N- at a higher temperature [1]. Correlatively, an increase in the pyrolyzing temperature with a large decrease in CEC was deemed to be the main reason for the decrease in Q_{max}, since ion exchange was considered to be the main mechanism in the NH_4^+ adsorption process [1]. In addition, elevated temperature could increase the hydrophobicity and aromaticity of the surface of biochars, thus discouraging the contact of biochar with hydrophilic NH_4^+ [1].

2.4.2 Metallic Oxides on Biochar

Biochars (modified and raw) comprises a variety of metal oxides, such as La_2O_3, Fe_2O_3, AlOOH, CaO, and MgO. Biochars made from peanut shells exhibited the highest Mg and Ca portions. The biochars made from wheat straws have a relatively high content of Ca, which could then lead to a stronger adsorbing capability for phosphate. This may be caused by the precipitation of phosphate ions or the surface deposition with the strong chemical bonds or the metal oxides *via* hydrogen bonds (weak bonds), respectively.

2.4.3 Solution pH

Solution pH is an extremely crucial parameter that could affect the adsorbing of phosphate and ammonium on the surface of biochars due to its influence on the capability of exchanging ions of biochar, variable charge on the biochar surface through deprotonation and protonation of surface functional groups, and the chemical form of the ions. In neutral or weak acidic solutions, ammonium could be chelated *via* exchanging of cations by displacing positively charged species on the biochars. The low ammonia removing rate at comparatively low pH (< 6) might be due to the H^+ ions and the intense competition of ammonium ions for effective adsorbing sites on the surface of biochars. On the other hand, under alkaline conditions, adsorbing process happens *via* electrostatic interactions between the ammonium and the negatively charged surface [3]. As the pH value increased, the protonation of negatively charged functional groups on the surface of biochars leads to electrostatic repulsion between biochars and NH_4^+.

It can be easily noticed that the rising pH had a negative effect on the process of removing anionic phosphate because of the pH-dependent speciation of phosphate (PO_4^{3-}, HPO_4^{2-}, and $H_2PO_4^-$) in the liquid phase [1]. The phosphate forms tend to have an increasing negative charge density with increasing pH. Although the biochars' point of zero charges (pH_{zpc}) is close to neutral pH, phosphate adsorption is favored in the acidic pH solutions. Protonation of functional groups on the surface of biochar at low pH under pH_{zpc} facilitates its electrostatic interaction and leads to an increase in positive surface charge, which is deemed to promote phosphate adsorption [3]. In addition, OH^- ions at a relatively higher pH might also compete with phosphate for adsorption sites on the surface of biochars.

2.4.4 Contact Time

The adsorption of N or P could rise with prolonged contact time until equilibrium is reached. This might be related to higher N or P ion transfer from the liquid medium to the higher sorbate kinetic energy (N/P), the active sites of biochars, and reduced boundary layer that would otherwise impede ion transfer from solution to the organic materials. When equilibrium is reached, the adsorption becomes equal to desorption, and the active sites of the biochars are fully occupied, where prolonged exposure has negligible or no effect on adsorption.

2.4.5 Ambient Temperature

The processes of adsorption are affected by the ambient temperatures of water environments because of heat exchange and physical diffusion processes, such as exothermic and endothermic reactions. Increasing the ambient temperatures will lead to a drop in the NH_4^+ adsorbing capability because the adsorption of NH_4^+ has been shown to be exothermic processes [1]. Nevertheless, as the temperature increased, the adsorption of NH_4^+ increased, indicating an endothermic process. A higher range of temperatures might also promote the diffusion of adsorbed NH_4^+ to the inner structure of biochars, resulting in increased adsorption [1]. The rising temperature could also facilitate the random thermal motions of ions, which may increase the probability of collisions between adsorption sites on biochar and phosphate [1].

2.4.6 Coexisting Ions

Sewage water contains not only relatively high concentrations of ammonium ions, nitrate, or phosphate, but also a variety of other coexisting ions, such as cations, arsenate (As(V)), chloride (Cl^-), sulfate (SO_4^{2-}), bicarbonates (HCO_3^-), and carbonates (CO_3^{2-}). Exchanging of cations is the main mechanism for the process of removing NH_4^+ from water-based solutions. Therefore, the existence of competing cations, such as Mg^{2+}, Ca^{2+}, K^+, and Na^+, may reduce the adsorption of NH_4^+ on biochars [1]. This is especially likely to happen for cations that have a higher affinity of binding for biochars than that of NH_4^+. The intensity of competition between other cations and NH_4^+ depends on the pH-dependent variable charge on the surface of biochars, as well as their relative electro-affinities for a given functional group. Divalent cations are generally stronger in the competition because they can occupy more adsorption sites on biochars and have greater charge density [1].

Adsorption of anions by biochar is regulated by the affinity and strength of positively charged sites created by metal oxides and metals at the surface of biochars, with the exception of specific covalent bonds prompted complexation. The anions' affinity to biochars increases with the decrease of hydrated ionic radius and the increase of biochars surface charge, indicating the existence of electrostatic interaction between anions and biochars. Anions such as sulfate and phosphate might compete with nitrate for adsorbing on biochars. More negatively charged anions are expected to take over more vacant adsorption sites on the surface of biochars, resulting in considerably reduced nitrate adsorbed by biochars. Compared with other anions, the net absorption of nitrate by biochar depends on the relative affinity of nitrate to the sites of adsorption. Anions such as HCO_3^-, NO_3^-, and Cl^- compete with PO_4^{3-}, thus reducing the adsorption of phosphate. This effect is the strongest when the three competing ions are mixed together in a solution. Anions such as NO_3^- and Cl^- cannot be precipitated with Mg in modified biochars, and the decrease in phosphate adsorption was due to the blocking or completion of the surface adsorption sites.

2.5 Nutrients Desorption from Biochar

2.5.1 Ammonium Desorption

Desorption solutions such as KCl are generally utilized for the determination of mineral nitrogen in soils. Based on the surface adsorption and van der Waals interaction, the weak interaction of NH_4-N with the biochars enables the easy leaching and uniform of adsorbates. The desorption of NH_4-N mainly occurs on the basis of ion exchange with K^+, which is verified by the maximum intensity at low pH (pH = 2) and the steady decrease of NH_4-N desorption with the rising pH of the solutions. It was also observed that desorption was slightly increased in a strong alkali environment since the OH^- ions existing in such an environment electrostatically can attract NH_4-N [3]. When pH reaches above 7, volatile NH_3 is converted from the NH_4-N ions, leading to incomplete recovery of NH_4-N. Nutrients that are physically adsorbed on the surface of biochars can be easily desorbed; in contrast, the opposite phenomenon can be witnessed in the case of chemisorption. Adsorption on the biochar made at a higher temperature typically usually occurs *via* physisorption, which gives explanation to the high desorption rates in these biochars.

2.5.2 Nitrate Desorption

Most reports showed that desorption of NO_3-N from raw biochar reaches very low values or becomes insignificant [3]. Due to the transfer of electrons in the aromatic structure of biochars, the reaction of nitrogen compounds is catalyzed by the biochars, leading to forming of other nitrogen forms than NO_3-N [3]. The biochars reduce NO_3-N, possibly to gaseous forms or nitrites, resulting in reduced desorption [3]. The increased desorption at a higher pH could be related to the creation of negative charges on the surface of biochars, which have a repulsive effect on the negatively charged NO_3–N [3]. In addition, a higher level of concentration of OH^- ions at a high pH leads to displacement of NO_3–N from the adsorption sites [3]. Physisorption mainly occurs on biochars made at a higher range of temperature (> 600 °C), a process that weakly binds NO_3–N ions, and therefore, this adsorbed nutrient can be more easily desorbed [3].

2.5.3 Phosphorous Desorption

The desorbing of PO_4–P clearly depends on the starting level of PO_4–P concentration utilized in the process of adsorption, which is reasonable since it verifies the desorbed nutrient amounts [3]. Nevertheless, desorption was not proportional to the adsorption of lower amounts of PO_4–P [3]. These desorption differences are related to the large difference between the water-based solution utilized as desorbate and the level of concentration of adsorbed ions. They are also related to the easier desorption of a variety of nutrients in successive adsorptive layers [3]. PO_4–P has poor solubility in an alkaline environment with increasing pH, and there is electrostatic repulsion between PO_4–P and OH^-. As a result, the amount of PO_4–P in the solutions desorbed from biochars decreases, which does not promote the ions leaching into the biochars. A more favorable option for the desorption of PO_4–P is the acidic environment [3], which is related to the high solubility of Mg–P such as $MgHPO_4$ and $Mg_3(PO_4)_2$ at a lower pH.

2.6 Nutrient-loaded Biochar as Potential Nutrient Suppliers

As shown in the above examples, the process of desorption of nutrients in biochar happens steadily [3]. However, the maximum nutrient released from the biochar's structure is relatively quick (from 24 to 72 hours) in most cases [3], which reduces the application of biochars as a steady and controlled-release fertilizer. The process of desorption of nutrients could be regulated to a certain extent by appropriate modifications, pH level, and the choice of substrate. Under ambient settings, regulating the moisture content or the level of pH could be problematic. Nevertheless, many investigations demonstrated the potential to repurpose biochars, which was began by the utilization of removing nutrients in water, and lately used as a type of fertilizer. However, a series of studies on environmental processes and desorption kinetics that may affect nutrient release are needed. The strength with which nutrients are bound by biochar is small, which introduces the risk of fast release of nutrients after this fertilizer is applied to the soil. As a result, the direct application of secondary biochars is problematic. At present, biochars might need extra processing steps to boost the strength of the interactions with the substrates, so that leaching of nutrients would occur in a more controlled manner and more steadily. The direct application of secondary biochars to soil should not be ruled out [3]. To the contrary, future investigations are required using more complex biochars and various organic materials, where the binding would be stronger. This would offer a slow and steady release of nutrients.

References

1 Zhang, M., Song, G., Gelardi, D.L., Huang, L., Khan, E., Masek, O., Parikh, S.J., Ok, Y.S., 2020. Evaluating biochar and its modifications for the removal of ammonium, nitrate, and phosphate in water. *Water Research*, 186, 116303.
2 Nasseh, I., Khodadadi, M., Khosravi, R., Beirami, A.R., Nasseh, N., 2016. Metronidazole removal methods from aquatic media: a systematic review. *Annals of Military and Health Sciences Research*, 4(4), 196–204.
3 Marcinczyk, M., Ok, Y.S., Oleszczuk, P., 2022. From waste to fertilizer: Nutrient recovery from wastewater by pristine and engineered biochars. *Chemosphere*, 306, 135310.
4 Wen, P., Wu, Z., Han, Y., Cravotto, G., Wang, J., and Ye, B.C., 2017. Microwave-assisted synthesis of a novel biochar-based slow-release nitrogen fertilizer with enhanced water-retention capacity. *ACS Sustainable Chemistry & Engineering*, 5(8), 7374–7382.
5 Song, X., Li, K., Ning, P., Wang, C., Sun, X., Tang, L., Ruan, H., Han, S., 2017. Surface characterization studies of walnut-shell biochar catalysts for simultaneously removing of organic sulfur from yellow phosphorus tail gas. *Applied Surface Science*, 425, 130–140.

3

Adsorption of Metals/Metalloids

Yuqing Sun[1] and Daniel C.W. Tsang[2]

[1] School of Agriculture, Sun Yat-Sen University, Guangzhou, Guangdong, China
[2] State Key Laboratory of Clean Energy Utilization, Zhejiang University, China

Heavy metals present an existential threat to environmental and human health. As a result, efficient and low-cost heavy metal removal technologies are urgently needed. Because it is extremely difficult to employ biodegradation and transformation methods to treat heavy metals, the adsorption method yield the most promising results in treatment in recent years. As a sustainable low-cost material that can effectively adsorb heavy metals, a variety of biochars have recently drawn much attention in the field because of their wide-ranging prospects in their applications. Although these new carbon-based materials have various advantages, compared with traditional activated carbon, their heavy metal adsorbing efficiency is relatively lower, which limits their practical applications. In addition, once the heavy metals are retained by biochar, desorption could be extremely difficult, and the materials that were used for absorption then become dangerous wastes, which could cost a great more capital and resources to be safely disposed of. As a result, it is very important to find a more suitable method for modifying biochar's surface, enhancing a variety of biochars' heavy metals removal abilities, and improving the reusability of biochars that are used for heavy metals absorption. This chapter evaluates and introduces the biochar modification methods covered in former studies, the mechanisms of removing heavy metals using a variety of materials, and the recycling potential of biochars that are used for the absorption of heavy metals.

Biochar Applications for Wastewater Treatment, First Edition. Edited by Daniel C.W. Tsang and Yuqing Sun.
© 2023 John Wiley & Sons, Inc. Published 2023 by John Wiley & Sons, Inc.

3.1 Metals/Metalloids in Wastewater

The last century has seen rapid global development and the rise of mega metropolitans, a large amount of heavy metals and metalloids entered groundwater and surface water through sewage water, and industrial waste discharges from battery manufacturing, chemical, mining, and metallurgy industries. Heavy metal pollution of water bodies presents an existential threat to the environment and human health. Heavy metals can be extremely carcinogenic and toxic. Also, when the maximum levels of tolerance are exceeded, the heavy metals could be fatally harmful to all living organisms in the natural environment. In addition, most heavy metals could not be easily biodegraded and would accumulate in living organisms, even at very low concentrations. On top of that, because heavy metals can coexist with other ions, they could easily form complexes with a variety of agents, such as ethylenediaminetetraacetic acid, aggravating the environmental risks and toxicity of heavy metals.

Currently, a variety of traditional technologies are utilized to treat heavy metal-contaminated wastewater to uphold ecological sustainability and public safety. The technologies include phytoremediation, flotation, complexation/chelation, ion exchange, membrane removal, and chemical precipitation. But, these technologies are economically inefficient and generate hazardous secondary wastes such as volatile organic compounds (VOCs) and persistent organic pollutants. As a result, they are either very expensive to adopt or pose a threat to the environment in spite of their effectiveness. Adsorbent materials have been established as an effective and convenient to use approach for removing heavy metals from polluted water and are generally employed for heavy metals remediation.

The most commonly utilized carbon-based adsorbent for sewage water and polluted water treatment is activated carbon. Because the activated carbon that is made from coal is very expensive, biochars made from organic materials have lately established themselves as a cost-efficient option for activated carbon with better or at least comparable performance for absorbing heavy metals. Biochars are various types of carbon-rich, highly aromatic, stable, insoluble, and porous solid products, produced by pyrolyzing organic raw materials at a high temperature (300 °C ~ 700 °C) under limited oxygen or oxygen-deprived conditions. Biochars could be produced with a wide variety of organic materials and residues, for example, tea residue, kitchen waste, distillery grains, sludge, sawdust, husk, and straw. With such abundant sources of raw materials for potential production, the costs of production and energy requirements for making biochars are comparably low.

In the field of wastewater treatment, the applications of biochars have been restricted by their poor efficiency in absorbing heavy metals, which is caused by the surface functional groups, fewer adsorption sites, limited specific surface area, and lower porosity of biochars compared with activated carbon. In addition, biochar is in powder form, which is difficult to recycle and recover, further hindering their possible real-world applications in treating waste or polluted water. On the other hand, once the heavy metals are retained by biochars, desorption could be extremely difficult. Also, the materials that were used for absorption then become dangerous wastes, which could become a threat to the environment and cost a great more capital and resources to be safely disposed of. As a result, it is urgently needed to find new methods to successfully modify the biochars' surface, refining their reusability, capacity for removing heavy metals in polluted water, and possible real-world application, as well as offering a helpful way for utilizing biomass materials in the wastewater.

3.2 Mechanisms of Biochar for Adsorption of Metals/Metalloids

The mechanism of heavy metal adsorption onto biochar might include precipitation, surface complexation, ion exchange, electrostatic interaction, and physical adsorption. Each of the mechanisms mentioned here plays a very distinctive role in heavy metal adsorption, depending on the properties of biochars itself, the solution's ionic environment, and the target metal ion. Solution pH is another crucial element, as it affects the complexation behavior of the functional groups, the biochars' surface charge, and the metal speciation [1]. In most cases, chemisorption, including precipitation, surface complexation, and ion exchange, plays a more critical part in removing the heavy metals in water-based solutions than physical adsorption and electrostatic interaction. Moreover, it is worth noting that there are multiple mechanisms available in specific water environments. For instance, surface complexation, electrostatic interaction, and ion exchange are closely linked to surface functional groups *via* covalent bonding, formation of binding sites, and electrostatic forces.

3.2.1 Physical Adsorption

The internal structure and surface of biochars include a mixed distribution of macropores, mesopores, and micro-pores, and biochar retains heavy metals in its surface and internal pore structure through physical adsorption. As a relatively weak process, physical adsorption involves the movement and diffusion of heavy metal pollutants into the biochars' porous structures, and then deposits on the biochars' surface without forming any chemical bonds. In most cases, the surface area, distribution of pores, and pore size of biochars could deeply affect the physical adsorption. The surface area of the biochars can be dramatically improved by increasing the number of micropores in the biochars, hence, promoting physical adsorption. On the other hand, increasing mesoporous could effectively improve the diffusion of contaminant, thereby, accelerating adsorption process. The choice of the organic raw materials and the method of synthesis, such as the carbonization or pyrolysis temperature is one of the basic factors that could determine the biochars' pore structures. As a result, the polarity and heterogeneity of the biochar's surface associated with the functional groups contribute to physical adsorption, especially the physical movement of heavy metals to the biochar surface driven by ionic dipole forces and electrostatic attraction.

In general, three aspects, the pore structure of biochar, its surface properties, and the nature of heavy metals, determine the efficiency of physical adsorption. It is a widespread phenomenon but unlikely to be the main mechanism for absorbing heavy metal pollutants. Biochar made with pyrolytic pinewood (at 700 °C) or switchgrass (at 300 °C) partially adsorbs Cu(II) and U(VI) *via* diffusion and physical adsorption. The Pb(II) adsorption on the biochar that is produced by hydrothermal liquefying organic raw materials is essentially a physical process of endothermic.

3.2.2 Electrostatic Interaction

Electrostatic interactions happen between negatively charged biochar surfaces and positively charged heavy metals, particularly in the presence of functional groups. Because it is a relatively weak process, electrostatic interactions only play a secondary role in the adsorption of heavy metals by biochars [1]. Since most biochars' surfaces have variable charges, the point of zero charges of the surface of biochars and solution pH could determine the prevalence of electrostatic interactions in the adsorption of heavy metals [1]. The charged interface between the solutions and the biochars largely depends on the surface groups' ionization.

The characteristic of the surface charges of biochars is one of the measures used to determine the strength between biochars and adsorbate *via* electrostatic adsorption, and the pH of the solution is crucial in absorbing the heavy metals in polluted water. The plot of the Zeta potential of biochars at different solution pH conditions could determine the point of zero charge (pH_{zpc}) of biochars. When the solution pH is greater than pH_{zpc}, the biochars' surface would be negatively charged because of deprotonation of the hydration surface of the biochars, increasing the electrostatic attraction between the heavy metals, such as Cu^{2+} and Zn^{2+}, and the surface of biochars, so that the heavy metals are strongly adsorbed. On the contrary, when the solution pH was less than pH_{zpc}, a positively charged surface was formed because of the protonation of the hydration surface of the biochars, which might be the reason for the lower adsorption capacity. In addition, H^+ in solution could compete with heavy metals for vacant adsorption sites of biochars.

3.2.3 Ion Exchange

Ion exchange between protons and heavy metals on the functional groups that contain oxygen such as hydroxyl and carboxyl groups is one of the main mechanisms for biochars to adsorb heavy metals. The ion exchange process' efficiency in the adsorption of heavy metals on the biochars' surface is mostly dependent on the surface functional groups' chemistry of the biochars and the size of ions of the heavy metal pollutants [1]. When ion exchange is the predominant mechanism, one important indicator of heavy metal adsorption is cation exchange capacity (CEC). The exchange of ions typically happens between divalent metals of M^{2+} and H^+ on the functional groups that contain oxygen. It is easy to conclude that solution pH can be a crucial factor affecting the exchange of ions. For instance, more protons (H^+) would be available to saturate metal binding sites at an acidic pH. On top of that, the functional groups that contain oxygen discharge H^+ after exchanging ions with heavy metal pollutants, leading to a drop in solution pH. The removal of Cu(II) in water by the biochars made by sludge is largely caused by exchanging the toxic heavy metal with the calcium ions on the surface of biochars. It was also found that ion exchange dominates the adsorption of heavy metal pollutants onto the surface of biochars made with canna.

3.2.4 Surface Complexation

Through the inner and outer spheres of surface complexation, complexes are formed, which have polyatomic structures, and the distinctive interactions of metal-functional groups play an influential part in the process of adsorbing heavy metals by biochars [1]. For instance, the heavy metal pollutants could be effectively combined with lactone, phenolic, and carboxyl functional groups in the biochars that are made at lower temperatures *via* complexation. The changes in hydroxyl (–OH) and/or carboxyl (–COOH) functional groups were revealed using FTIR imaging, which was caused by the complexation with Cd^{2+} and Pb^{2+} ions prior to and after adsorption. The adsorption of Cr(VI) on the biochars that were treated by H_3PO_4 adsorb Cr(VI) *via* the formation of inner spherical complexes with the functional groups that contain oxygen on the biochars' surface.

3.2.5 Precipitation

Precipitation happens either on a surface or in solution with the solid products' formation during the process of adsorption. Mineral components from subsequent additions or raw materials, such as SiO_3^{2-}, PO_4^{3-}, and CO_3^{2-}, play a crucial part in the precipitation similarly, which helps enhance the adsorption capacity of the biochars and the efficiency of the adsorption sites for heavy metals [1]. For instance, removing Pb(II) from water-based solutions by biochars might be governed by a mechanism of precipitation that converts lead ions into minerals such as hydrocerrusite ($Pb_3(CO_3)_2(OH)_2$) and cerussite ($PbCO_3$) on the surface of biochars. Although precipitation is one of the main mechanisms for removing heavy metals in polluted water, other mechanisms usually work synergistically with precipitation, including surface complexation, electrostatic interaction, and ion exchange. Heavy metal precipitation onto biochars might possess faster kinetics than other mechanisms. Also, surface functional groups have merely little influence on the mechanism of heavy metal precipitation onto biochars. However, surface functional groups can implicitly stimulate the process of precipitation by affecting other mechanisms of adsorption. For instance, surface functional groups could facilitate multiple mechanisms involved in the adsorption process of Cr(III) by biochars produced with urban sludge and stimulate the process of surface precipitation to produce $Cr(OH)_3$ on the surface of biochars.

3.2.6 Reduction

Reduction is deemed to be the main mechanism for variable valence metals such as Pb^{2+} and Cr^{6+}. For example, in reference to nZVI-modified biochars, two leading nZVI reduction mechanisms were detected during the process of removing variable valence metals: (1) nZVI directly reduces the metals; (2) the process of metal adsorption initially happened on the nZVI's distinct core-shell structures, and subsequently, the heavy metals adsorbed are reduced by nZVI producing Fe^{2+}/Fe^{3+}.

3.3 Modified Biochar for Adsorption of Metals/Metalloids

To improve the efficiency of biochars for removing heavy metals, a variety of methods are utilized to alter the biochars' properties, including changing the content of functional groups, the porous structures, and the specific surface area [2].

3.3.1 Biochar/Layered Double Hydroxide Composites

Removing heavy metals efficiently has also been extensively recounted utilizing various biochar/LDH composites. The presence of numerous metal cations and functional groups, as well as the high porosity of biochars, facilitates better elimination of toxic heavy metals by biochar/LDH compared to other modified biochars [1]. The maximum capacity of adsorption of biochar/LDH was 35.6, 331, 74.1, and 682 mg/g for Cd(II), Cr(VI), Cu(II), and Pb(II), respectively [3].

Adsorption of divalent cations such as Pb(II), Cd(II), and Cu(II) have been reported at pH values in the range of 3 to 6. This is because of the existence of enough surface hydroxyl groups on biochar/LDH to effectively neutralize hydrogen ions and facilitate Pb(II) surface precipitation, by forming $Pb_3(CO_3)_2(OH)_2$. Lower pH conditions (< 3) significantly inhibit the Pb(II) adsorption due to destruction of LDH structure as an abundance of hydrogen ions. In addition, the buffering effect of LDH on biochar matrix inhibits the effect of hydrogen ions and promote better Cu(II) removal at acidic pH range (3–5). Similarly, for anionic metals ions such as Cr(VI) and As(V), the optimum removal was reported at pH 2. This is due to the highly protonated surface of biochar/LDH leading to a strong electrostatic attraction for As(V) and Cr(VI) species mostly in the fashion of $H_2AsO_4^-$ and $Cr_2O_7^{2-}$ at pH 3–7.

The equilibrium time for removing heavy metal pollutants using the majority of biochar/LDH composites is observed to be 60–300 min. Biochar/LDH composites have also exhibited preferential uptake of hazardous heavy metals with the existence of competing ions. This might be caused by the strong affinity of surface hydroxyl groups on biochar/LDH towards Pb(II) and Cu(II) in multicomponent systems, and higher selectivity and lower free energy of Cu(II) compared to other coexisting cations.

3.3.2 Magnetic Biochar Composites

Heavy metals exist in environmental media in a variety of forms, and they can be divided into cationic and anionic categories in accordance with their various properties. The arsenic, such as As(V) and As(III), and Cr(VI) are anionic heavy metals. In general, the synergistic effect among biochars and Fe improves the adsorbing capacity of pollutants of biochar composites that are magnetic. First, Fe–O and the oxygen-containing functional groups of biochars could absorb heavy metal pollutants *via* ion exchange and surface complexation. Second, the biochar composites that have magnetic properties exhibit a more negatively charged surface than plain biochars, as a result, facilitating the cationic pollutants' adsorption and leading to the processes of removing cationic pollutants *via* complexation, electrostatic attraction, and chemical precipitation.

The adsorbing capacity of biochars with magnetic properties for Cr(VI) fluctuates from 8.35 mg/g to 220 mg/g, once again illustrating that the magnetic biochars' performance could be greatly influenced by the organic raw materials and various methods of production. The mechanism of using magnetic biochars to remove Cr(VI) comprises co-precipitation, complexation with functional groups, reduction, ion exchange, and electrostatic adsorption. The adsorbing capacity of As(III) and As(V) on different biochars that have magnetic properties can also be fluctuating, ranging from 1.63 ~ 10.1 mg/g and 1.31 ~ 45.8 mg/g, respectively. The mechanism of magnetic biochar for arsenic removal largely comprises complexation and electrostatic adsorption with a variety of functional groups, in which iron oxide plays a vital part in removing arsenic.

The pollutants removed from cationic heavy metals by biochars with magnetic properties are primarily Hg(II), Sn(II), Sb(II), Ni(II), Pb(II), Cu(II), and Cd(II) [1]. Also, their removal efficiency might be affected by their distinctive chemical and physical characteristics. The adsorbing capacities of biochars with magnetic properties for Cu(II) fluctuate from 11.0 to 85.9 mg/g, for Cd(II) from 6.34 to 198 mg/g, and for Pb(II) from 4.96 to 476 mg/g [4]. In the same way, it could be concluded from the various studies that the adsorbing amounts of biochars with magnetic properties for Hg(II), Sn(II), Sb(II), and Ni(II) are 0.953, 0.344, 25.0, 111, and 47.9 mg/g, respectively. The mechanism of removing cationic heavy metal pollutants using biochars with magnetic properties includes all or at least some of the following activities: (1) co-deposition; (2) hydrogen bonding; (3) internal spherical complexation; (4) π-π interaction; (5) surface complexation; (6) ion exchange; and (7) electrostatic adsorption.

3.3.3 Biochar-Supported nZVI Composites

According to their unique characteristics, heavy metals of a variety of kinds could be captured by nZVI *via* a range of mechanisms such as co-precipitation, precipitation, oxidation, adsorption, and reduction. In addition, multiple mechanisms might be involved for each heavy metal. The reduction of heavy metal pollutants with redox potentials above −0.44 V could be one of the most critical reaction mechanisms, including heavy metals such as Ag, Hg, Pt, Pd, Cu, Co, Se, Ni, Pb, U, As, and Cr. For heavy metals with higher redox potentials, the reduction reaction with ZVI happens preferentially; for example, in the presence of Pb^{2+}, the reduction reaction of Cr^{6+} happens preferentially. In addition, hydrogen ions might be consumed by the passivation of nZVI, leading to a rise in pH, which could be conducive to precipitation processes of Zn, Co, Cd, Pb, and Cu. The formation of iron oxides and the release of free iron ions during the rusting process could initiate processes involving the mechanisms of adsorption (Ba, Zn, Co, Se, Ni, Pb, Cd, U, As, and Cr) and co-precipitation (Se, Ni, As, and Cr).

The existence of biochars improves the ability of nZVI/biochar composites to remove heavy metal pollutants. Biochars could also boost the removing capacity of nZVI for Zn^{2+}, Cu^{2+}, Pb^{2+}, Cr^{6+}, and As(V). A number of mechanisms have been suggested for employing biochars to remove heavy metal pollutants, such as π–π electron donor–acceptor interaction, complexation with surface functional groups, ion exchange, and electrostatic attraction. The effectiveness of biochars in removing heavy metal pollutants can be attributed to four aspects:

1) The biochars' negative charges that are produced during deprotonation could quickly combine with cationic form of heavy metals through electrostatic

attraction, such as Cu, Cd, and Pb. Due to the higher point of zero charge (pH_{pzc}) value of biochars, nZVI/biochar composites possess greater pH_{pzc} values (e.g., 8.4–11.6 for nZVI/biochar) than nZVI (8.3) [5], which, as a result, promotes electrostatic attractions with anionic heavy metals, causing the creation of inner-sphere complexes.

2) The O atoms in the Fe–O groups of nZVI and the O–H and C=O groups of biochars in nZVI/biochar composites can procedure coordination complexes with heavy metal ions through donating free electron pairs [5]. Several studies have validated the surface complexation of Zn^{2+}, Cu^{2+}, and Pb^{2+} by biochar [5].
3) Electron transfer from Fe^0 to Fe^{3+} is promoted by biochars, leading to more Fe^{2+} being created, which co-precipitates with Cr^{6+}.
4) Biochars and heavy metals could also be involved in redox reactions. In addition, biochars, particularly those made with municipal waste sludge and manure of livestock, are also augmented with phosphate and carbonate groups. They could produce precipitates with cationic heavy metals such as Pb [5]. As a result, biochar in nZVI/biochar systems could offer more active surface sites for removing the anionic and cationic forms of heavy metals [5].

A synergistic effect between biochars and nZVI is also detected in the process of removing heavy metal pollutants. The dramatic improvement accomplished with nZVI/biochar composites could be attributed to three aspects:

1) The raised dispersion and reduced size of nZVI crystallite improved the nZVI's reactivity on the biochars.
2) Biochar components of the composites could produce multidentate complexes with As(V) [5].
3) Biochar facilitates electron transfer from ZVI to As(V), for example, in the biochar matrices that are produced with switchgrass and red oak, the ZVI was passivated to FeOOH, which eventually combines with As to create FeAsOOH [5]. In actual fact, a number of mechanisms might happen at the same time; for example, adsorption of As(V) *via* surface complexation with FeOOH and biochars can happen simultaneously with coprecipitation and intraparticle diffusion with Fe^{3+} [5].

3.3.4 Comparison of Different Modification Methods for Metals/Metalloids

As a result of a comparative analysis of the adsorption capacities and adsorption mechanisms of a variety of heavy metal pollutants by altered biochars, the most appropriate method of modification is chosen to enhance the adsorption capacities of biochars for objective contaminants. The biochar-supported nanoscale metal oxide and inorganic metal composite substances, such as alumina, manganese oxide, and iron oxide, are very efficient at removing Pb(II), as Pb(II) can produce stable inner sphere complexes with these metal oxides and metals. On the other hand, in terms of Cd(II), alkali activation or manganese modification biochar has more affinity with Cd(II) in water-based solutions than the biochars that are modified with iron, as the surface proton content would be high and such kind of biochars tend to be acidic, preventing Cd(II) that is positively charged from contacting [2]. Alkali modified or manganese biochar generally has specific surface alkalinity, which is conducive to improving the electrostatic attractions with cations. N-doped biochar shows better selectivity to Cu(II), because of the exceedingly high stability constant of nitrogen-Cu(II) complexes [6]. On top of that, the biochars that are modified with metal oxides or metals could effectively eradicate Cu(II) ions in sewage water and municipal sludge. With a mild reduction reactivity, the zero-valent nanosize iron could lead to the highly efficient removing performance of Cr(VI) adsorption, particularly in acidic solutions [2]. The altered biochars that are positively charged could easily attract Cr(VI) anions [2]. Also, the Cr(VI) absorbed on the surface of biochars oxidizes Fe^0 to Fe(II) and reduces itself to Cr(III), and then the ions of Cr(III) precipitate as Cr(III) hydroxides [2]. As a valuable approach, sulfur impregnation could improve the removing process of Hg(II) due to the creation of HgS precipitates, and the modified biochars' functional groups that contain oxygen could similarly further the adsorption of Hg(II) [2]. Primarily existing as anions in low pH water environments, the biochars that are modified with metals or metal oxides can effectively remove As(V) from water solutions because of the electrostatic attraction between the metal or metal oxide particles on the anions and the composites [2]. The biochars that are modified with iron absorbs As(V) by producing iron arsenates.

Different heavy metal removal mechanisms rely on varying approaches to modification. As a result, when the approaches of modification are properly combined, the biochars will have sufficient heavy metal removing capacities, and the targeted heavy metal pollutants will be more sturdily kept in the biochars *via* the improved removing mechanisms and chemisorption. For instance, utilizing CO_2 as the pyrolyzing atmosphere could create a more established porous structure and lead to more oxygen containing functional groups introduced to the biochars' surface compared with a nitrogen pyrolysis atmosphere. On top of that, when CO_2 is utilized in place of nitrogen during the pyrolyzing procedure, the organic feedstock could be modified with nano-metals or other substances. As a result, the produced composite materials possess the combined benefits of nanomodification and gas modification. Ferromanganese dioxides are generally utilized for removing contaminants because of their unique ligand affinity with heavy metals, magnetism, and moderately sound structures. Applying different biochar modification techniques at the same time can provide more desirable outcomes than utilizing only one ingredient such as manganese oxide or iron. Furthermore, combining microwave assisted heating with one of the other methods of modification might lead to the resultant biochars with a broader range of effective chemical and physical mechanisms for heavy metal removal. Nevertheless, current studies on combinatorial methods of modification are limited, thus our research would only serve as an initial hypothesis that needs more attention and further investigations.

3.4 Biochar Recycling after Adsorption of Metals/Metalloids

Modified and unmodified biochars all possess a good affinity for heavy metal pollutants. They could discharge adsorbed heavy metals as resulting contaminants when heavy metal-laden biochars are abandoned in the ecosystem, particularly under acidic conditions. Moreover, regardless of whether the absorbed heavy metals undergo transformation and solidification, they could turn into possible sources of contamination and would be released in due course. It is not yet clear if the heavy metal-laden biochars would remain in the environment for many years, and whether they will turn into dangerous pollutants and become an even greater threat to the natural ecosystem. It may require further investigations because finding a safe way to treat heavy metal contaminated waste materials has always been a challenging but imperative task.

To evaluate biochars' reutilizing ability and further grasp the process of adsorption of biochars, it is vitally important to establish the process of desorbing heavy metals. Research has shown that adsorbed biochars saturated with heavy metal pollutants can be eluted using strong acids such as H_2SO_4, HCl, and HNO_3 or strong bases, for example, NaOH, for multiple washes. And then the treated biochars can be recycled for removing more heavy metals.

Absorbed metals and biochars complement one another in the pyrolyzing process, and the supplemented metals improve the biochars' catalytic performance, in which the biochars play multiple parts in the catalytic process. Biochar can increase the transformation of tar in the pyrolyzing process and can transform metals with high valence into metals with low valence, refining the properties of catalysis.

References

1. Li, A., Deng, H., Jiang, Y., Ye, C., Yu, B., Zhou, X., Ma, A., 2020. Superefficient removal of heavy metals from wastewater by Mg-loaded biochars: Adsorption characteristics and removal mechanisms. *Langmuir*, 36(31), 9160–9174.
2. Zhang, A., Li, X., Xing, J., Xu, G., 2020. Adsorption of potentially toxic elements in water by modified biochar: A review. *Journal of Environmental Chemical Engineering*, 8(4), 104196.

3 Salh, B., Denizli, A., Piskin, E., 1996. Congo red-attached poly (EGDMA-HEMA) microbeads for removal of heavy metal ions. *Separation Science and Technology*, 31(5), 715–727.

4 Francis, S. and Varshney, L., 2005. Studies on radiation synthesis of PVA/EDTA hydrogels. *Radiation Physics & Chemistry*, 74(5), 310–316.

5 Wang, S., Zhao, M., Zhou, M., Li, Y.C., Wang, J., Gao, B., Sato, S., Feng, K., Yin, W., Igalavithana, A.D., Oleszczuk, P., Wang, X., Ok, Y.S., 2019. Biochar-supported nZVI (nZVI/BC) for contaminant removal from soil and water: A critical review. *Journal of Hazardous Materials*, 373, 820–834.

6 Han, B., Weatherley, A.J., Mumford, K., Bolan, N., He, J., Stevens, G.W., Chen, D., 2021. Modification of naturally abundant resources for remediation of potentially toxic elements: A review. *Journal of Hazardous Materials*, 421, 126755.

4

Adsorption of PPCPs

Yuqing Sun[1] and Daniel C.W. Tsang[2]

[1] *School of Agriculture, Sun Yat-Sen University, Guangzhou, Guangdong, China*
[2] *State Key Laboratory of Clean Energy Utilization, Zhejiang University, China*

The widespread and inappropriate use of pharmaceutical and personal care products (PPCPs) has resulted in severe water pollution over the last several decades, with severe impacts on ecosystems and human beings [1]. As a result, an efficient and ecologically conscious technique that could remove PPCPs and a variety of harmful chemicals from the water systems has long been urgently needed. Technology focused on adsorption mechanisms could be a suitable candidate to tackle this problem. In this chapter, we first elaborate on the hazards of PPCPs and the ecological incidents caused by the use of them. We then comprehensively examine the latest advance in the removal of PPCPs from water utilizing biochars. A variety of mechanisms for adsorbing, such as pore filling, hydrogen bonding, and electrostatic interactions, are examined in detail. This chapter concludes that modification methods to the biochars' surface and the biochar production processes, such as pyrolysis temperature change, have a major influence on the performance of removing pollutants. The composition of the polluted water and the pH of the solution to be treated are also crucial to the efficiency of removing PPCPs and numerous chemicals.

4.1 PPCPs in Wastewater

PPCPs include all personal care products and pharmaceutical drugs, such as agents of diagnostics, nonsteroidal anti-inflammatory (NSAIDs), pain relievers, antibiotics, skin care products, sunscreens, perfumes, soaps, toothpaste, and hair shampoos. They are generally utilized to cure illnesses and better the quality of our everyday life. Because these products are deeply integrated into the fabric of our daily experiences, the PPCPs used by humans are in astronomical amounts and usually discharged into wastewater treatment plants. During routine treatment procedures, PPCPs could be transformed into metabolites and various byproducts, which are highly likely to be then released into the ecosystem in numerous ways. The advance in analytical techniques has made detecting PPCPs at the trace level become possible.

Although only a few thousand ng/L of PPCPs are detected in the environment – which is also the record concentration [2] – they are characterized by refractory biodegradation, refractory biological activity, and pseudo-persistence. The presence of PPCPs, even with such a low concentration, could pose a serious and lasting threat to natural water systems and human well-being. PPCPs from water systems enter animals, plants, and people through respiration, skin contact, or diet. Contact with these chemicals can damage the reproductive systems and the kidney of animals such as livestock and water-living creatures. These PPCPs could also result in convulsions, vomiting, dyspnea, acute pulmonary edema, coma, nausea, jaundice, and even sleep cycle changes in humans.

Antibiotic abuse has grown to be a looming threat to the well-being of humans, causing the generation of antibiotic-resistant genes (ARG) and the developments of antibiotic-resistant bacteria (ARB) and finally compromise environmental safety and the health of people. Prolonged contact with antibiotics considerably hinders the growth of water-living creatures. In addition, numerous types of drug-resistant bacteria that evolve through the overuse of antibiotics could also be introduced to humans via our daily diets. Their high-level cytotoxicity can affect cells' survival and growth, and the reproduction of them will cause interference with human hormone secretion, impairment of human immunity, gastrointestinal side effects, and even the deteriorated reproductive mechanism of humans. In a number of instances, they would accumulate via the food chain because of their biorefractory nature and high log K_{ow}. Not to mention that the parent PPCPs could even derive intermediates that are significantly more toxic.

A number of techniques are reported to be effective for removing PPCPs – such as adsorption, chlorination, reverse osmosis, advanced oxidation, ultrasonic treatment, anaerobic ammonium oxidation, biologically activated sludge, sedimentation, flocculation, and coagulation. On the other hand, the removal effect of biologically activated sludge, filtration, sedimentation, and coagulation is poor.

The reverse osmosis method is ineffective due to a large amount of energy and raw materials required in the process of treatment. In addition, the high cost and low salinity hamper the wide usage of oxidation by chemicals. After a close examination of all the above techniques, adsorption has the advantages of multiple functions, low cost, less secondary pollution, and good operability, and is deemed to be the most suitable method for removing PPCPs.

Among a variety of studied materials for removing PPCPs, including carbon-based, cyclodextrin-based adsorbents, zeolite, resin, and clay mineral adsorbents, the materials that are carbon-based have proved to be the most efficient because of their low cost nature, high activity of reaction, large specific surface area (SSA), and abundant oxygen-containing functional groups such as epoxy, carbonyl, carboxyl, and hydroxyl groups [2].

Biochars are a group of carbon-based materials made by thermochemical treatment of a variety of organic raw materials, including corn stalks, leaves, rice husks, cotton stalks, and wood chips, under limited-oxygen or oxygen-deprived atmosphere without additional energy-demanding steps for activation. The pores' sizes and surface area are influenced by activation methods, heating temperature, and the type of organic raw materials. On top of that, the properties below make the materials more effective than other adsorbents: (1) variable distribution of the particle sizes, large specific surface area, and low bulk density; and (2) abundant raw materials and low preparation cost.

4.2 Biochar Mechanisms for PPCPs Adsorption

In the treatment of wastewater-containing PPCPs, removal techniques that are based on biological, chemical, and physical mechanisms are offered by using biochars:

- Surface complexation, ion exchange, electrostatic interaction, and utilization of hydrogen bonds with functional groups and the van der Waals forces to adsorb PPCPs to the surface of biochars
- Absorbing PPCPs through hydrophobic partitioning
- Using micro-organisms to biodegrade at the surface of biochars
- Intraparticle diffusion and pore filling

4.2.1 π-π Interaction

As an important weak interaction, the π-π interaction mainly happens in the aromatic ring system [2]. The biochars' functional groups play the role of acceptors or donors of π-electron, whereas the pollutants' benzene rings play the role of the counter-part donors or acceptors of π-electron [2]. The alkaline-acid modified magnetic biochars are used for adsorbing the tetracycline, and the main force is discovered to be the π-π interactions [3]. It was speculated that tetracycline plays the role of a donor of π-electron, whereas the functionalized biochars are considered to act as an acceptor of π-electron.

4.2.2 Hydrogen Bonding

The electrostatic forces between the partially negative and highly electronegative atom B and hydrogen nucleus on the strong polar bond (A–H) are called the hydrogen bond [2]. The biochars' functional groups that contain oxygen could bind with oxhydryl through hydrogen bonding. The prepared biochar-montmorillonite composite is employed in the process of adsorbing atenolol [4]. A range of hydroxyl groups of montmorillonites and oxygen-containing functional groups of biochars contributed to the easy formation of hydrogen bonds with atenolol. The synthesized magnetic biochar-montmorillonite composites demonstrated that the OTC could bond with the oxygen-containing functional groups on the adsorbents' surface through hydrogen [5].

4.2.3 Electrostatic Interaction

Because polluted water has a range of ionizable functional groups at various pH levels, electrostatic interaction is generally assessed at different pH levels. For instance, when the water-based solution's pH is below A, tetracycline (TC) will be positively charged. In addition, when the solution's pH is from A to B, TC will be uncharged or negatively charged, and when the solution's pH is above B, TC will be negatively charged [2]. At the same time, the pH value of the solution affects the surface charge of the biochars. When the solution's pH is under the pH_{pzc} (pH_{pzc} > A), the biochars' surface will be positively charged [2]. As a result, when the solution's pH is under A, the biochars' surface and TC will be positively charged [2]. As a consequence, electrostatic repulsion comes into play. In addition, the biochars' surface that is positively charged will draw both negatively and neutrally charged TC because of the electrostatic attractions when the solution' pH ranges from A to pH_{pzc} [2]. Nevertheless, when the solution's pH is above pH_{pzc}, electrostatic repulsions come into play again because of negatively charged biochars and TC.

4.2.4 Other Mechanisms

One of the key mechanisms is deemed to be pore filling. On the other hand, it occurs on biochars that contain micro-pores all the time. Molecular size is an important element in the process of diffusion, which is considered to have more vital roles than the PPCPs' active sites. Pore filling's essential mechanism is physical absorption. On top of that, another more usual mechanism of adsorption is believed to be the ion exchange reactions. The montmorillonite, which is covered on the magnetic biochar composites, facilitates the process of adsorption *via* exchange reactions of the cation [5].

4.3 Factors Affecting PPCPs Adsorption by Biochar

4.3.1 Pyrolysis Temperature

The biochars that are prepared at different pyrolyzing temperatures might possess a variety of adsorption mechanisms for removing PPCPs. At a lower pyrolyzing temperature (250–450 ºC), the elevated partitioning, because of the residues of uncarbonized raw organic materials, can raise the PPCPs removing rate, and there are more organic functional groups on the surface of biochars. At a higher pyrolyzing temperature (550–900 ºC), the larger specific surface area and the elevated aromaticity are the main causes that promote the biochars' efficiency in removing PPCPs [1]. As a result, the biochars that are made with the same organic raw materials at different pyrolyzing temperatures would possess notably distinctive adsorption capacities for the targeted PPCPs.

4.3.2 Biochar Surface Modification

The untreated biochars have the disadvantages of a limited range of adsorption and insufficient adsorbing capacity. As a result, modifying biochars to expand their applicability in the processes of polluted water treatment is very much necessary. An assortment of methods that can modify biochars, such as magnetic, physical, and chemical methods, are required to successfully address such issues. After carefully planned modification procedures, the treated biochars would possess the properties of improved pore volume, surface area, surface charge, and surface hydrophilicity / hydrophobicity of biochars, which can enhance their adsorbing capacities for organic contaminants.

Two types of modification methods for biochars are proven to be very successful in enhancing the PPCPs removing rate: (a) biochars with acid or alkali treatment to enhance the profusion of oxygenated functional groups on the surface of biochars; and (b) impregnation of nano-materials, such as Ag and Fe, to the biochars composites [1]. For instance, potassium hydroxide (KOH) is generally utilized for removing tetraccline during acid or alkali treatments [6]. By doing so, the more oxygenated functional groups created on the surface of biochars and the resulting higher surface area and will lead to improved adsorbing capacities of the biochars.

The biochars with magnetic properties have drawn much interest from the public because of the advantages of reducing the cost of recycling biochars and the separation effect verification. The biochars that are loaded with magnetic materials have the benefit of relatively larger surface area, smaller size of particles, and more surface organic functional groups, which will help show better adsorption capacities than pristine biochars. The adsorption mechanism of the Fe-biochar for chlortetracycline mostly depends on interparticle diffusion and hydrogen bondings [7].

After the pretreatment using clay minerals, the resultant biochars would possess an increased number of oxygenated surface functional groups and ultra-small silica grains on the biochar substrate's surface [8]. In the end, the oxygenated surface functional groups and the silica grains could play the role of adsorption sites for pollutants in water-based solutions, greatly improving the adsorption efficiency of biochars. If the nature of the contaminants is ionic and in the case of electrostatic repulsion, the biochars modified by impregnation with clay minerals may exacerbate the issue [9]. For the biochar composites that are pretreated with clay, the biochar will provide space for the clay particle distribution, thereby enhancing the adsorbing capacities of the pristine biochars [10]. On top of that, the active sites could be enhanced by introducing clay minerals, resulting in higher efficiency of adsorption and better interactions between the biochar composites and ionizable PPCPs molecules.

The adsorption properties of both magnetic-biochar composites and clay mineral-biochar composites are more robust than those of the biochars in the pristine state [2]. On top of that, we found that the biochars that are loaded with iron display much better sorption capacities than clay-laden biochars. But, according to the previous studies, the ratio of contribution of each component to the sorption capacities of the composites was unclear, and their sorption mechanisms were not studied in depth.

4.3.3 Properties of PPCPs

The performance of adsorption can be strongly influenced by the adsorbates and the adsorbents. PPCPs, which are the targeted adsorbates, have the characteristics of rich molecular chemical structures and low concentration. They also have the shortcomings of high ecological risk, low acute toxicity, difficulty in targeted removal, and coexistence with other contaminants. PPCPs in the ecosystem are responsible for a variety of pollutants with distinctive properties, such as amphoteric, hydrophilicity, and hydrophobicity. The amount of PPCPs that the adsorbents could adsorb increases with the hydrophobicity of the adsorbates. In contrast, the adsorption capacities decrease as the hydrophilicity of PPCPs increases. In natural waters, because PPCPs are usually mixed with organic substances such as humic acids, their competitive sorption will have a noticeable influence on the removal efficiency of PPCPs. Even though the distinctive properties of PPCPs as adsorbates could influence the performance of adsorption, adsorption is still appropriate and more cost-efficient than other approaches [2].

4.3.4 Environmental pH

The effect of pH on the adsorption of biochars for removing PPCPs relies on the pH_{pzc} of the biochars and the pK_a of the target PPCP [1]. The maximum adsorption capacity of an individual PPCP is realized when the minimum repulsion between and the biochar surface and the compound happens (pK_a of a specific compound < solution pH < pH_{pzc} of biochar) [1]. Generally, acidic pH is favorable for removing PPCPs because of the lack of deprotonation on the biochar surface at low pH [1].

4.3.5 Wastewater Composition

The complex composition of polluted water could considerably influence the removal process of PPCPs by biochars, primarily because of the competition for sorption sites. For instance, since both triclosan and phosphate exist as anions in wastewater, the elevated phosphate content in the polluted water might reduce the adsorption of triclosan by biochars because of the competition for adsorption sites [11]. Conversely, it has been reported that the adsorption of heavy metals such as Cd^{2+} increases the positive charge on the biochars' surface, thereby raising the adsorption of sulfamethoxazole on biochar [12]. It has been reported that high ionic strength and high salinity (e.g., electrolyte ions SO_4^{2-}, Cl^-, K^+, Mg^{2+}, Ca^{2+}, and Na^+) in polluted water can improve the adsorption of sulfamethoxazole and ibuprofen to biochars, while elevated levels of humic acid and carbonates negatively affect drug adsorption because of competition for active adsorption sites [13].

References

1 Shahraki, Z., Mao, X., 2022. Biochar application in biofiltration systems to remove nutrients, pathogens, and pharmaceutical and personal care products from wastewater. *Journal of Environmental Quality*, 51(2), 129–151.
2 Wang, T., He, J., Lu, J., Zhou, Y., Wang, Z., Zhou, Y., 2021. Adsorptive removal of PPCPs from aqueous solution using carbon-based composites: A review. *Chinese Chemical Letters*, 33(8), 3585–3593.
3 Zhou, Y., He, Y., He, Y., Liu, X., Xu, B., Yu, J., Dai, C., Huang, A., Pang, Y., Luo, L., 2018. Analyses of tetracycline adsorption on alkali-acid modified magnetic biochar: Site energy distribution consideration. *Science of the Total Environment*, 650, 2260–2266.

4 Fu, C., Zhang, H., Xia, M., Lei, W., Wang, F., 2019. The single/co-adsorption characteristics and microscopic adsorption mechanism of biochar-montmorillonite composite adsorbent for pharmaceutical emerging organic contaminant atenolol and lead ions. *Ecotoxicology and Environmental Safety*, 187, 109763.

5 Liang, G., Wang, Z., Yang, X., Qin, T., Xie, X., Zhao, J., Li, S., 2019. Efficient removal of oxytetracycline from aqueous solution using magnetic montmorillonite-biochar composite prepared by one step pyrolysis. *Science of the Total Environment*, 695, 133800.

6 Sophia, A., Lima, E., 2018. Removal of emerging contaminants from the environment by adsorption. *Ecotoxicology and Environmental Safety*, 150, 1–17.

7 Wei, J., Liu, Y., Li, J., Zhu, Y., Yu, H., Peng, Z., 2019. Adsorption and co-adsorption of tetracycline and doxycycline by one-step synthesized iron loaded sludge biochar. *Chemosphere*, 236, 124254–124254.

8 Li, Y., Wang, Z., Xie, X., Zhu, J., Li, R., Qin, T., 2017. Removal of norfloxacin from aqueous solution by clay-biochar composite prepared from potato stem and natural attapulgite. *Colloids and Surfaces A-Physicochemical and Engineering Aspects*, 514, 126–136.

9 Ashiq, A., Adassooriya, N.M., Sarkar, B., Rajapaksha, A.U., Ok, Y.S., Vithanage, M., 2019. Municipal solid waste biochar-bentonite composite for the removal of antibiotic ciprofloxacin from aqueous media. *Journal of Environmental Management*, 236, 428–435.

10 Premarathna, K.S.D., Rajapaksha, A.U., Adassoriya, N., Sarkar, B., Sirimuthu, N.M.S., Cooray, A., Ok, Y.S., Vithanage, M., 2019. Clay-biochar composites for sorptive removal of tetracycline antibiotic in aqueous media. *Journal of Environmental Management*, 238, 315–322.

11 Kimbell, L.K., Tong, Y., Mayer, B.K., McNamara, P.J., 2018. Biosolids-derived biochar for triclosan removal from wastewater. *Environmental Engineering Science*, 35(6), 513–524.

12 Han, X., Liang, C., Li, T., Wang, K., Huang, H., Yang, X., 2013. Simultaneous removal of cadmium and sulfamethoxazole from aqueous solution by rice straw biochar. *Journal of Zhejiang University SCIENCE B*, 14(7), 640–649.

13 Lin, L., Jiang, W., Xu, P., 2017. Comparative study on pharmaceuticals adsorption in reclaimed water desalination concentrate using biochar: Impact of salts and organic matter. *Science of the Total Environment*, 601, 857–864.

5

Stormwater Biofiltration Media

Jingyi Gao[1], Yuqing Sun[2], and Daniel C.W. Tsang[3]

[1] EIT Institute for Advanced Study, Ningbo, Zhejiang, China
[2] School of Agriculture, Sun Yat-Sen University, Guangzhou, Guangdong, China
[3] State Key Laboratory of Clean Energy Utilization, Zhejiang University, China

With the growth of the global population, the process of urbanization has accelerated. Urban construction has led to an increase in the impervious area of roads and hardening of the underlying soil surface, resulting in an increase in peak flow, the runoff coefficient, and the nonpoint source pollution load (Vijayaraghavan et al., 2021; J. Wang et al., 2022). Various pollutants such as suspended solids, nutrients, heavy metals, organic substances, and microorganisms are released into surface water from point and nonpoint sources in large quantities, resulting in serious threats to water quality (Kong et al., 2021). Therefore, before the polluted surface runoff flows into the waterway, the strategy of controlling pollutants in the runoff close to the source has become the core stormwater management and control measures in various countries, and these management strategies are given some different terms in different countries, such as low impact development (LID) and best management practice for water pollution (BMP) in Canada and the United States, low impact urban design and development (LIUDD) in New Zealand, sustainable drainage systems (SuDS) in the United Kingdom, water-sensitive urban design (WSUD) in Australia, Sponge City in China, and green infrastructure (GI) in Europe (Eckart et al., 2017).

Biochar Applications for Wastewater Treatment, First Edition. Edited by Daniel C.W. Tsang and Yuqing Sun.
© 2023 John Wiley & Sons, Inc. Published 2023 by John Wiley & Sons, Inc.

5.1 Introduction

Biofiltration or bioretention – also called rain gardens, swales, or bioswales – is one of the most effective and environmentally friendly methods used in urban stormwater management practices (Bratieres et al., 2008). It not only has the ecological benefits of increasing biodiversity, reducing stormwater runoff, and making a contribution to groundwater recharge and the water cycle as a whole, but also has social benefits of making people understand nature and generate ideas to protect the natural environment (Church, 2015). Due to its flexibility in size, location, configuration, and appearance, as well as its ideal removal of pollutants (Dagenais et al., 2018), biofiltration has been used for stormwater transport and treatment in several countries and is considered one of the promising stormwater control measures (SCMs) (Bratieres et al., 2008).

The main components that make up the biofiltration system are the water ponding depression, followed by the filtration media, with vegetation growing in the filtration media, drainage material/module, and underdrain (Vijayaraghavan et al., 2021). Since the substrate determines the health and growth of plants and determines permeability, pollutant removal potential, and stability of the bioretention system, the overall performance of the biofiltration system depends on the characteristics of the substrate (Fassman-Beck et al., 2015). The appropriate bioretention media, therefore, has a significant impact on the effectiveness of the bioretention system. Ideal biofiltration media should have characteristics of high infiltration rate, high water-holding capacity and air-filled porosity, high sorption capacity, high stability, less leaching, minimal organic constituents, and good anchorage and support a wide variety of vegetation and microbial communities (Vijayaraghavan et al., 2021). It is often difficult to select a single filtration media to meet these properties simultaneously, so for efficient removal of various stormwater runoff pollutants, it is common practice to combine traditional biofiltration substrates (mixtures of sand and loam, sandy loam, or loamy sand) with the desired properties mix of suitable materials (Liu et al., 2014), such as fly ash, vermiculite, perlite, zeolites, biological waste material, activated carbon (AC), biochar, and water treatment residual (W. He et al., 2022; Wang et al., 2019; Xiong et al., 2019; Yan et al., 2016; Zuo et al., 2019). The structure of the biofiltration system using biochar-modified medium is shown in Figure 5.1.

In recent years, the application of pyrolytic carbonaceous materials in rainwater treatment has received extensive attention, due to its porous structure, high specific surface area, functional groups, and hydrophobic properties (Nabiul Afrooz and Boehm, 2017; Ulrich et al., 2015). Biochar, which is a pyrolyzed biomass produced from waste biomass, has become a more competitive filter medium for rainwater treatment due to its low cost and green production method (Lehmann and

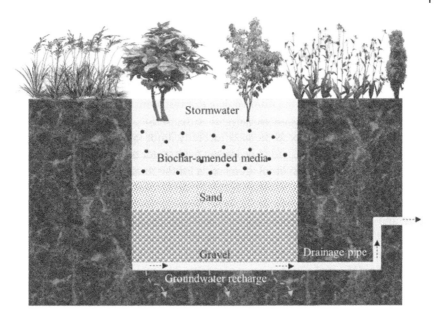

Figure 5.1 Biofiltration system using biochar amended media (Basanta Kumar Biswal et al., 2022 / with permission from Elsevier).

Joseph, 2015). Biochar has been proven to have good removal effects on heavy metals, organic pollutants, nutrients, and microorganisms, and has been widely used in wastewater treatment, soil remediation, greenhouse gas emission reduction, and carbon fixation (H. Li et al., 2020; Liu et al., 2019; Pandey et al., 2022; Sun et al., 2022; Yuan et al., 2022). In terms of soil improvement, biochar has obvious advantages due to its excellent physical and chemical properties and has been used for a long time to improve soil hydraulic properties and remove various pollutants in rainwater (Liang et al., 2021). In addition, adding biochar to bioretention packing can improve the water retention capacity of the facility and the hydraulic conductivity of the bioretention pond, delay the flood peak time, reduce the flood peak flow, and reduce the occurrence of overflow (Xiao et al., 2016; Xiong et al., 2022).

This chapter introduces the types of stormwater pollutants and the physical and chemical properties of biochar and explains the removal mechanism and application advantages of biochar added to biofiltration media for key pollutants in stormwater, such as metals/metalloids, organic pollutants, nutrients, and microbial pollutants. At the same time, the current limitations of biochar as an additive for biofiltration media and possible future research and development directions are described.

5.2 Common Pollutants in Stormwater

Common stormwater pollutants include chemical pollutants and microbial pollutants. Chemical pollutants are mainly composed of suspended solids, nutrients, heavy metals, and organic chemicals. Total suspended solids (TSS) are due to road wear, particulate emissions from vehicle exhaust, construction materials, debris generated on roads, atmospheric particulate deposition, etc (Mullaney and Lucke, 2014). Elevated solids levels increase turbidity, limit light penetration, limit the growth of aquatic plants, and may destroy habitats for benthic organisms such as fish, as well as provide accumulation media for the transport and storage of other pollutants (US EPA, 2013). Nutrients mainly come from natural processes such as atmospheric deposition, man-made processes such as fertilizers used in agriculture and horticulture, animal manure, and car cleaners (Wijeyawardana et al., 2022). Nitrogen (N) and phosphorus (P) in urban runoff can cause plant and algal blooms in rivers and streams, reduce oxygen levels, and damage fish and other aquatic life (Ma et al., 2021; Tian et al., 2019). The main sources of heavy metals are tires, road asphalt, motor oil, grease, and metal plantings washed into urban stormwater runoff and subsequently discharged into surface and groundwater sources (Reddy et al., 2014). Heavy metals are toxic, nondegradable, and bioaccumulative, and in plants and animals exposed to heavy metal-contaminated water and soil, metals can accumulate in their tissues, posing a hazard to both public health and the environment (Wijeyawardana et al., 2022). Table 5.1 shows the removal of TSS, nutrients, heavy metals and microbial pollutants in rainwater by biofiltration system.

The use of herbicides/insecticides and industrial chemicals, petroleum-derived chemicals has led to the production of organic chemicals in stormwater (Pamuru et al., 2022). Typical organic pollutants are polycyclic aromatic hydrocarbons (PAHs), polychlorinated biphenyls (PCBs), pesticides, and a few volatile organic compounds (Cao et al., 2019), which are stable and persist in the environment longer, less water-soluble, and more toxic (Convention, 2008). Harmful microorganisms in stormwater come from an improperly handled pet and livestock manure and some unreasonably constructed household septic tank systems, which can cause gastrointestinal diseases and threaten human health (Shen et al., 2018). These types of pollutants and their concentrations are not fixed in different regions and at different times, because they vary with factors such as population density, land use, geological topography, rainwater duration, and intensity in the watershed (Pamuru et al., 2022). Table 5.2 shows a summary of major stormwater pollutants concentrations from the National Stormwater Quality Database (NSQD).

Table 5.1 Summary statistics of stormwater in the biofiltration system from the National Stormwater Quality Database (NSQD) (Clary et al., 2017). NSQD is an urban stormwater runoff characterization database that was initially developed in 2001 and updated periodically through 2015 (Pitt et al., 2018).

Pollutants		Median Influent/Effluent Concentration		
		In	Out	Unit
TSS		44	10	mg/L
Nutrients	TN	1.06	0.89	
	NOx as N	0.360	0.441	
	Ammonia as N	0.300	0.0500	
	TP	0.190	0.240	
Heavy metals	As	1.31	1.60	µg/L
	Ca	0.130	0.0825	
	Cr	4.00	0.738	
	Cu	13.1	7.13	
	Pb	5.70	0.932	
	Zn	62.0	12.8	
	Fe	556	595	
	Ni	4.20	2.80	
Microbial Pollutants	E. coli	275	158	MPN/100 mL
	Enterococcus	586	218	

Table 5.2 Summary of statistics national stormwater pollutants concentrations from NSQD. *Source:* INT'L STORMWATER BMP DBASE [WWW Document], n.d; Pitt et al, 2018.

		Mean	Stdev	Median	Min	Max	Unit
	TSS	133	260	58	0.11	4800	mg/L
	COD	77	91	52	1	1674	
	Oil and grease	10.349	80.813	2.000	0.060	2980.000	
	Ph	7.3	0.8	7.3	3.4	10.7	
Microcial pollutants	Fecal coliforms (FC)	55151	282910	3700	1	5230000	#/100 Ml
Nutrients	Total Kjeldahl nitrogen (TKN)	2.0	3.5	1.4	175.0	175.0	mg/L
	Ammonia	0.6	1.0	0.3	0.0	11.9	
	Nitrites plus nitrates	0.9	1.3	0.6	0.0	24.7	
	Total phosphorus (TP)	0.380	0.683	0.236	0.007	21.200	

(Continued)

Table 5.2 (Continued)

		Mean	Stdev	Median	Min	Max	Unit
Metals	Cadmium	1.45	5.45	0.25	0.04	105	µg/L
	Chromium	7.1	13.5	2.0	0.5	200.0	
	Copper	26.5	54.6	11.3	0.3	1360	
	Lead	24.4	60.6	6.0	0.0	1200	
	Nickel	7.2	14.7	1.0	1.0	226	
	Zinc	160	356	82	0.4	14700	
Organic chemicals	Total petroleum hydrocarbon	3.9	4.4		0.3	37.5	mg/L
	Benzene	84.7	79.3		2	160	µg/L
	2-Chloroethylvinylether	3.38	2.60		0.2	36.85	
	Chloroform	74.82	159.42		1.7	651	
	Dichlorobromoethane	0.848	0.474		0.41	2.35	
	1,1-Dichloroethane	0.630	0.057		0.59	0.67	
	1,2-Dichloroethane	1.519	3.640		0.05	26.7	
	Ethylbenzene	0.825	0.963		0.2	2.8	
	Methylchloride	5.150	4.080		0.5	12	
	Methylenechloride	12.226	9.374		0.2	80	
	Tetrachloroethylene	1.520	1.001		0.4	3.3	
	Toluene	1.49	2.13		0.2	8.6	
	1,1,1-Trichloroethane	2.4	1.98		0.2	5	

5.3 Biochar for Biofiltration Media

5.3.1 Production of Biochar

The type of feedstocks used to produce biochar and its production conditions will affect the physicochemical properties of biochar (Li et al., 2021). Biochar is a kind of pyrolysis carbon-rich materials generated from plant/waste biomass, such as crop residues, wood residues, food waste digestate, yard waste, sewage sludge, which is obtained under anaerobic or anoxic atmosphere through thermochemical processes of pyrolysis, gasification, hydrothermal carbonization, and microwave-assisted pyrolysis (M. He et al., 2022; Jin et al., 2016; Seo et al., 2022). The most common method for producing biochar is pyrolysis at a temperature of 250–850 °C, which is divided into slow pyrolysis and fast pyrolysis according to heating rate and residence time (Tripathi et al., 2016). Fast pyrolysis yields higher bio-oil, while slow pyrolysis has higher biochar yield (Qian et al., 2015).

5.3.2 Physicochemical Properties of Biochar

The adsorption capacity of biochar is affected by its physicochemical properties. The physical properties of biochar include surface area, porosity, surface functional groups, acidity, alkalinity, etc. At higher pyrolysis temperatures, biochar would have more aromatic carbon, higher surface area, and micropore volume, but lower yield and average pore size. An increase in temperature results in a decrease in H, O, H/C, O/C, (O + N)/C, and functional groups, while an increase in carbon, ash, pH, conductivity, and surface roughness (Seo et al., 2022). But in some cases, biochar produced at high temperatures may exhibit lower surface area and porosity. At too high pyrolysis temperature (> 900 °C), due to the expansion and coalescence of continuous pores, as well as the softening and melting of ash, the plugging of pores and the reduction of the number of pores will result (J. Li et al., 2020). According to previous research, compared with wood or plant-derived biochar, non-wood-derived biochar may contain fewer aromatic groups, but more aliphatic groups, higher ash content, and nutrient concentration (Subedi et al., 2016). Lignin-rich biomass tends to produce biochar with a macroporous structure, while cellulose-rich biomass is more likely to produce biochar with a predominantly microporous structure (Li et al., 2017). At the same time, the ash composition of biochar also differed significantly among different feedstocks. Biochar based on biosolids and animal manure has a higher ash content, while biochar produced from herbaceous feedstocks (e.g., corn, sugar beet, sweet potato) has a lower ash content (Li et al., 2019). Cation exchange capacity (CEC) is one of the important chemical properties of biochar. Higher CEC means stronger adsorption of cations, and higher ash content may help to improve CEC and enhance the adsorption capacity of specific metals (Hopkins and Hawboldt, 2020). In conclusion, the physical and chemical properties of biochar are closely related to the feedstocks and production conditions (such as pyrolysis temperature and residence time) of biochar, so to achieve the removal of specified pollutants, the design of the biochar production process is optional.

5.4 Removal of Pollutants in Biochar-Based Biofiltration Systems

5.4.1 Metals/Metalloids

Heavy metal levels in urban stormwater have the potential to exceed recommended thresholds in guidelines for recreational and drinking water use (Ma et al., 2016). Although the concentration of a single heavy metal does not necessarily pose a direct risk to human health, there are many types of heavy metals contained in urban runoff, such as copper, lead, zinc, chromium, cadmium, iron, manganese, aluminum, mercury, and nickel, among which Cu, Pb, and Zn are the most common heavy metals and major contributors to the toxicity of urban runoff (Clary et al., 2017). Biochar has been proven to be a green sustainable material for heavy metal removal from stormwater, used as an additional adsorption medium to existing filtration media in bioretention systems. Currently, five mechanisms have been proposed to control the adsorption of metals by biochar from aqueous solutions, namely complexation, cation exchange, precipitation, electrostatic interaction, and chemical reduction, but the adsorption mechanism is different for each target metal (Li et al., 2017; Yang et al., 2019). Researchers have conducted batch and column-based removal studies of key metal pollutants from stormwater using biochar in recent years (Table 5.3).

Su et al. (2022) used alkali-modified biochar, and the comparative area of KBC produced is significantly larger than that of BC. At the same time, the ash content, pH value, and surface charged groups of biochar are increased, and it has higher aromaticity. Through the adsorption of Zn, it is found that KBC (89.0–97.5%) has higher adsorption rate and adsorption rate than BC (62.5%), and can inhibit the secondary release of Zn. The adsorption of five heavy metals, Cd, Cu, Ni, Pb, and Zn in simulated rainwater by biochar and iron-coated biochar (FeBC) was discussed by Esfandiar et al. (2022), and it was concluded that BC had good effect on Pb, Cr, and Cu, but the removal effect of Ni, Zn, and Cd is poor, and FeBC is not effective for metal removal. The reason for this result is that the ability of biochar to remove metals is related to the presence of oxygen-containing functional groups on its surface (Okaikue-Woodi et al., 2020). The BC in this study has fewer oxygen-containing groups and has a higher aromatic structure, even with a higher specific surface area; the complexing ability of metal ions is also weak. Sun et al. (2020) compared the removal efficiency of Cr, Cu, Cd, Pb, Ni, and Zn between waste wood-derived biochar and sulfuric acid-treated biochar, and found that the average removal performance of sulfuric acid-treated biochar is lower than untreated biochar. The removal of heavy metals (Cu, Cd, and Zn) by nanoscale zero-valent iron-modified pine-based biochar (named BC-nZVI) was explored by Hasan et al. (2020). It was found that the removal effects of BC and BC-nZVI on Cu were ideal, both greater than 99%, but the adsorption efficiency of BC-nZVI on Cd and Zn was better than that of BC in both single-metal solution and mixed-metal synthetic rainwater. In the column experiments, the removal efficiencies of BC-nZVI and BC were 91.7% and 42.6% for Cd, and 94.2% and 42.2% for Zn, respectively. It was concluded that the adsorption of metals on BC-nZVI occurs mainly through chemical reduction and surface complexation, demonstrating that the removal of metals is mainly controlled by the chemisorption process rather than physical adsorption. Generally speaking, adsorbents with high aromatics and hydrophobic structures are not suitable for removing inorganic pollutants such as metals, and adsorbents with high specific surface area may not have good metal removal capacity. According to previous studies, materials from

Table 5.3 Removal of metals/metalloids by biochar.

Biochar source	Modification method	Pollutants	Removal efficiency	References
Enteromorpha prolifera	KOH-BC (KBC)[a]	Zn	89.0–97.5%	(Su et al., 2022)
	BC		62.5%	
Conifer softwood	BC[b]	Cd, Cu, Ni, Pb, Zn	Cd (80%), Cu, Ni, Pb (80%), Zn	(Spahr et al., 2022)
Pinewood	BC	Cd, Cr, Cu, Ni, Pb, Zn	Cu (~100%), Pb (~100%), Cr (~100%), Ni (~40%), Cd (~25%), Zn (~20%)	(Esfandiar et al., 2022)
	FeBC[c]		Ineffective	
Forestry wood scraps	H_2SO_4-BC[d]	Cu, Cd, Ni, Zn	Cu (89%), Cd (26.6%), Ni (31.9%), Zn (33%)	(Sun et al., 2020)
	BC		Cu (96.8%), Cd (60.7%), Ni (57.1%), Zn (80.3%)	
Pinewood biomass	BC-nZVI[e]	Cu, Cd, Zn	Cu (>99%), Cd (91.7%), Zn (94.2%)	(Hasan et al., 2020)
	BC		Cu (>99%), Cd (42.6%), Zn (42.2%)	
Pine wood chips	BC	Cu, Zn	Cu (48.5–59.2%), Zn (52.7–78.0%)	(Jalizi et al., 2020)
Wood chips	BC	Pb, Cu, Zn, Cd	Pb (4%), Cu (1%), Zn (−11%), Cd (5%)	(Cairns et al., 2020)
	BC amended with wood ash		Pb (100%), Cu (99%), Zn (98%), Cd (100%)	
	BC amended with basaltic rock dust		Pb (100%), Cu (74%), Zn (52%), Cd (29%)	
Mountain Crest Gardens	BC	Cd, Cu, Ni, Pb, Zn	Cd (~100%), Cu (~100%), Ni (~100%), Pb (~100%), Zn (54%)	(Ashoori et al., 2019a)

[a] KOH-BC: Biochar modified by 10% KOH (w/v).
[b] BC: Unmodified biochar.
[c] BC-Fe: Biochar modified by Fe.
[d] BC-H_2SO_4: Biochar modified by 10% H_2SO_4 (v/v).
[e] BC-nZVI: Biochar modified by nanoscale zerovalent iron composite.

agricultural biomass are more efficient than wood biomass at effectively removing metals (Zhao et al., 2019). Taken together, biochar treated by suitable modification method will have better adsorption capacity for metals, and the biomass raw material, pyrolysis temperature and modification method of biochar will largely determine its effectiveness.

5.4.2 Nutrient

Nutrients in stormwater include nitrogen and phosphorus. In the biofiltration system, nitrogen-containing stormwater is first absorbed by the soil and then removed by a series of microorganisms through the three steps of ammoniation, nitrification, and denitrification (Xiong et al., 2022). Due to the porosity of biochar, abundant functional groups and surface sites on the surface, ammonium or nitrate can be adsorbed by electrostatic attraction (Sui et al., 2021). The incorporation of biochar into biofiltration media can enhance the anaerobic conditions for denitrification by increasing water retention or nutrient adsorption, thereby increasing nitrogen removal efficiency (Subramaniam et al., 2016). In addition, biochar can affect the growth of microorganisms in bioretention, thereby enhancing nitrification and denitrification (Chen et al., 2020). The denitrifying bacteria in the bioretention pond can use the huge surface area of the biochar as a growth site and use the carbon source provided by the biochar to enhance the denitrification of nitrogen and improve the denitrification effect (Wu et al., 2017). Removal and immobilization of dissolved phosphorus in bioretention ponds mainly rely on adsorption and precipitation (Liu and Davis, 2014), and the addition of biochar can enhance these processes. In biochar-based biofiltration media, the dissolved phosphorus can be removed by anion exchange due to the electrostatic interaction between the surface charge of biochar and phosphate (Marvin et al., 2020). The effective removal of phosphorus in the biofiltration sysytem is affected by the surface charge and pH value of the biochar. When the pH value of the urban runoff entering the bioretention pond is less than pH_{PZC} (the zero charge point of the biochar), the surface of the biochar is positively charged, which will cause electrostatic adsorption of anions and enhance phosphorus removal (Xiong et al., 2022), while if the interaction between biochar and phosphate changes from electrostatic attraction to electrostatic repulsion at higher pH (alkaline medium), phosphorus removal is weakened (Wang et al., 2016).

Table 5.4 summarizes some studies on the removal of nutrients (N and P) from stormwater using biochar-based biofiltration systems. Wang et al. (2021) used bamboo leaf and panda manure as raw materials to produce biochar BLBC and PMBC, respectively, in which PMBC was modified to obtain nZVZ-PMBC and nZVZ-CMC-PMBC, the adsorption capacity of four materials for TP The order is BLBC < PMBC < nZVZ-PMBC < nZVZ-CMC-PMBC. The maximum adsorption capacity of nZVZ-CMC-PMBC composite for TP (154.30 mg·g^{-1}) was 2.73 times that of PMBC and 1.7 times that of nZVZ-PMBC. In addition, the adsorption performance of NH_4^+-N was ranked as BLBC < nZVZ-PMBC < PMBC < nZVZ-CMC-PMBC. The adsorption capacity of nZVZ-CMC-PMBC composite was still the largest (40.31 mg·g^{-1}). A bi-layer biochar woodchip-modified bioretention system was designed by (Peng et al., 2022), and the nitrogen removal efficiency of the

Table 5.4 Removal of nutrients by biochar.

Biochar feedstocks	Modification method	Pollutants	Adsorption capacity/ Removal efficency	Reference
Panda manure biochar (PMBC)	nZVZ-CMC-PMBC[a]	NH_4^+-N	40.31 mg·g^{-1}	(Wang et al., 2021)
		TP	154.30 mg·g^{-1}	
Pine wood	BC[b]	NO_3-N	46.8 ± 21.1%	(Peng et al., 2022)
		TN	58.0 ± 10.8%	
Wood	BC	TN	44.59%-47.55%	(Rahman et al., 2020)
Pine wood	BC	NO_3-N	>99%	(Alam and Anwar, 2020)
Soil Reef™	Biochar/ZVI[c]	NO_3-N	30.6-95.7%	(Tian et al., 2019)
Cassava straw	Mg-BC[d]	NH_4^+-N	24.04 mg·g^{-1}	(Jiang et al., 2019)
Banana straw		TP	31.15mg·g^{-1}	
Rice husk	BC	NH_4^+-N	>95%	(Xiong et al., 2019)
		NO_3-N	>90%	
	Fe-BC[e]	NH_4^+-N	98.3%	
		NO_3-N	60-80%	
River sediment	BC	TN	12-54%	(Sang et al., 2019)
		TP	28-70%	

[a] nZVZ-CMC-PMBC: Modification of PMBC by $ZnCl_2$ and CMC.
[b] BC: Unmodified biochar.
[c] Biochar/ZVI: Biochar modified by anoscale zerovalent iron composite.
[d] Mg-BC: $MgCl_2$-modified biochar.
[e] Fe-BC: $FeCl_3$-modified biochar (Iron/carbon mass ratio: 0.70).

improved system (from 46.8 ± 21.1% to 73.6 ± 6.5%) was about five times higher than conventional system (only −38.6 ± 18.4% and 24.4 ± 6.0%), which proved the biochar wood chip modification system greatly expanded the reaction volume and enhanced the denitrification capacity. Similar results were obtained (Rahman et al., 2020), and it was found that the common bioretention medium, sand has better porosity but poor water retention, while the pore structure of biochar increased storage of water, which promotes nitrification and denitrification, leading to higher TN removal. Zhou et al. (2022) used Fe_3O_4 modified sludge-activated carbon to prepare BSF, and the removal efficiency of TP was higher than that of unmodified biochar (BC) due to the iron ions supported on its surface. The filtration effect of BSF on ammonia nitrogen (NH_4^+-N) is better than that on TP, indicating that the advantages of BSF as a medium in bioretention systems are mainly reflected in the filtration of NH_4^+-N. In a study by Tian et al. (2019), iron-supported biochar (zero-valent iron-modified biochar) was also fabricated to enhance nitrogen removal, which showed higher water retention, promoted anoxic conditions and longer residence time, and provided electrons for microbial denitrification, resulting in higher nitrate (NO_3-N) removal efficieny (30.6–95.7%). Jiang et al. (2019) produced biochar from six different feedstocks (banana straw, corn straw, Chinese fir straw, cassava straw, taro straw, and Camellia oleifera shell), and used $MgCl_2$ as modifider. Among them, cassava straw biiocahr (CSB) and banana straw biochar (BSB) had the highest adsorption capacities for NH_4^+-N and TP, which were 24.04 mg·g^{-1} and 31.15 mg·g^{-1}, respectively, and the loading of Mg enhanced the electrostatic attraction between biochar and $H_2PO_4^-$ in solution, and the adsorption capicity of $NH4^+$-N and TP was strongly correlated with Mg.

5.4.3 Organic Chemicals

Organic pollutants in stormwater include hydrocarbons, corrosion inhibitors, antifreeze, pesticides, and herbicides. Polycyclic aromatic hydrocarbons (PAHs), polychlorinated biphenyls (PCBs), pharmaceuticals and personal care products (PPCPs), and other trace organic pollutants (TrOCs) are the main organic chemicals (Furen et al., 2022; Markiewicz et al., 2020; Spahr et al., 2020; Ulrich et al., 2015). Bioretention facilities remove organic pollutants in stormwater runoff mainly through media adsorption, sedimentation, filtration, plant absorption, and biodegradation (Björklund and Li, 2017). In conventional stormwater treatment systems (such as sedimentation and biofiltration), particles are used to attach and remove many organic pollutants, but a large proportion of hydrophobic organic pollutants exist in the dissolved and colloidal phases of rainwater, and organic compounds pass through (Esfandiar et al., 2021). The colloids are carried, making it easier to pass through soil pores or filter materials (Flanagan et al., 2018). Therefore, traditional sand-based biofiltration media cannot retain and degrade most organic pollutants in urban runoff (Tsang et al., 2019). According to previous studies, activated carbon (AC) produced from sewage sludge can achieve effective removal of hydrophobic organic compounds (HOCs) such as PAHs, phthalates, and alkylphenols (Björklund and Li, 2017). Therefore, the addition of more economical biochar is considered to be very potential in the treatment of organic pollutants in rainwater.

Table 5.5 summarizes various examples of the application of biochar to the removal of organic chemicals from stormwater. In the study by Esfandiar et al. (2021), biochar (BC) and iron-coated biochar (FeBC) were used as adsorbents to remove the four most common PAHs – pyrene (PYR), phenanthrene (PHE), acenaphthene (ACY), and naphthalene (NAP) – from real stormwater. FeBC was shown to be an efficient adsorbent for the removal of NAP from water, and the

Table 5.5 Removal of organic chemicals by biochar.

Biochar feedstocks	Biochar type	Pollutants	Removal efficency	Partition coefficient (K_d) (L/kg)	References
Pinewood	BC	NAP	About 100%	40,520	(Esfandiar et al., 2021)
		PHE		170,234	
		PYR		393,670	
		ACY		108,240	
	FeBC	NAP		28,167	
		PHE		52,360	
		PYR		125,706	
		ACY		45,492	
Pinewood	BC	2,4-D	Over 95%		(Ray et al., 2019)
		TCEP			
		ATR			
		DIU			
		FPR			
		PFOA			
		PFOS			
Pinewood	BN-biochar[a]	2,4-D			(Ulrich et al., 2015)
		Atrazine		3.80E	
		Benzotriazole		4.34E	
		Diuron		4.76E	
		Fipronil		4.11E	
		Oryzalin		4.00E	
		TCPP		3.57E	
	MCG-biochar[b]	2,4-D		4.95E	
		Atrazine		5.87E	
		Benzotriazole		5.75E	
		Diuron		6.18E	
		Fipronil		5.92E	
		Oryzalin		5.89E	
		TCPP		5.67E	
Pinewood	MCG-biochar	TOrCs	>99%		(Ulrich et al., 2017a)
Pinewood	BC	Sulfamethoxazole (SMX)	75%		(Shimabuku et al., 2016)

(Continued)

Table 5.5 (Continued)

Biochar feedstocks	Biochar type	Pollutants	Removal efficency	Partition coefficient (K_d) (L/kg)	References
Pinewood	MCG-biochar	Atrazine	About 100%[c]	5.17E+5	(Ashoori et al., 2019b)
		2,4-D		1.45E+5	
		Diuron		4.21E+6	
		TCEP		4.30E+5	
		Fipronil		4.06E+5	
		1H-Benzotriazole		1.40E+6	
Pinewood	BC	PCBs	54–96%		(Shinneman, 2019)
Wood dust	BC	BPA	43.6–100%		(Lu and Chen, 2018)

[a] BN-biochar: A pyrolysis biochar.
[b] MCG-biochar: A gasification biochar.
[c] Removal percentage of mass after the 4-month challenge tests using synthetic stormwater spiked with 50 μg L^{-1} of fipronil, diuron, benzotriazole, atrazine, 2,4-D, and TCEP at 250 pore volumes of weekly spiking.

iron oxide coating could enhance the adsorption performance of biochar (Yi et al., 2020). It was found that the PHE adsorption partition coefficient (K_d): BC (170,234 L/kg) > FeBC (52,360 L/kg), confirming the positive effect of aromaticity (H/C) on PAHs adsorption as well as the negative effect of the polarity [(O + N) /C], while there is a weak correlation between the surface area (SA) of the material and its K_d value for the adsorption of PAHs.

Ray et al. (2019) applied biochar obtained by pine wood gasification to adsorb eight representative trace organic compounds in urban runoff, namely atrazine (ATR), 2,4-dichlorophenoxyacetic acid (2,4-D), Diuron (DIU), tris(2-chloroethyl) phosphate (TCEP), fipronil (FPR), perfluorooctanoic acid (PFOA), and perfluorooctane sulfonic acid (PFOS). The results showed that biochar removed more than 95% of all trace organics within 5 minutes at all initial concentrations, because the adsorption mechanism of TrOCs onto biochar was surface interactions (e.g., electrostatics, hydrogen bonds, π electrons) and physical interactions, such as pore filling mechanisms. It was also shown that the polyaromatic hydrophobic biochar surface is well suited for organic compound adsorption, and the large degree of microporosity and surface area of biochar can facilitate removal.

Ashoori et al. (2019b) designed a biochar-modified woodchip bioreactor and calculated that breakthrough of trace organics would occur between 10,000 and 32,000 pore volumes. Under ideal conditions, the reactor can be used for decades, assuming no other failures, demonstrating the cost-effectiveness of biochar. Bisphenol A (BPA) in stormwater was removed by combining biochar with a biofilter in the study by Lu and Chen (2018), which found that the biochar produced under the pyrolysis conditions of 700 °C has a higher specific surface area and pore volume than the biochar produced under the pyrolysis conditions of 300–500 ºC, which can remove BPA in synthetic rainwater more efficiently. The average BPA removal efficiency of the biofilter modified with biochar was up to 141 times higher than that of the real stormwater containing 200 μg/L BPA at an HLR of 40 cm/h without biochar. Meanwhile, the growth of plants was promoted.

Biochar properties, such as surface area, particle size, and pore size distribution, can affect the adsorption kinetics of organic pollutants. The decrease in pore size will increase the adsorption capacity and rate, but the organic carbon will block the pores more easily, so the smaller mesopore volume will lead to slower adsorption kinetics (Ulrich et al., 2015). Therefore, by controlling the particle size and pore size of biochar, the adsorption of organic pollutants will be improved. Besides, dissolved organic carbon (DOC) is also a key factor affecting the performance of biochar-based biofilters in removing organic pollutants (Ulrich et al., 2017b). Natural organic matter (NOM) in stormwater negatively affects the removal of organic pollutants in biochar-enhanced biofilter systems because NOM competes with organic pollutants and blocks the micropores of biochar, resulting in reduced availability of adsorption sites (Reguyal and Sarmah, 2018; Shimabuku et al., 2016). The presence of NOM may reduce the removal of TrOCs due to adverse reactions caused by sorbent fouling. However, the degree of scaling depends on the micropore size of the biochar and the molecular size of the NOM (Biswal et al., 2022; Boehm et al., 2020).

5.5 Microplastic in Urban Runoff

Microplastics (MPs) are plastics smaller than 5 mm in size and are considered a serious threat and challenge to terrestrial and aquatic ecosystems due to their widespread presence and environmental persistence in the natural environment (Ahmed et al., 2022). MP in the aquatic environment poses a health threat to more than 100 marine species (Perumal and Muthuramalingam, 2022), while MP can negatively affect organic matter decomposition and soil aggregate formation if it enters the soil, directly or indirectly affect soil function (de Souza Machado et al., 2018; Palansooriya et al., 2022). The main pathways of MP pollution transfer to the water environment include atmospheric deposition, discharge from wastewater treatment plants, and urban runoff (Fang et al., 2021; Werbowski et al., 2021). Sources of MP from urban runoff include tire wear, road wear, plastic waste, artificial turf, paint, construction materials, footwear, fibers used in textiles, pre-production granules and powders, industrial waste, and landfill leachate (e.g., Smyth et al., 2021). Current stormwater treatment technologies are mainly used to remove total suspended solids, heavy metals, organic matter, and nutrients from stormwater (Valenca et al., 2021). Biofiltration systems have been reported as a potential strategy for reducing microplastics, effectively filtering microplastics, and preventing their spread to the downstream environment (Gilbreath et al., 2019; Pankkonen, 2020). According to previous studies, sand filtration systems have limited ability to remove microplastics, but the addition of biochar improves the retention of MPs in de-sand filtration systems, and biochar has a strong potential to remove 10 μm microplastic spheres (Hsieh et al., 2022; Wang et al., 2020). Selected biochars from different raw materials and added them to the sand column, and found that the woodchip-derived biochar exhibited the highest MPs retention performance, followed by the cellulose-derived biochar, followed by lignin-derived biochar. Also adding biochar modification into the biofiltration system can achieve better microplastic removal, adding biochar and Fe_3O_4-biochar to quartz sand reduces the transport of plastic particles in porous media and increases plastic particles retention, and a very small fraction released from the column (Tong et al., 2020). The removal mechanism of biochar for microplastics in rainwater runoff is mainly electrostatic attraction and adsorption, while the sand filtration system is mainly through the interception of small filter pores and the adhesion of sand surface (C. Wang et al., 2022), and the combination of the two methods can lead to excellent removal (Chen et al., 2021). However, the current use of biochar for microplastics removal is mostly focused on the effluent treatment of sewage treatment plants, and it is rarely used in biofiltration systems. Moreover, the interaction between MPs and co-pollutants in urban runoff is not yet clear, and the aging, fragmentation, and biodegradation processes of microplastics in biofiltration systems need further research.

5.6 Challenge and Perspective

The application of biochar to biofiltration media to enhance the lifetime of their systems is currently a knowledge gap. Most of the current work focuses on predicting the media depletion time of pollutants removed by abiotic processes, with the lifespan of traditional biofilters limited to 10–15 years based on the accumulation of nondegradable metals. According to data collected on artificial and wild weathered biochar (Spokas et al., 2014), biochar will disappear through physical decomposition of the biochar structure. BC materials with high oxygen:carbon (O:C) ratios dissolve rapidly when exposed to drying and rewetting/saturation cycles (i.e., relaxation). The soil particle stability of biochar needs to be further examined in the future, which is an important mechanism for extending the life

5.6 Challenge and Perspective

of biochar. The value of regeneration and reuse of biochar should also be explored and developed according to the characteristics of biochar, so as to prolong its service life and increase its economic benefits.

In the study by Ulrich et al. (2017a), biochar-modified biofilters could maintain TOrC removal for more than 10 years (i.e., it is possible to maintain TOrC retention throughout the biofilter life cycle). Whether biodegradation can mineralize TOrC in biochar-modified biofilters and whether TOrC accumulates and leaches from depleted media is to be determined. Therefore, the fate and removal of organic pollutants after retention by biochar-improved biological filtration systems is still unclear, but their impact on the environment and public health is great, and more research in this area should be conducted in the future.

At present, most of the researches on adding biochar to biofiltration media are in laboratory scenarios, selecting synthetic stormwater or real stormwater for pollutant removal experiments (column experiments, batch experiments, etc.), however, there are few reports on the application of biochar to stormwater in practical scenarios. Biochar is effective as a bioretention filler modifier for unit-scale pollutant removal, but studies on watershed-scale performance are insufficient (Boehm et al., 2020). The unstable/intermittent flow of urban runoff, unpredictable weather conditions, impervious areas to be treated, different pollutants of concern, competition among pollutants and other complex factors can affect the effectiveness of treatment at the watershed scale. Therefore, future field and watershed-scale studies should be conducted to validate the results of laboratory column experiments or batch experiments, with a focus on long-term performance.

Although biochar has an advantage over activated carbon in terms of cost (the average cost of activated carbon and biochar is $5.6/kg and $5/kg, respectively) (Alhashimi and Aktas, 2017), its cost varies widely depending on feedstock, transportation, pyrolysis operation, and other factors, and is estimated to be between $350 and $18,000 per metric ton (Boehm et al., 2020). At present, although the modified bio-carbon designed on the laboratory scale has a good removal effect on pollutants, it also increases the cost. In practical application, a balance between effect and price is required. The feedstocks and production methods of biochar have an important impact on the removal of pollutants. It is meaningful to develop more types of feedstocks, such as various biomass-rich wastes, and at the same time achieve stable and efficient removal of pollutants by improving production conditions, which can realize waste recycling, make it more economical, and promote environmental sustainability. It is hoped that more types of raw materials can be explored in future research, such as various biomass-rich wastes and stable, efficient removal of pollutants by improving production conditions, so as to realize waste recycling and make it more economically beneficial and then promote the sustainable development of the environment.

5.7 Conclusion

Biochar-enhanced biofilters show great promise at laboratory and mesoscale for removal of common pollutants (heavy metals, nutrients, organic chemicals) from stormwater, and also on novel pollutants, microplastics good removal prospects. Biochar is a carbon-negative material that reduces greenhouse gas emissions. And the use of biomass waste to produce biochar makes the garbage usable, which has far-reaching environmental protection significance. Compared with activated carbon, the use of biochar in stormwater biofiltration has better economic benefits. The types of feedstocks and production methods (pyrolysis temperature, modification methods, etc.) of biochar will affect the physicochemical properties of biochar. The physicochemical properties of biochar, including surface area, pore size distribution, pH value, functional groups, elemental composition, and so on, are of great significance to the type and ability of removing pollutants. It is the future pursuit to design a biofiltration media with excellent performance and low price, and actually apply it to the biofiltration systems of the city to ensure its good service lifetime and to have the possibility of regeneration and reuse.

References

Ahmed, R., Hamid, A.K., Krebsbach, S.A., He, J., Wang, D., 2022. Critical review of microplastics removal from the environment. *Chemosphere*, 293, 133557. https://doi.org/10.1016/j.chemosphere.2022.133557.

Alam, M.Z. and Anwar, A.H.M.F., 2020. Nutrients adsorption onto biochar and alum sludge for treating stormwater. *Journal of Water and Environment Technology*, 18, 132–146. https://doi.org/10.2965/jwet.19-077.

Alhashimi, H.A. and Aktas, C.B., 2017. Life cycle environmental and economic performance of biochar compared with activated carbon: A meta-analysis. *Resources, Conservation and Recycling*, 118, 13–26. https://doi.org/10.1016/j.resconrec.2016.11.016.

Ashoori, N., Teixido, M., Spahr, S., LeFevre, G.H., Sedlak, D.L., Luthy, R.G., 2019a. Evaluation of pilot-scale biochar-amended woodchip bioreactors to remove nitrate, metals, and trace organic contaminants from urban stormwater runoff. *Water Research*, 154, 1–11. https://doi.org/10.1016/j.watres.2019.01.040.

Ashoori, N., Teixido, M., Spahr, S., LeFevre, G.H., Sedlak, D.L., Luthy, R.G., 2019b. Evaluation of pilot-scale biochar-amended woodchip bioreactors to remove nitrate, metals, and trace organic contaminants from urban stormwater runoff. *Water Research*, 154, 1–11. https://doi.org/10.1016/j.watres.2019.01.040.

Biswal, B.K., Vijayaraghavan, K., Tsen-Tieng, D.L., Balasubramanian, R., 2022. Biochar-based bioretention systems for removal of chemical and microbial pollutants from stormwater: A critical review. *Journal of Hazardous Materials*, 422, 126886. https://doi.org/10.1016/j.jhazmat.2021.126886.

Björklund, K. and Li, L., 2017. Removal of organic contaminants in bioretention medium amended with activated carbon from sewage sludge. *Environmental Science and Pollution Research*, 24, 19167–19180. https://doi.org/10.1007/s11356-017-9508-1.

Boehm, A.B., Bell, C.D., Fitzgerald, N.J.M., Gallo, E., Higgins, C.P., Hogue, T.S., Luthy, R.G., Portmann, A.C., Ulrich, B.A., Wolfand, J.M., 2020. Biochar-augmented biofilters to improve pollutant removal from stormwater – Can they improve receiving water quality? *Environmental Science: Water Research and Technology*, 6, 1520–1537. https://doi.org/10.1039/D0EW00027B.

Bratieres, K., Fletcher, T.D., Deletic, A., Zinger, Y., 2008. Nutrient and sediment removal by stormwater biofilters: A large-scale design optimisation study. *Water Research*, 42, 3930–3940. https://doi.org/10.1016/j.watres.2008.06.009.

Cairns, S., Robertson, I., Sigmund, G., Street-Perrott, A., 2020. The removal of lead, copper, zinc and cadmium from aqueous solution by biochar and amended biochars. *Environmental Science and Pollution Research*, 27, 21702–21715. https://doi.org/10.1007/s11356-020-08706-3.

Cao, S., Capozzi, S.L., Kjellerup, B.V., Davis, A.P., 2019. Polychlorinated biphenyls in stormwater sediments: Relationships with land use and particle characteristics. *Water Research*, 163, 114865. https://doi.org/10.1016/j.watres.2019.114865.

Chen, Y., Li, T., Hu, H., Ao, H., Xiong, X., Shi, H., Wu, C., 2021. Transport and fate of microplastics in constructed wetlands: A microcosm study. *Journal of Hazardous Materials*, 415, 125615. https://doi.org/10.1016/j.jhazmat.2021.125615.

Chen, Y., Shao, Z., Kong, Z., Gu, L., Fang, J., Chai, H., 2020. Study of pyrite based autotrophic denitrification system for low-carbon source stormwater treatment. *Journal of Water Process Engineering*, 37, 101414. https://doi.org/10.1016/j.jwpe.2020.101414.

Church, S.P., 2015. Exploring green streets and rain gardens as instances of small scale nature and environmental learning tools. *Landscape and Urban Planning*, 134, 229–240. https://doi.org/10.1016/j.landurbplan.2014.10.021.

Clary, J., Jones, J., Leisenring, M., Hobson, P., Strecker, E., 2017. International stormwater BMP database 2016 summary statistics. *Water Environment & Reuse Foundation*. https://static1.squarespace.com/static/5f8dbde10268ab224c895ad7/t/5fbd3c237ad3fe66120f69ea/1606237239545/2016_BMPDBSummaryStatistics_03-SW-1COh.pdf

Convention, S., 2008. *The 12 initial POPs under the Stockholm convention*. SCPOP Geneva, Switzerland.

de Souza Machado, A.A., Lau, C.W., Till, J., Kloas, W., Lehmann, A., Becker, R., Rillig, M.C., 2018. Impacts of microplastics on the soil biophysical environment. *Environmental Science and Technology*, 52, 9656–9665. https://doi.org/10.1021/acs.est.8b02212.

Eckart, K., McPhee, Z., Bolisetti, T., 2017. Performance and implementation of low impact development – A review. *Science of the Total Environment*, 607–608, 413–432. https://doi.org/10.1016/j.scitotenv.2017.06.254.

Esfandiar, N., Suri, R., McKenzie, E.R., 2021. Simultaneous removal of multiple polycyclic aromatic hydrocarbons (PAHs) from urban stormwater using low-cost agricultural/industrial byproducts as sorbents. *Chemosphere*, 274, 129812. https://doi.org/10.1016/j.chemosphere.2021.129812.

Esfandiar, N., Suri, R., McKenzie, E.R., 2022. Competitive sorption of Cd, Cr, Cu, Ni, Pb and Zn from stormwater runoff by five low-cost sorbents; Effects of co-contaminants, humic acid, salinity and pH. *Journal of Hazardous Materials*, 423, 126938. https://doi.org/10.1016/j.jhazmat.2021.126938.

Fang, Q., Niu, S., Yu, J., 2021. Characterising microplastic pollution in sediments from urban water systems using the diversity index. *Journal of Cleaner Production*, 318, 128537. https://doi.org/10.1016/j.jclepro.2021.128537.

Fassman-Beck, E., Wang, S., Simcock, R., Liu, R., 2015. Assessing the effects of bioretention's engineered media composition and compaction on hydraulic conductivity and water holding capacity. *Journal of Sustainable Water in the Built Environment*, 1, 04015003. https://doi.org/10.1061/JSWBAY.0000799.

Flanagan, K., Branchu, P., Boudahmane, L., Caupos, E., Demare, D., Deshayes, S., Dubois, P., Meffray, L., Partibane, C., Saad, M., Gromaire, M.-C., 2018. Field performance of two biofiltration systems treating micropollutants from road runoff. *Water Research*, 145, 562–578. https://doi.org/10.1016/j.watres.2018.08.064.

Furén, R., Flanagan, K., Winston, R.J., Tirpak, R.A., Dorsey, J.D., Viklander, M., Blecken, G.-T., 2022. Occurrence, concentration, and distribution of 38 organic micropollutants in the filter material of 12 stormwater bioretention facilities. *Science of the Total Environment*, 846, 157372. https://doi.org/10.1016/j.scitotenv.2022.157372.

Gilbreath, A., McKee, L., Shimabuku, I., Lin, D., Werbowski, L.M., Zhu, X., Grbic, J., Rochman, C., 2019. Multiyear water quality performance and mass accumulation of PCBs, mercury, methylmercury, copper, and microplastics in a bioretention rain garden. *Journal of Sustainable Water in the Built Environment*, 5, 04019004.

Hasan, M.S., Geza, M., Vasquez, R., Chilkoor, G., Gadhamshetty, V., 2020. Enhanced heavy metal removal from synthetic stormwater using nanoscale zerovalent iron-modified biochar. *Water, Air and Soil Pollution*, 231, 220. https://doi.org/10.1007/s11270-020-04588-w.

He, M., Zhu, X., Dutta, S., Khanal, S.K., Lee, K.T., Masek, O., Tsang, D.C.W., 2022. Catalytic co-hydrothermal carbonization of food waste digestate and yard waste for energy application and nutrient recovery. *Bioresource Technology*, 344, 126395. https://doi.org/10.1016/j.biortech.2021.126395.

He, W., Lin, X., Shi, Z., Yu, J., Ke, S., Lu, X., Deng, Z., Wu, Y., Wang, L., He, Q., Ma, J., 2022. Nutrient removal performance and microbial community analysis of amended bioretention column for rainwater runoff treatment. *Journal of Cleaner Production*, 374, 133974. https://doi.org/10.1016/j.jclepro.2022.133974.

Hopkins, D., Hawboldt, K., 2020. Biochar for the removal of metals from solution: A review of lignocellulosic and novel marine feedstocks. *Journal of Environmental Chemical Engineering*, 8, 103975. https://doi.org/10.1016/j.jece.2020.103975.

Hsieh, L., He, L., Zhang, M., Lv, W., Yang, K., Tong, M., 2022. Addition of biochar as thin preamble layer into sand filtration columns could improve the microplastics removal from water. *Water Research*, 221, 118783. https://doi.org/10.1016/j.watres.2022.118783.

Jalizi, S., Ashley, K., Chan, C.C.V., 2020. Restoration of an urban creek water quality using sand and biochar filtration galleries, in: Moore, J., Attia, S., Abdel-Kader, A., Narasimhan, A., (Eds.), *Ecocities Now: Building the Bridge to Socially Just and Ecologically Sustainable Cities*. Springer International Publishing, Cham, 161–173. https://doi.org/10.1007/978-3-030-58399-6_11.

Jiang, Y.-H., Li, A.-Y., Deng, H., Ye, C.-H., Wu, Y.-Q., Linmu, Y.-D., Hang, H.-L., 2019. Characteristics of nitrogen and phosphorus adsorption by Mg-loaded biochar from different feedstocks. *Bioresource Technology*, 276, 183–189. https://doi.org/10.1016/j.biortech.2018.12.079.

Jin, J., Li, Y., Zhang, J., Wu, S., Cao, Y., Liang, P., Zhang, J., Wong, M.H., Wang, M., Shan, S., Christie, P., 2016. Influence of pyrolysis temperature on properties and environmental safety of heavy metals in biochars derived from municipal sewage sludge. *Journal of Hazardous Materials*, 320, 417–426. https://doi.org/10.1016/j.jhazmat.2016.08.050.

Kong, Z., Shao, Z., Shen, Y., Zhang, X., Chen, M., Yuan, Y., Li, G., Wei, Y., Hu, X., Huang, Y., He, Q., Chai, H., 2021. Comprehensive evaluation of stormwater pollutants characteristics, purification process and environmental impact after low impact development practices. *Journal of Cleaner Production*, 278, 123509. https://doi.org/10.1016/j.jclepro.2020.123509.

Lehmann, J. and Joseph, S., 2015. *Biochar for Environmental Management: Science, Technology and Implementation*. Routledge.

Li, H., Dong, X., da Silva, E.B., de Oliveira, L.M., Chen, Y., Ma, L.Q., 2017a. Mechanisms of metal sorption by biochars: Biochar characteristics and modifications. *Chemosphere*, 178, 466–478. https://doi.org/10.1016/j.chemosphere.2017.03.072.

Li, H., Li, Y., Xu, Y., Lu, X., 2020. Biochar phosphorus fertilizer effects on soil phosphorus availability. *Chemosphere*, 244, 125471. https://doi.org/10.1016/j.chemosphere.2019.125471.

Li, J., Li, M., Wang, S., Yang, X., Liu, F., Liu, X., 2020. Key role of pore size in Cr(VI) removal by the composites of 3-dimentional mesoporous silica nanospheres wrapped with polyaniline. *Science of the Total Environment*, 729, 139009. https://doi.org/10.1016/j.scitotenv.2020.139009.

Li, S., Harris, S., Anandhi, A., Chen, G., 2019. Predicting biochar properties and functions based on feedstock and pyrolysis temperature: A review and data syntheses. *Journal of Cleaner Production*, 215, 890–902. https://doi.org/10.1016/j.jclepro.2019.01.106.

Li, Y., Yu, H., Liu, L., Yu, H., 2021. Application of co-pyrolysis biochar for the adsorption and immobilization of heavy metals in contaminated environmental substrates. *Journal of Hazardous Materials*, 420, 126655. https://doi.org/10.1016/j.jhazmat.2021.126655.

Liang, J., Li, Y., Si, B., Wang, Y., Chen, X., Wang, X., Chen, H., Wang, H., Zhang, F., Bai, Y., Biswas, A., 2021. Optimizing biochar application to improve soil physical and hydraulic properties in saline-alkali soils. *Science of the Total Environment*, 771, 144802. https://doi.org/10.1016/j.scitotenv.2020.144802.

Liu, J. and Davis, A.P., 2014. Phosphorus speciation and treatment using enhanced phosphorus removal bioretention. *Environmental Science and Technology*, 48, 607–614. https://doi.org/10.1021/es404022b.

Liu, J., Sample, D.J., Bell, C., Guan, Y., 2014. Review and research needs of bioretention used for the treatment of urban stormwater. *Water*, 6, 1069–1099. https://doi.org/10.3390/w6041069.

Liu, X., Mao, P., Li, L., Ma, J., 2019. Impact of biochar application on yield-scaled greenhouse gas intensity: A meta-analysis. *Science of the Total Environment*, 656, 969–976. https://doi.org/10.1016/j.scitotenv.2018.11.396.

Lu, L. and Chen, B., 2018. Enhanced bisphenol A removal from stormwater in biochar-amended biofilters: Combined with batch sorption and fixed-bed column studies. *Environmental Pollution*, 243, 1539–1549. https://doi.org/10.1016/j.envpol.2018.09.097.

Ma, Y., Egodawatta, P., McGree, J., Liu, A., Goonetilleke, A., 2016. Human health risk assessment of heavy metals in urban stormwater. *Science of the Total Environment*, 557–558, 764–772. https://doi.org/10.1016/j.scitotenv.2016.03.067.

Ma, Y., Wang, S., Zhang, X., Shen, Z., 2021. Transport process and source contribution of nitrogen in stormwater runoff from urban catchments. *Environmental Pollution*, 289, 117824. https://doi.org/10.1016/j.envpol.2021.117824.

Markiewicz, A., Strömvall, A.-M., Björklund, K., 2020. Alternative sorption filter materials effectively remove non-particulate organic pollutants from stormwater. *Science of the Total Environment*, 730, 139059. https://doi.org/10.1016/j.scitotenv.2020.139059.

Marvin, J.T., Passeport, E., Drake, J., 2020. *State-of-the-art review of phosphorus sorption amendments in bioretention media: A systematic literature review.* https://doi.org/10.1061/JSWBAY.0000893.

Mullaney, J. and Lucke, T., 2014. Practical review of pervious pavement designs. *CLEAN – Soil Air Water*, 42, 111–124. https://doi.org/10.1002/clen.201300118.

Nabiul Afrooz, A.R.M. and Boehm, A.B., 2017. Effects of submerged zone, media aging, and antecedent dry period on the performance of biochar-amended biofilters in removing fecal indicators and nutrients from natural stormwater. *Ecological Engineering*, 102, 320–330. https://doi.org/10.1016/j.ecoleng.2017.02.053.

Okaikue-Woodi, F.E.K., Cherukumilli, K., Ray, J.R., 2020. A critical review of contaminant removal by conventional and emerging media for urban stormwater treatment in the United States. *Water Research*, 187, 116434. https://doi.org/10.1016/j.watres.2020.116434.

Palansooriya, K.N., Sang, M.K., Igalavithana, A.D., Zhang, M., Hou, D., Oleszczuk, P., Sung, J., Ok, Y.S., 2022. Biochar alters chemical and microbial properties of microplastic-contaminated soil. *Environmental Research*, 209, 112807. https://doi.org/10.1016/j.envres.2022.112807.

Pamuru, S.T., Forgione, E., Croft, K., Kjellerup, B.V., Davis, A.P., 2022. Chemical characterization of urban stormwater: Traditional and emerging contaminants. *Science of the Total Environment*, 813, 151887. https://doi.org/10.1016/j.scitotenv.2021.151887.

Pandey, B., Suthar, S., Chand, N., 2022. Effect of biochar amendment on metal mobility, phytotoxicity, soil enzymes, and metal-uptakes by wheat (Triticum aestivum) in contaminated soils. *Chemosphere*, 307, 135889. https://doi.org/10.1016/j.chemosphere.2022.135889.

Pankkonen, P., 2020. *Urban stormwater microplastics–characteristics and removal using a developed filtration system.*

Peng, Y., Deng, S., Kong, Z., Yuan, Y., Long, H., Fang, J., Ma, H., Shao, Z., He, Q., Chai, H., 2022. Biochar and woodchip amended bioreactor extending reactive volume for enhanced denitrification in stormwater runoff. *Journal of Water Process Engineering*, 46, 102541. https://doi.org/10.1016/j.jwpe.2021.102541.

Perumal, K. and Muthuramalingam, S., 2022. Global sources, abundance, size, and distribution of microplastics in marine sediments - A critical review. *Estuarine, Coastal and Shelf Science*, 264, 107702. https://doi.org/10.1016/j.ecss.2021.107702.

Pitt, R., Maestre, A., Clary, J., 2018. *The National Stormwater Quality Database (NSQD, version 4.02)* https://yosemite.epa.gov/oa/eab_web_docket.nsf/Attachments%20By%20ParentFilingId/0579A9DB19C5446F85257CBA00593372/$FILE/Attachment%202%20-%20Pitt%2C%20National%20Stormwater%20Quality%20Database%20Version%203.1.pdf.

Qian, K., Kumar, A., Zhang, H., Bellmer, D., Huhnke, R., 2015. Recent advances in utilization of biochar. *Renewable and Sustainable Energy Reviews*, 42, 1055–1064. https://doi.org/10.1016/j.rser.2014.10.074.

Rahman, M.Y.A., Nachabe, M.H., Ergas, S.J., 2020. Biochar amendment of stormwater bioretention systems for nitrogen and escherichia coli removal: Effect of hydraulic loading rates and antecedent dry periods. *Bioresource Technology*, 310, 123428. https://doi.org/10.1016/j.biortech.2020.123428.

Ray, J.R., Shabtai, I.A., Teixidó, M., Mishael, Y.G., Sedlak, D.L., 2019. Polymer-clay composite geomedia for sorptive removal of trace organic compounds and metals in urban stormwater. *Water Research*, 157, 454–462. https://doi.org/10.1016/j.watres.2019.03.097.

Reddy, K.R., Xie, T., Dastgheibi, S., 2014. Removal of heavy metals from urban stormwater runoff using different filter materials. *Journal of Environmental Chemical Engineering*, 2, 282–292. https://doi.org/10.1016/j.jece.2013.12.020.

Reguyal, F. and Sarmah, A.K., 2018. Adsorption of sulfamethoxazole by magnetic biochar: Effects of pH, ionic strength, natural organic matter and 17α-ethinylestradiol. *Science of the Total Environment*, 628–629, 722–730. https://doi.org/10.1016/j.scitotenv.2018.01.323.

Sang, M., Huang, M., Zhang, W., Che, W., Sun, H., 2019. A pilot bioretention system with commercial activated carbon and river sediment-derived biochar for enhanced nutrient removal from stormwater. *Water Science and Technology*, 80, 707–716. https://doi.org/10.2166/wst.2019.310.

Seo, J.Y., Tokmurzin, D., Lee, D., Lee, S.H., Seo, M.W., Park, Y.-K., 2022. Production of biochar from crop residues and its application for biofuel production processes – An overview. *Bioresource Technology*, 361, 127740. https://doi.org/10.1016/j.biortech.2022.127740.

Shen, P., Deletic, A., Urich, C., Chandrasena, G.I., McCarthy, D.T., 2018. Stormwater biofilter treatment model for faecal microorganisms. *Science of the Total Environment*, 630, 992–1002. https://doi.org/10.1016/j.scitotenv.2018.02.193.

Shimabuku, K.K., Kearns, J.P., Martinez, J.E., Mahoney, R.B., Moreno-Vasquez, L., Summers, R.S., 2016. Biochar sorbents for sulfamethoxazole removal from surface water, stormwater, and wastewater effluent. *Water Research*, 96, 236–245. https://doi.org/10.1016/j.watres.2016.03.049.

Shinneman, J., 2019. *A laboratory investigation of PCB removal from stormwater using biochar-amended engineered soils in columns*.

Smyth, K., Drake, J., Li, Y., Rochman, C., Van Seters, T., Passeport, E., 2021. Bioretention cells remove microplastics from urban stormwater. *Water Research*, 191, 116785. https://doi.org/10.1016/j.watres.2020.116785.

Spahr, S., Teixidó, M., Gall, S.S., Pritchard, J.C., Hagemann, N., Helmreich, B., Luthy, R.G., 2022. Performance of biochars for the elimination of trace organic contaminants and metals from urban stormwater. *Environmental Science: Water Research and Technology*, 8, 1287–1299. https://doi.org/10.1039/D1EW00857A.

Spahr, S., Teixidó, M., Sedlak, D.L., Luthy, R.G., 2020. Hydrophilic trace organic contaminants in urban stormwater: Occurrence, toxicological relevance, and the need to enhance green stormwater infrastructure. *Environmental Science: Water Research and Technology*, 6, 15–44. https://doi.org/10.1039/C9EW00674E.

Su, Z., Sun, P., Chen, Y., Liu, J., Li, J., Zheng, T., Yang, S., 2022. The influence of alkali-modified biochar on the removal and release of Zn in bioretention systems: Adsorption and immobilization mechanism. *Environmental Pollution*, 310, 119874. https://doi.org/10.1016/j.envpol.2022.119874.

Subedi, R., Taupe, N., Pelissetti, S., Petruzzelli, L., Bertora, C., Leahy, J.J., Grignani, C., 2016. Greenhouse gas emissions and soil properties following amendment with manure-derived biochars: Influence of pyrolysis temperature and feedstock type. *Journal of Environmental Management*, 166, 73–83. https://doi.org/10.1016/j.jenvman.2015.10.007.

Subramaniam, D., Mather, P., Russell, S., Rajapakse, J., 2016. Dynamics of nitrate-nitrogen removal in experimental stormwater biofilters under intermittent wetting and drying. *Journal of Environmental Engineering*, 142, 04015090. https://doi.org/10.1061/(ASCE)EE.1943-7870.0001043.

Sui, M., Li, Y., Jiang, Y., Wang, L., Zhang, W., Sathishkumar, K., Zakaria, H., 2021. Sediment-based biochar facilitates highly efficient nitrate removal: Physicochemical properties, biological responses and potential mechanism. *Chemical Engineering Journal*, 405, 126645. https://doi.org/10.1016/j.cej.2020.126645.

Sun, Y., Chen, S.S., Lau, A.Y.T., Tsang, D.C.W., Mohanty, S.K., Bhatnagar, A., Rinklebe, J., Lin, K.-Y.A., Ok, Y.S., 2020. Waste-derived compost and biochar amendments for stormwater treatment in bioretention column: Co-transport of metals and colloids. *Journal of Hazardous Materials*, 383, 121243. https://doi.org/10.1016/j.jhazmat.2019.121243.

Sun, Y., Zhang, Q., Clark, J.H., Graham, N.J.D., Hou, D., Ok, Y.S., Tsang, D.C.W., 2022. Tailoring wood waste biochar as a reusable microwave absorbent for pollutant removal: Structure-property-performance relationship and iron-carbon interaction. *Bioresource Technology*, 362, 127838. https://doi.org/10.1016/j.biortech.2022.127838.

Tian, J., Jin, J., Chiu, P.C., Cha, D.K., Guo, M., Imhoff, P.T., 2019. A pilot-scale, bi-layer bioretention system with biochar and zero-valent iron for enhanced nitrate removal from stormwater. *Water Research*, 148, 378–387. https://doi.org/10.1016/j.watres.2018.10.030.

Tong, M., He, L., Rong, H., Li, M., Kim, H., 2020. Transport behaviors of plastic particles in saturated quartz sand without and with biochar/Fe_3O_4-biochar amendment. *Water Research*, 169, 115284. https://doi.org/10.1016/j.watres.2019.115284.

Tripathi, M., Sahu, J.N., Ganesan, P., 2016. Effect of process parameters on production of biochar from biomass waste through pyrolysis: A review. *Renewable and Sustainable Energy Reviews*, 55, 467–481. https://doi.org/10.1016/j.rser.2015.10.122.

Tsang, D.C.W., Yu, I.K.M., Xiong, X., 2019. Chapter 18 - Novel application of biochar in stormwater harvesting, in: Ok, Y.S., Tsang, D.C.W., Bolan, N., Novak, J.M., Eds.,

Biochar from Biomass and Waste. Elsevier, 319–347. https://doi.org/10.1016/B978-0-12-811729-3.00018-2.

Ulrich, B.A., Im, E.A., Werner, D., Higgins, C.P., 2015. Biochar and activated carbon for enhanced trace organic contaminant retention in stormwater infiltration systems. *Environmental Science and Technology*, 49, 6222–6230. https://doi.org/10.1021/acs.est.5b00376.

Ulrich, B.A., Loehnert, M., Higgins, C.P., 2017a. Improved contaminant removal in vegetated stormwater biofilters amended with biochar. *Environmental Science: Water Research and Technology*, 3, 726–734. https://doi.org/10.1039/C7EW00070G.

Ulrich, B.A., Vignola, M., Edgehouse, K., Werner, D., Higgins, C.P., 2017b. Organic carbon amendments for enhanced biological attenuation of trace organic contaminants in biochar-amended stormwater biofilters. *Environmental Science and Technology*, 51, 9184–9193. https://doi.org/10.1021/acs.est.7b01164.

US EPA, O., 2013. *Residual risk report to Congress 1999 [WWW Document]*. https://www.epa.gov/fera/residual-risk-report-congress-1999 (accessed 11.3.22).

Valenca, R., Le, H., Zu, Y., Dittrich, T.M., Tsang, D.C.W., Datta, R., Sarkar, D., Mohanty, S.K., 2021. Nitrate removal uncertainty in stormwater control measures: Is the design or climate a culprit? *Water Research*, 190, 116781. https://doi.org/10.1016/j.watres.2020.116781.

Vijayaraghavan, K., Biswal, B.K., Adam, M.G., Soh, S.H., Tsen-Tieng, D.L., Davis, A.P., Chew, S.H., Tan, P.Y., Babovic, V., Balasubramanian, R., 2021. Bioretention systems for stormwater management: Recent advances and future prospects. *Journal of Environmental Management*, 292, 112766. https://doi.org/10.1016/j.jenvman.2021.112766.

Wang, C., O'Connor, D., Wang, L., Wu, W.-M., Luo, J., Hou, D., 2022. Microplastics in urban runoff: Global occurrence and fate. *Water Research*, 225, 119129. https://doi.org/10.1016/j.watres.2022.119129.

Wang, J., Meng, Q., Zou, Y., Qi, Q., Tan, K., Santamouris, M., He, B.-J., 2022. Performance synergism of pervious pavement on stormwater management and urban heat island mitigation: A review of its benefits, key parameters, and co-benefits approach. *Water Research*, 221, 118755. https://doi.org/10.1016/j.watres.2022.118755.

Wang, M., Hu, S., Wang, Q., Liang, Y., Liu, C., Xu, H., Ye, Q., 2021. Enhanced nitrogen and phosphorus adsorption performance and stabilization by novel panda manure biochar modified by CMC stabilized nZVZ composite in aqueous solution: Mechanisms and application potential. *Journal of Cleaner Production*, 291, 125221. https://doi.org/10.1016/j.jclepro.2020.125221.

Wang, Z., Sedighi, M., Lea-Langton, A., 2020. Filtration of microplastic spheres by biochar: Removal efficiency and immobilisation mechanisms. *Water Research*, 184, 116165. https://doi.org/10.1016/j.watres.2020.116165.

Wang, Z., Shen, D., Shen, F., Li, T., 2016. Phosphate adsorption on lanthanum loaded biochar. *Chemosphere*, 150, 1–7. https://doi.org/10.1016/j.chemosphere.2016.02.004.

Wang, Z., Tan, K., Cai, J., Hou, S., Wang, Y., Jiang, P., Liang, M., 2019. Silica oxide encapsulated natural zeolite for high efficiency removal of low concentration heavy metals in water. *Colloids and Surfaces A: Physicochemical and Engineering Aspects*, 561, 388–394. https://doi.org/10.1016/j.colsurfa.2018.10.065.

Wijeyawardana, P., Nanayakkara, N., Gunasekara, C., Karunarathna, A., Law, D., Pramanik, B.K., 2022. Improvement of heavy metal removal from urban runoff using modified pervious concrete. *Science of the Total Environment*, 815, 152936. https://doi.org/10.1016/j.scitotenv.2022.152936.

Wu, B., Fang, L., Fortner, J.D., Guan, X., Lo, I.M.C., 2017. Highly efficient and selective phosphate removal from wastewater by magnetically recoverable La(OH)$_3$/Fe$_3$O$_4$ nanocomposites. *Water Research*, 126, 179–188. https://doi.org/10.1016/j.watres.2017.09.034.

Xiao, Q., Zhu, L., Shen, Y., Li, S., 2016. Sensitivity of soil water retention and availability to biochar addition in rainfed semi-arid farmland during a three-year field experiment. *Field Crops Research*, 196, 284–293. https://doi.org/10.1016/j.fcr.2016.07.014.

Xiong, J., Liang, L., Shi, W., Li, Z., Zhang, Z., Li, X., Liu, Y., 2022. Application of biochar in modification of fillers in bioretention cells: A review. *Ecological Engineering*, 181, 106689. https://doi.org/10.1016/j.ecoleng.2022.106689.

Xiong, J., Ren, S., He, Y., Wang, X.C., Bai, X., Wang, J., Dzakpasu, M., 2019. Bioretention cell incorporating Fe-biochar and saturated zones for enhanced stormwater runoff treatment. *Chemosphere*, 237, 124424. https://doi.org/10.1016/j.chemosphere.2019.124424.

Yan, Q., Davis, A.P., James, B.R., 2016. Enhanced organic phosphorus sorption from urban stormwater using modified bioretention media: Batch studies. *Journal of Environmental Engineering*, 142, 04016001. https://doi.org/10.1061/(ASCE)EE.1943-7870.0001073.

Yang, X., Wan, Y., Zheng, Y., He, F., Yu, Z., Huang, J., Wang, H., Ok, Y.S., Jiang, Y., Gao, B., 2019. Surface functional groups of carbon-based adsorbents and their roles in the removal of heavy metals from aqueous solutions: A critical review. *Chemical Engineering Journal*, 366, 608–621. https://doi.org/10.1016/j.cej.2019.02.119.

Yi, Y., Huang, Z., Lu, B., Xian, J., Tsang, E.P., Cheng, W., Fang, J., Fang, Z., 2020. Magnetic biochar for environmental remediation: A review. *Bioresource Technology*, 298, 122468. https://doi.org/10.1016/j.biortech.2019.122468.

Yuan, X., Wang, J., Deng, S., Dissanayake, P.D., Wang, S., You, S., Yip, A.C.K., Li, S., Jeong, Y., Tsang, D.C.W., Ok, Y.S., 2022. Sustainable food waste management: Synthesizing engineered biochar for CO$_2$ capture. *ACS Sustainable Chemistry and Engineering*, 10, 13026–13036. https://doi.org/10.1021/acssuschemeng.2c03029.

Zhao, J., Shen, X.-J., Domene, X., Alcañiz, J.-M., Liao, X., Palet, C., 2019. Comparison of biochars derived from different types of feedstock and their potential for heavy metal removal in multiple-metal solutions. *Scientific Reports*, 9, 9869. https://doi.org/10.1038/s41598-019-46234-4.

Zhou, J., Fan, X., Zhang, D., Tang, Y., Wang, X., Yuan, Z., Zhang, H., Zhang, J., 2022. Potential exploration of Fe_3O_4/biochar from sludge as the media of bioretention system and its comparison with conventional media. *Environmental Science and Pollution Research*, 29, 37906–37918. https://doi.org/10.1007/s11356-021-17334-4.

Zuo, X., Guo, Z., Wu, X., Yu, J., 2019. Diversity and metabolism effects of microorganisms in bioretention systems with sand, soil and fly ash. *Science of the Total Environment*, 676, 447–454. https://doi.org/10.1016/j.scitotenv.2019.04.340.

Dagenais, D., Brisson, J., Fletcher, T.D., 2018. The role of plants in bioretention systems; does the science underpin current guidance? Ecological Engineering 120, 532–545. https://doi.org/10.1016/j.ecoleng.2018.07.007.

Werbowski, L.M., Gilbreath, A.N., Munno, K., Zhu, X., Grbic, J., Wu, T., Sutton, R., Sedlak, M.D., Deshpande, A.D., Rochman, C.M., 2021. Urban Stormwater Runoff: A Major Pathway for Anthropogenic Particles, Black Rubbery Fragments, and Other Types of Microplastics to Urban Receiving Waters. *ACS EST Water* 1, 1420–1428. https://doi.org/10.1021/acsestwater.1c00017.

Spokas, K.A., Novak, J.M., Masiello, C.A., Johnson, M.G., Colosky, E.C., Ippolito, J.A., Trigo, C., 2014. Physical Disintegration of Biochar: An Overlooked Process. *Environ. Sci. Technol. Lett.* 1, 326–332. https://doi.org/10.1021/ez500199t.

6

Biochar Solution for Anaerobic Digestion

Yanfei Tang[1], Wenjing Tian[2], and Daniel C.W. Tsang[3]

[1] *College of Environmental Science and Engineering, Tongji University, Shanghai, China*
[2] *Institute of Environment and Ecology, Chongqing University, Chongqing, China*
[3] *State Key Laboratory of Clean Energy Utilization, Zhejiang University, China*

6.1 Introduction

The role of biochar (BC) as an additive for enhancing anaerobic digestion (AD) performance is attracting increasing attention. AD is one of the most promising waste management technologies for pollution control as well as the recovery of energy and nutrient from biomass. AD is a continuous biochemical process containing stages of hydrolysis, acidogenesis/acetogenesis, and methanogenesis. Issues such as low-flammable biogas (i.e., hydrogen, methane) production and process instability are often encountered in AD due to (1) ineffective hydrolysis-acidification, thus resulting in fewer available fermentative intermediates (e.g., volatile fatty acids (VFAs), alcohols, hydrogen (H_2), and formate), and (2) restricted microbial syntrophic interaction between fermentative bacteria and methanogens due to the inefficient interspecies electron transfer (IET). In addition, the productions of VFAs and H_2 essentially rely on the syntrophic metabolisms between fermentative bacteria and acidogenic bacteria. It was reported that eco-compatible biochar can serve as an effective additive for AD. It has at least five positive effects on AD (Chiappero et al., 2020):

1) Leaching minerals for buffering and nutrients
2) Adsorbing inhibitory compounds
3) Supporting microbial colonization
4) Enriching functional microorganisms

Biochar Applications for Wastewater Treatment, First Edition. Edited by Daniel C.W. Tsang and Yuqing Sun.
© 2023 John Wiley & Sons, Inc. Published 2023 by John Wiley & Sons, Inc.

5) Introducing direct interspecies electron transfer (DIET), whose electron transfer efficiency is 1000 times greater than that of IET

In terms of inducing DIET, BC may work through both surface redox active moieties and its battery behavior, thus avoiding diffusion limitation of electron carriers (i.e., H_2/formate) via electrical pili (e-pili) or other biological accessory structures (i.e., c-type cytochromes). The occurrence of DIET induced by BC provided a microcosmic (electron level) explanation for the mechanism of BC solution on AD development, and opened up a new research direction for further AD optimization and utilization. However, the sequence of the importance of BC's multiple roles needs further study, and more solid microbiome evidence is required to determine whether BC really does rapidly establish DIET.

In this chapter, the effects of BC on altering AD performance are first summarized, and then the relevant mechanisms are discussed. Lastly, this chapter presents the challenges and prospects of BC integration with AD, aiming to provide inspiration for future research and application of BC in AD.

6.2 Application of BC as an Additive in Anaerobic Digestion

6.2.1 pH Buffering

The volatile fatty acids (VFAs) produced as intermediates in AD tend to reduce the pH. Before the VFAs convert to methane and carbon dioxide via syntrophic reactions between acetogens and methanogens microorganisms, VFAs accumulation may occur when VFAs production rate exceeds the consumption rate, especially in case of high organic loads of easily biodegradable wastes. The serious pH drop would result in instability and even the failure of AD. The buffer capacity is determined by the alkalinity of AD, mainly in the form of carbon dioxide, bicarbonate, and ammonia. Several methods have been proposed to overcome this technological challenge; so far, the most diffused solution to improve buffer capacity implemented both in lab and full scale is co-digestion because of the high ammonia content in manure or sewage sludge or substrate.

However, it remains a huge challenge to develop simple, permanent, and cost-effective methods to improve buffer capacity of AD systems. Nowadays, BC addition can be regarded as an attractive alternative to the aforementioned methods since it can be produced via cost-effective and environmentally friendly approaches, moreover, its physiochemical properties can be matched with the operational conditions.

BC's buffer capacity mainly depends on two factors:

1) *Functional groups.* Rapid accumulation of VFAs during AD process results in a medium with low pH value in which some functional groups of BC like amine

adsorb H⁺ and accept electron. This phenomenon could mitigate the sudden pH drop.
2) *Inorganic materials.* Ash portion of BC contains inorganic materials such Ca, K, Mg, Na, Al, Fe, Si and S. Among them, alkali and alkaline earth metals (AAEMs) are responsible for alkalinity of BC via reaction (6.1) (Ca and C_xH_xCOOH are selected as representative of AAEMs and VFAs, respectively) (Chiappero et al., 2020):

$$CaCO_3 + 2C_xH_xCOOH \rightarrow [C_xH_xCOOH]2Ca + H_2O + CO_2 \qquad (6.1)$$

Most studies have suggested that the alkalinity of BC due to AAEMs in ash fraction could effectively contribute to the buffering capacity of AD against VFAs inhibition. As Jang et al. (2018) reported, during the AD of dry dairy manure, BC from high-nutrient dairy manure (9.1% Ca, 3.6% Mg, 1.3% N, 0.14% P) and alkalinity potential might lead to a lower total VFAs concentration and increased methane production. Wei et al. (2020) found enhanced methane production and solids removal by adding BC from corn stover (rich of alkaline earth metals) to AD of primary sludge. Moreover, extra dosages of trace metals may be required for an effective activity of methanogens without VFAs accumulation. Ambaye et al. (2020) found that the BC addition enhanced methane production and VFAs degradation and speculated that BC could provide adequate concentrations of trace metals for AD stability.

6.2.2 Adsorption of Inhibitors

In general, inhibition is defined as the predominant cause of reduction of methane yields and instability of AD system. A substance can be regarded as an "inhibitor" if it creates an adverse shift in the microbial community or arrests bacterial growth (Chen et al., 2008). The direct inhibitors can be excessive metals (Cu^{2+}, Zn^{2+}, Cr^{3+}, Cd, Ni, Pb^{4+}, Hg^{2+}, Na^+, K^+, Mg^{2+}, Ca^{2+}, Al^{3+}), refractory organic compounds and that with biological toxicity (chlorophenols, halogenated aliphatics, pesticides, antibiotics, lignocellulose hydrolysate). The indirect inhibitors generally include high content of VFAs, long-chain fatty acids, hydrogen (for AD system to produce methane), ammonium and sulphides (Fagbohungbe et al., 2017). Based on several studies focusing on BC impacts on AD and ammonia inhibition, it seems that BC is promising for mitigating ammonia inhibition, resulting in reduced lag phase and enhanced methane yields with respect to control reactors. Mumme et al. (2014) suggested that BC with the feedstock of paper sludge and wheat husks could relieve mild ammonia inhibition (2.1 g ammonia-N, kg^{-1}). Su et al. revealed the BC addition could alleviated the inhibition in case of upon 1500 mg L^{-1} ammonia-N in AD of food waste (FW). One of the best performances of BC solution for ammonia inhibition (up to 7 g-N L^{-1}) has been reported by Lu et al. (2016) (Figure 6.1). However, the details of this mechanisms are still not fully

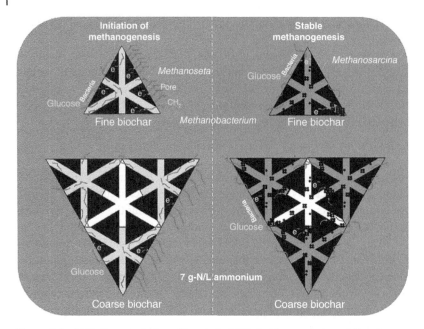

Figure 6.1 With the permission of Lu et al., 2016 / with permission of Elsevier.

clear, and the main hypotheses contain the physicochemical factors of cation exchange capacity, chemical and/or physical adsorption capacity, and surface functional groups; and also the microbial aspects of the promotion of direct interspecies electron transfer (DIET) and the immobilization of microorganisms.

6.2.3 Effects on Microbial Growth and Activities

6.2.3.1 Microbial Colonization and Functional Microbes Enrichment

Immobilization and acclimation of balanced microbial consortia on various support media are commonly adopted for enhancing AD, thanks to the intensification of syntrophic conversion relationships. Wang et al. (2018a) suggested that BC could act as an inert core for microbial aggregation, resulting in a higher microbial growth rate and accelerating sludge granulation. Cooney et al. (2016) confirmed the role of BC on accelerating biofilm formation during the startup of a packed bed anaerobic digester at pilot scale. This is notable, given that, in a relatively short time, the AD system reached stable and good performances. Likewise, Bu et al. (2021) reported that BC boosted hydrogen production by 317.1% mainly through stimulating bacterial growth and intensifying synergistic effect of sugars degradation/hydrogen production between *Lachnospiraceae*, *Clostridiaceae*, and *Pseudomonadaceae* (Figure 6.2).

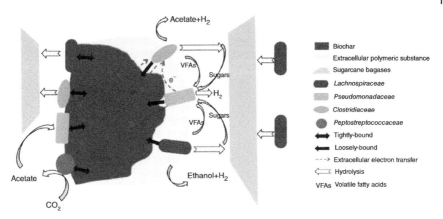

Figure 6.2 With the permission of Bu et al., 2021 / with permission of Elsevier.

For microbial immobilization, high SSA, TPV, and superficial hydrophobicity seemed to be important properties of BC. Interestingly, in terms of porosity structure and particle size, this should be proper since some authors reported that macropores can help the attachment of bacterial cells, while some authors suggest that bacteria could access more easily fine particles than coarse particles (Lu et al., 2016). Functionally, the attachment and colonization of microbial populations on BC can limit the risks of wash-out, accelerate the acclimation of microbes during substrate-induced inhibition, reduce the distance between syntrophic bacteria and methanogens, and facilitate interspecies electron transfer and exchanges of VFAs or other metabolites. Li et al. (2018) found that methanogens distinctly survived under acidic threatening with BC addition in co-digestion of food waste (FW) and waste activated sludge (WAS). Further, the immobilization of microbes can relieve the ammonia inhibition (Lu et al., 2016). One more reason for attenuating ammonia inhibition may be that the blockage of BC's pores due to the initial colonization and possible subsequent aggregation of *Methanosarcina* would protect cells located deeper in the pores from inhibitors. In addition, the colonization of porous materials by microbes can make the dominant species more resistant and more rapidly acclimatized to substrate-induced inhibition. For instance, magnetic BC favored the enrichment of acido/acetogens and methanogens absorbed on its surface shortening the microbial contact distance, thus VFAs produced by acido/acetogen bacteria could be more quickly transported to methanogens than in control digesters in AD of OFMSW in batch tests. The distance of less than 1 μm has been reported to be essential for the oxidation of VFAs and hydrogen production (Fagbohungbe et al., 2017; Stams, 1994).

The relevant anaerobic bacteria and archaea enriched in BC amended digestate are reported in Table 6.1, together with BC properties, substrate and inoculum

Table 6.1 A summary of representative studies reporting the selective enrichment of bacteria and archaea as well as relevant effect mechanisms by biochar addition during hydrogen-and methane-producing AD processes.

Biochar			Anaerobic digestion				
Feedstock	Production		Substrate	Effect mechanism	Enriched bacteria	Enriched archae	Reference
Fallen leaves	Pyrolysis: 300, 450, 600 °C (10 °C/min); 2h holding time; sieve to < 150 μm		Fermentative medium with carbon source of pretreated sugarcane bagasse (PSCB)	Improving critical enzymatic activities, manipulating the ratio of NADH/NAD+ and enhancing electron transfer efficiency	Lachnospiraceae, Pseudomonadaceae, Clostridiaceae, Peptostreptococcaceae	No data	(Bu et al., 2021)
Sawdust/ wheat bran/ peanut shell/ sewage sludge	Pyrolysis: 300/ 500/700 °C (15 °C min^{-1}); 1h holding time, ground to 0.25–1 mm		Food wastes and dewatered sewage sludge	Strong pH buffering capacity of biochar keep the stable pH during the VFAs accumulating and H$_2$ producing process	Some elongated rod-shaped bacteria were ammended on biochar	No data	(Wang et al., 2018d)
Pinewood	Pyrolysis: 650, 900 °C ; 20 min holding time; stainless steel ring milled to a size fraction of 3.5–25.9 μm)		White bread	Minerals in biochar increased VFAs and Clostridium butyricum proportion. Iron species in biochar play a key role in promoting hydrogen production	Clostridium butyricum	No data	

Material	Preparation	Substrate	Mechanism	Bacteria	Archaea	Reference
Rice straw and $FeCl_3 \cdot 6H_2O$	10 g rice straw and 10 g $FeCl_3 \cdot 6H_2O$ were mixed via ball milling at 300 rpm/min for 2 min, followed by a pause for 20 min (repeated 6 times) → 850 °C for 2h (5 °C/min) (namely MSBC)	Waste activated sludge	High electron transfer (ET) capability of MSBC with an intrinsic graphene-oxide-like structure and Fe species changed the extracellular ET pathway from indirect interspecies hydrogen transfer (IIHT) to direct interspecies electron transfer (DIET)	g_norank_c_Aceothermiia, Thermovirga	Methanoaseta, norank_o_Methanomicrobiales	(Liu et al., 2022)
Iron, magnesium, and chitosan (CTS)	Four modified-biochar: iron-loaded bagasse carbon (Fe + C), chitosan bagasse carbon (CTS + C), iron-chitosan bagasse carbon (Fe + C + CTS) and iron-magnesium-chitosan bagasse carbon (Fe + Mg + C + CTS)	Food waste was taken from a university canteen	Modified biochar with iron enhanced bacterial and archaeal energy metabolism, carbohydrate metabolism and membrane transport by the microorganisms	Anaerolineaceae distinguishly enriched under the biochar containing Fe to maintain the structural integrity of the anaerobic granular sludge	Methanothrix (benefit for direct interspecies electron transfer (DIET), may be related to the highest iron content in the chitosan-biochar)	(Su et al., 2021)
Corn straw	Pyrolysis; 550 °C, pure nitrogen atmosphere, sieve to a particle size of 0.5–1 mm	Sewage sludge and food waste	nZVI-BC facilitated the growth of hydrogentrophic methanogens mainly through the enhancement of DIET between bacteria and methanogens, and the enrichment of hydrogenotrophic methanogens	Firmicutes	Hydrogenotrophic methanogens of Methanobacterium and Methanobacteriaceae	(Zhang and Wang, 2021)

(Continued)

Table 6.1 (Continued)

Biochar			Anaerobic digestion				
Feedstock	Production		Substrate	Effect mechanism	Enriched bacteria	Enriched archae	Reference
Pinewood	Pyrolysis; 600 °C, pure nitrogen atmosphere, sieve to a particle size of 0.5–1 mm		Synthetic wastewater with butyrate and propionate	Under the enrichment conditions employed, DIET was only effective with biochar as a conduit, suggest that establishing DIET that relies on direct biological electrical connections may be more difficult with propionate or butyrate as the electron donor than has previously been observed for DIET with ethanol	*Geobacter* *Thermanaerovibrio* *Syntrophomonas*	*Methanosaeta* *Methanosarcina* *Methanospirillum*	(Zhao et al., 2016)
Fruitwood	Pyrolysis: 800–900 °C; sieve to 2–5 mm, 0.5–1 mm and 75–150 μm in size		Synthetic wastewater with three total ammonia nitrogen (TAN) "stress level" of 0.26, 3.5, 7.0 g/L	Hydrogenotrophic *Methanobacterium* was actively resistant to ammonium. However, acetoclastic *Methanosaeta* can survive at VFAs concentrations up to 60–80 mmol-C/L by improved affinity to conductive biochar. The selection of appropriate biochar particles sizes was important in facilitating the initial colonization of microbial cells	Enterobacteriaceae	*Methanobacterium* *Methanosarcina* *Methanosaeta*	(Lu et al., 2016)

used in AD tests. Many bacterial species were found in reactors supplemented with BC, none of them identified as more recurrent. Among archaea, most studies identified *methanosaeta, methanosarcina, methanobacterium,* and *methanolinea* species in BC amended reactors. Different studies [ref] investigated the spatial distribution of bacteria and archaea by dividing sludge samples into different fractions, from suspended to attached to BC. Lu et al. (2016) postulated an explanation for the spatial distribution of methanogens into BC pores by their cell morphology and dimension. The short fibrous form of Methanosaeta (0.8–7 μm in length) could explain its attachment into internal and external pores, while the long fibrous form of Methanobacterium (1.2–120 μm in length) could limit its penetration into BC pores [44, 166].

6.2.3.2 Direct Interspecies Electron Transfer (DIET) Acceleration

Many studies suggest that BC addition might improve electron transfer between anaerobic bacteria and archaea closely attached to BC surface. The overall AD efficiency depends on effective syntrophic interactions between bacteria and methanogens exchanging electrons to satisfy their energy requirements (Martins et al., 2018), happening through various routes:

- Indirect interspecies electron transfer (IIET) via soluble (i.e., hydrogen, formate, acetate) (Morris et al., 2013) and insoluble (humic substances) (Lovley et al., 1999) compounds
- Direct interspecies electron transfer (DIET) via electrical conductive pili, membrane-bound electron transport proteins, and conductive materials (i.e., magnetite, BC, granular activated carbon, carbon cloth) (Lovley, 2017)

In the IIET, hydrogen and formate act as electron shuttles between syntrophic-producing bacteria and consuming-methanogens (Martins et al., 2018). Diffusion regulates the transfer efficiency of a metabolite between microorganisms, as defined by Fick's law (Stams and Plugge, 2009): the shorter the distance, the higher the flux of metabolites between microbes. Therefore, if anaerobic bacteria and methanogenic archaea form compact structures likely acting as an organ, the cells aggregate and the rate of interspecies hydrogen transfer can be enhanced. However, the diffusion ways of soluble metabolites as energy and information transfer mediates are relatively slow and the hydrogen IIET is well agreed as a bottleneck in methane production.

On the other hand, DIET is superior at the formation of an electric current between electron-donating and electron-accepting microorganisms without the mediation of electron shuttles. Although DIET has only been observed in defined cocultures, it is likely to play an important role in mixed inocula with conductive materials, given that methane production of the bioreactors with conductive materials can be greatly strengthened with specific microbes enriched on the additives. Conductive materials such as magnetite, GAC, and BC were shown to successfully establish and effectively mediate DIET between syntrophic partners.

With the addition of conductive biochar, many studies focused on microbial community changes and justified the enhancement of AD activity by means of the improvement of hydrogen and formate interspecies transfer mechanisms or, more surprisingly, by DIET. However, most of these findings are only based on indirect observations, such as the enrichment of bacterial and archaeal species able to participate to DIET as potential ways. Liu et al. (2022) synthesized a magnetic-straw-based biochar and reported that the MSBC had an intrinsic graphene-oxide-like structure and high electron transfer (ET) capability, which could shift extracellular pathway from indirect interspecies hydrogen transfer to DIET via reinforcing methanogenesis pathway by creating a favorable environment for acetoclastic methanogens, which is usually fitted with DIET (Figure 6.3). Zhao [51] found the selective enrichment on BC of *Geobacter* and *Methanosaeta* during AD of synthetic wastewater with butyrate and propionate in UASB reactors. They suggested that butyrate and propionate could be degraded via DIET in the presence of a conductive material, and they found an abundance of *Syntrophomonas* and *Smithella*, stating that the metabolism via interspecies H_2 transfer for butyrate and propionate degradation was probably present. Wang et al. (2018c) revealed that BC addition brought about the enrichment of *Anaerolineaceae* and *Methanosaeta*, typical microorganisms for DIET, thus increasing methane production rate and shortening the lag phase during mesophilic AD of dewatered WAS and FW.

Aside from physical properties such as SSA and porosity, functional microbial colonization, the electrochemical characteristics of BC may be crucial in the promotion of electron transfer. The electrical conductivity (EC) of digestate has been

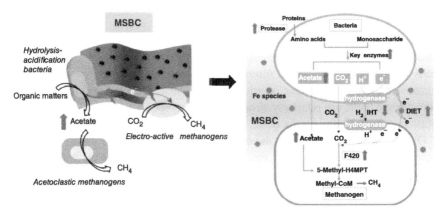

Figure 6.3 With the permission of Bu et al., 2021 / with permission of Elsevier.

well investigated for DIET in AD in literatures, and which could increase in presence of BC. However, the digestate EC seemed unrelated to the conductivity of BC, which may rather depend on the metabolism and composition of microbial species. The capability of BC in enhancing DIET appeared to be comparable to that of GAC, even if the EC of BC was roughly 1000 times lower. Wang et al. (2018b) suggested that BC from sawdust may act as a temporary electron acceptor for VFAs oxidation during thermophilic AD. Further, for better investigating whether BC from sawdust might substitute hydrogen as an electron acceptor in syntrophic oxidation of VFAs, they conducted a series of batch experiments with butyrate or propionate as substrates in which methanogenesis was inhibited. In comparison, in control AD the butyrate and propionate were not degraded while the experimental AD with BC addition stimulated their oxidation as well as the production of acetate, supporting the hypothesis of electron-accepting capacity of BC in the syntrophic process.

The proper design is important to clarify which role of BC (physical properties, i.e., SSA and porosity, vs. electrical properties) on methane production is more crucial. Cruz Viggi et al. (2017) introduced two controls without BC and with nonconductive silica sand for AD of FW. Notably, they found that VFAs degradation and methane generation were faster in the case of BC amended reactors than both the control reactors, indicating the predominant influence of the electrical properties of BCs.

6.3 Effects of BC on Digestate Quality

Anaerobic digestate has been considered as a soil improver because it is rich in nutrients. Challenges related to digestate management have recently grown in association with EU regulation on phytotoxic factors of ammonia, volatile organic acids, phenolic compounds, heavy metals, PAHs, and PCBs. Nowadays, most technologies available to exploit digestate as soil improver are based on mechanical or physical (e.g., mechanical dewatering, semipermeable membranes and evaporation) or chemical (e.g., ammonia stripping and nutrients adjustment) processes.

Therefore, one of the key aspects that should be considered when supplying additives to AD process is their effects on the quality of digestate for subsequent uses, particularly as soil improver or conditioner. Although literature has not fully explored the fate and properties of digestate with BC addition to land applications, some potential benefits of BC amendment can be identified as follows. In addition, BC remaining in digestate after AD acts as a nutrient retention helper and catalyst. It mitigates leaching of heavy metals and pollutants via physical and chemical absorption of organics (e.g., humic acids and/or other micromolecular biostimulants), phosphate, ammonium, nitrate, nitrite, metals, and CO_2 (Andrey Bagreev 2001; Fagbohungbe et al., 2017). The effects on digestate quality can be related to BC features such as SSA, surface functional groups, ash content, and presence of metals.

6.4 Conclusions and Perspectives

This chapter addressed three key issues related to the comprehension of BC role in enhancing AD processes:

1) *BC properties on AD performances and their ability to counteract AD challenges.* The primary BC features are SSA, porous structure and distribution, nature of surface functional groups (related to CEC and adsorption capacity, buffer capacity, ability to immobilize microbial communities), elemental composition, and ash content. However, some microscopic mechanisms (e.g., BC role of ammonium adsorption and BC influence on microbial redox activities) are still not fully understood. Besides, other challenges are related to digestate management. For instance, the influence of BC at high dosages on the rheological properties of the digestate should be explored, as well as eventual leaching of pollutants in the environment in case of digestate application as soil improver.
2) *Investigation of optimal BC production chain (i.e., feedstock-pyrolysis-activation) to achieve the desired features.* Lignocellulosic biomasses, slow pyrolysis, and physical activation seem to be a good combination in general, while other feedstocks and/or chemical activation should be evaluated for specific needs and tailor-made applications. Nevertheless, a systematic measurement of the correlations linking BC physiochemical characteristics and AD performances, carefully exploring one by one the BC roles mentioned in this chapter, is greatly needed for a deeper understanding.
3) *Economic and environmental advantages connected to BC use in AD processes.* For this issue, there is a lack of literature related to BC use in AD processes; therefore, only general statements can be formulated. BC production can be cheaper and its use has less environmental impacts than conventional AD improvers (e.g., physiochemical pretreatments and GAC addition). The integration of AD and pyrolysis achieved economic feasibility and positive environmental behaviors when compared with nonintegrated processes. Future studies are expected to investigate the optimization of technical, economic, and environmental performances of BC production chain and its integration in AD systems.

References

Ambaye, T.G., Rene, E.R., Dupont, C., Wongrod, S., van Hullebusch, E.D., 2020. Anaerobic digestion of fruit waste mixed with sewage sludge digestate biochar: Influence on biomethane production. *Frontiers in Energy Research*, 8, 31.

Andrey Bagreev, T.J.B. and Locke, D.C., 2001. Pore structure and surface chemistry of adsorbents obtained by pyrolysis of sewage sludge-derived fertilizer. *Carbon*, 39, 1971–1979.

Bu, J., Wei, H.L., Wang, Y.T., Cheng, J.R., Zhu, M.J., 2021. Biochar boosts dark fermentative H_2 production from sugarcane bagasse by selective enrichment/colonization of functional bacteria and enhancing extracellular electron transfer. *Water Research*, 202, 117440.

Chen, Y., Cheng, J.J., Creamer, K.S., 2008. Inhibition of anaerobic digestion process: A review. *Bioresour Technology*, 99(10), 4044–4064.

Chiappero, M., Norouzi, O., Hu, M., Demichelis, F., Berruti, F., Di Maria, F., Mašek, O., Fiore, S., 2020. Review of biochar role as additive in anaerobic digestion processes. *Renewable and Sustainable Energy Reviews*, 131, 110037.

Cooney, M.J., Lewis, K., Harris, K., Zhang, Q., Yan, T., 2016. Start up performance of biochar packed bed anaerobic digesters. *Journal of Water Process Engineering*, 9, e7–e13.

Cruz Viggi, C., Simonetti, S., Palma, E., Pagliaccia, P., Braguglia, C., Fazi, S., Baronti, S., Navarra, M.A., Pettiti, I., Koch, C., Harnisch, F., Aulenta, F., 2017. Enhancing methane production from food waste fermentate using biochar: The added value of electrochemical testing in pre-selecting the most effective type of biochar. *Biotechnol Biofuels*, 10, 303.

Fagbohungbe, M.O., Herbert, B.M., Hurst, L., Ibeto, C.N., Li, H., Usmani, S.Q., Semple, K.T., 2017. The challenges of anaerobic digestion and the role of biochar in optimizing anaerobic digestion. *Waste Management*, 61, 236–249.

Jang, H.M., Choi, Y.K., Kan, E., 2018. Effects of dairy manure-derived biochar on psychrophilic, mesophilic and thermophilic anaerobic digestions of dairy manure. *Bioresource Technology*, 250, 927–931.

Li, Q., Xu, M., Wang, G., Chen, R., Qiao, W., Wang, X., 2018. Biochar assisted thermophilic codigestion of food waste and waste activated sludge under high feedstock to seed sludge ratio in batch experiment. *Bioresource Technology*, 249, 1009–1016.

Liu, H., Xu, Y., Li, L., Yuan, S., Geng, H., Tang, Y., Dai, X., 2022. A novel green composite conductive material enhancing anaerobic digestion of waste activated sludge via improving electron transfer and metabolic activity. *Water Research*, 220, 118687.

Lovley, D.R., 2017. Happy together: Microbial communities that hook up to swap electrons. *ISME Journal*, 11(2), 327–336.

Lovley, D.R., Fraga, J.L., Coates, J.D., Blunt-Harris, E.L., 1999. Humics as an electron donor for anaerobic respiration. *Environmental Microbiology*, 1(1), 89–98.

Lu, F., Luo, C., Shao, L., He, P., 2016. Biochar alleviates combined stress of ammonium and acids by firstly enriching Methanosaeta and then Methanosarcina. *Water Research*, 90, 34–43.

Martins, G., Salvador, A.F., Pereira, L., Alves, M.M., 2018. Methane production and conductive materials: A critical review. *Environmental Science & Technology*, 52(18), 10241–10253.

Morris, B.E., Henneberger, R., Huber, H., Moissl-Eichinger, C., 2013. Microbial syntrophy: Interaction for the common good. *FEMS Microbiology Reviews*, 37(3), 384–406.

Mumme, J., Srocke, F., Heeg, K., Werner, M., 2014. Use of biochars in anaerobic digestion. *Bioresource Technology*, 164, 189–197.

Stams, A.J. and Plugge, C.M., 2009. Electron transfer in syntrophic communities of anaerobic bacteria and archaea. *Nat Reviews Microbiology*, 7(8), 568–577.

Stams, A.J.M., 1994. Metabolic interactions between anaerobic bacteria in methanogenic environments. *Antonie Van Leeuwenhoek*, 66(1), 271–294.

Su, C., Tao, A., Zhao, L., Wang, P., Wang, A., Huang, X. and Chen, M., 2021. Roles of modified biochar in the performance, sludge characteristics, and microbial community features of anaerobic reactor for treatment food waste. *Science of the Total Environment*, 770, 144668.

Wang, C., Liu, Y., Gao, X., Chen, H., Xu, X., Zhu, L., 2018a. Role of biochar in the granulation of anaerobic sludge and improvement of electron transfer characteristics. *Bioresource Technology*, 268, 28–35.

Wang, G., Li, Q., Gao, X., Wang, X.C., 2018b. Sawdust-derived biochar much mitigates VFAs accumulation and improves microbial activities to enhance methane production in thermophilic anaerobic digestion. *ACS Sustainable Chemistry & Engineering*, 7(2), 2141–2150.

Wang, G., Li, Q., Gao, X., Wang, X.C., 2018c. Synergetic promotion of syntrophic methane production from anaerobic digestion of complex organic wastes by biochar: Performance and associated mechanisms. *Bioresource Technology*, 250, 812–820.

Wang, G., Li, Q., Dzakpasu, M., Gao, X., Yuwen, C. and Wang, X.C., 2018d. Impacts of different biochar types on hydrogen production promotion during fermentative co-digestion of food wastes and dewatered sewage sludge. *Waste Management*, 80, 73–80.

Wei, W., Guo, W., Ngo, H.H., Mannina, G., Wang, D., Chen, X., Liu, Y., Peng, L., Ni, B.J., 2020. Enhanced high-quality biomethane production from anaerobic digestion of primary sludge by corn stover biochar. *Bioresource Technology*, 306, 123159.

Zhang, M. and Wang, Y., 2021. Impact of biochar supported nano zero-valent iron on anaerobic co-digestion of sewage sludge and food waste: Methane production, performance stability and microbial community structure. *Bioresource Technology*, 340, 125715.

Zhao, Z., Zhang, Y., Holmes, D.E., Dang, Y., Woodard, T.L., Nevin, K.P. and Lovley, D.R., 2016. Potential enhancement of direct interspecies electron transfer for syntrophic metabolism of propionate and butyrate with biochar in up-flow anaerobic sludge blanket reactors. *Bioresource Technology*, 209, 148–156.

7

Biochar-Assisted Anaerobic Ammonium Oxidation

Wenjing Tian[1], Yanfei Tang[2], Dongdong Ge[3], and Daniel C.W. Tsang[4]

[1] Institute of Environment and Ecology, Chongqing University, Chongqing, China
[2] College of Environmental Science and Engineering, Tongji University, Shanghai, China
[3] School of Environmental Science & Engineering, Shanghai Jiao Tong University, Shanghai, China
[4] State Key Laboratory of Clean Energy Utilization, Zhejiang University, China

7.1 Overview of Anaerobic Ammonium Oxidation

7.1.1 Introduction

Anaerobic ammonium oxidation (anammox) is a chemoautotrophic biological process, during which nitrates and nitrites can accept electrons generated by the ammonium oxidation, allowing the conversion of ammonium to nitrogen, as shown in Eq. (1) and Figure 7.1 (Chen et al., 2022a). Anammox process take a vital place in marine nitrogen circulation, and it is estimated that over 50% of the nitrogen in the marine ecosystem is generated from the anammox process. As a sustainable and eco-friendly biological nitrogen removal technology (Zhang et al., 2022a), anammox has been gaining extensively interest in the field of wastewater treatment since it was first discovered in a denitrification fluidized bed reactor in 1995 (Mulder et al., 1995). Compared to traditional nitrification-denitrification technology, anammox has plentiful low-carbon and cost-effective advantages of low oxygen demand, no consumption of organic carbon sources, less sludge production, and smaller space requirements (Gao et al., 2023; Zhang et al., 2022b). In addition, Arora et al. (2019) reported that their anammox-based wastewater treatment system could achieve the high nitrogen removal rate (NRR) of 81% along with the decrement of 68% concomitant in green gases (GHGs) emissions (Arora et al., 2019). Han et al. (2020) reported that on account of the decreased requirements in organic addition, the combination of partial nitrification and anammox could save $3.5 million in operational costs annually. Therefore, anammox has been wildly applied in the treatment of wastewater with high ammonia nitrogen and low C/N ratio, such as sludge digestate, landfill leachate, coking wastewater, feed processing wastewater, pharmaceutical wastewater, and other kinds of

Biochar Applications for Wastewater Treatment, First Edition. Edited by Daniel C.W. Tsang and Yuqing Sun.
© 2023 John Wiley & Sons, Inc. Published 2023 by John Wiley & Sons, Inc.

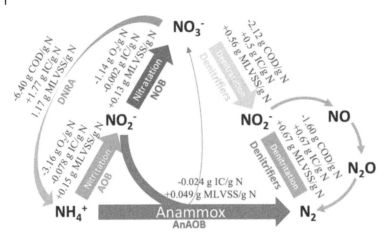

Figure 7.1 Nitrogen circulation with permission from ref (Chen et al., 2022 / with permission of Elsevier) Copyright (2020) Elsevier. (AnAOB refers to anammox bacteria; AOB refers to ammonium oxidizing bacteria; NOB refers to nitrite oxidizing bacteria; DNRA refers to dissimilatory nitrate reduction to ammonium; COD refers to chemical oxygen demand; IC refers to inorganic carbon; MLVSS refers to mixed liquor volatile suspended solids).

wastewater. According to the reports, the world's first full-scale anammox reactor was started up in the Rotterdam-Dokhaven Sewage Treatment Plant in the Netherlands in 2002 (Van der Star et al., 2007), and currently there are more than 200 practical anammox projects in operation worldwide (Cao et al., 2017).

$$NH_4^+ + 1.32\ NO_2^- + 0.066\ HCO_3^- + 0.13\ H^+ \rightarrow 1.02\ N_2 + 0.26\ NO \qquad (7.1)$$

The operation of anammox process is highly dependent on the activity of anammox bacteria (AnAOB), which are irregular globular Gram-negative bacteria with an average diameter of 0.8 μm to 1.1 μm and appear to the naked eye as a distinctive red color (Kang et al., 2018; Li et al., 2022a). At present, six genera of AnAOB have been identified by the molecular biological techniques: *Candidatus Brocadia, Candidatus Scalindua, Candidatus Kuenenia, Candidatus Anammoximicrobium, Candidatus Anammoxoglobus,* and *Candidatus Jettenia,* respectively (Ren et al., 2022). All of them belong to the phylum *Planctomycetes.* Although AnAOB could be wildly found in the anoxic natural environment (e.g., marine sediment, soil, lake wetland, and other oxygen minimum zones) (Ding et al., 2022; Karthäuser et al., 2021; Negi et al., 2022; Rios-Del Toro et al., 2017), there is no successful case of pure culture of AnAOB on a laboratory scale because of its sensitive growth characteristics (Ji et al., 2022). In addition, the widespread inhibitors in the nitrogen-rich wastewater put obstacles in the stable implementation of anammox process. Therefore, the large-scale application and further industrialization of anammox technology are currently subject to many constraints in the biological process.

7.1.2 Constraints

7.1.2.1 Inhibitors in Wastewater

In recent decades, pervasive anthropogenic activities sparked off excessive discharge of hazardous pollutants into the environment, so the wastewater collected in the sewage treatment plants commonly contains a lot of toxic substances for AnAOB, including ammonia, nitrite, heavy metals, phosphate, sulfide, salts, and some organic matter (e.g., aldehydes, phenols, and antibiotics) (Jin et al., 2012).

According to the earlier studies, when the anammox sludge was exposed to the ambient concentration of 38 mg/L free ammonia (FA) for a short period, its activity was dramatically reduced by half (Fernández et al., 2012). Further increases in exposure time to 281 days, the reactor performance became extremely unstable with the FA levels more than 35–40 mg/L (Fernández et al., 2012). The inhibition of high FA level is attributed to its superior diffusion capacity through cytomembrane, which could change the intracellular pH values and even cause cell apoptosis (Kadam and Boone, 1996). Besides, previous literature pointed out that the existence of heavy metals is another inhibition for anammox process, which could enter the cell due to the structural similarity, further displacing the binding sites of essential metals to nucleic acid and enzyme (Zhang et al., 2022a). Moreover, Zhang et al. claimed that the anammox reactors would collapse quickly as exposed to 3 mg/L of Zn^{2+} and Cu^{2+} (Zhang et al., 2022a).

7.1.2.2 Difficulties in AnAOB Enrichment

AnAOB are susceptible to the variations of external environment, including pH changes, temperature fluctuation, and oxygen exposure (Chen et al., 2022b). Their suitable pH range is narrow and basically needs to stay neutral (Meng et al., 2022). As a kind of strictly anaerobic chemoautotrophic bacteria, AnAOB has low tolerance to the dissolved oxygen (DO), whose activity could be inhibited at a low DO level (Yang et al., 2022), so it is necessary to strictly control the operational parameters of anammox process. But the complexity of sewage feed and the changes in seasonal temperature add to the intractability of parameter control (Li et al., 2022b; Rong et al., 2022). Meanwhile, AnAOB have a very low growth rate along with a long doubling time of around 10–14 days (Chen et al., 2022a). Under such conditions, it is technically difficult to enrich AnAOB; therefore, adding the startup and stabilization time for the anammox system, which could range from tens to hundreds of days (Zhang et al., 2021). Besides, the nitrogen-rich wastewater itself usually contains a lot of organic matter, which is conducive to the dominance of heterotrophic microorganisms and further forms dilution effects on autotrophs such as AnAOB (Lou et al., 2022). In addition, AnAOB-rich inoculum that can be used for autotrophic denitrification of municipal sewage is extremely scarce worldwide due to the technical barriers (Ji et al., 2022). Hence, there still remains one of the biggest challenges for the anammox process – how to rapidly improve the in-situ enrichment rate and biological interception of AnAOB in bioreactors.

7.1.2.3 Interactions among Different Microbial Groups

As illustrated in Figure 7.1, the anammox-based process is a representative multi-substrate and multi-culture bioprocess. There are four major microbial groups involved, including NOB, AOB, AnAOB, and heterotrophic bacteria (Cao et al., 2017). Both competitive and synergistic relationships exist among different microbial groups. On the one hand, AOB competes with AnAOB for ammonium. Heterotrophic denitrifiers, NOB, and AnAOB fight for nitrites. It should be noted that no matter whether NOB, AOB, or heterotrophic bacteria, their tolerance to environmental stress and growth rate were superior to AnAOB (Chen et al., 2023; Rong et al., 2022). For example, the growth of AnAOB would be significantly inhibited at low and moderate temperature from around 10–30 °C, while NOB and AOB could still stay active at relatively high levels (Cao et al., 2017). As a result, AnAOB is easily at a weak position in substrate competition, especially when the external environment changes, which could severely reduce the nitrogen removal efficiency and even cause the failure of anammox bioreactors. On the other hand, heterotrophic denitrifiers are responsible for accepting electrons from ammonia oxidation, thus allowing the stable metabolism of AnAOB (Cui et al., 2021). Previous studies pointed out that one of the intrinsic limitations of the anammox process is rooted in the difficulties to maintain the dynamic growth balance of AnAOB and other microbial groups, which requires systematic regulation and the assistance of exogenous additives (Ren et al., 2022; Zhang et al., 2022b).

7.2 Roles of Biochar in Promoting Anammox

In terms of the abovementioned challenges, biochar, a kind of formless carbonaceous material, has been deemed to a promising candidate to promote anammox robustness against environmental stress, due to their inherent merits, including favorable adsorption capacity, biological compatibility, highly developed pore structure, and rich redox-active sites (as shown in the Figure 7.2 and Table 7.1) (He et al., 2022; Xu et al., 2021). In recent years, the redox properties of biochar on mediating the electron transfer chain involved in biological process has raised extensive interest. Besides, the waste-derived and facile synthesis properties of biochar satisfy the low-carbon pursuit. Several studies have pointed out that the porosity of biochar endows its superior adsorption capacity for harmful substances for AnAOB, like heavy metals, antibiotics, sulfite, phosphate, and acids (Xie et al., 2022a). In addition, biochar surface could provide suitable habitats for

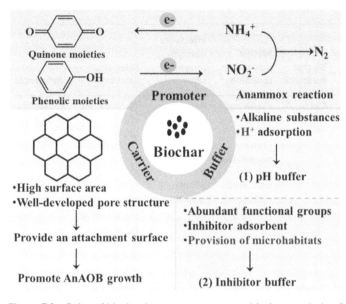

Figure 7.2 Roles of biochar in anammox system with the permission from ref (Zhang et al., 2022 / with permission of Elsevier).

Table 7.1 Application of biochar as exogenous additives in the anammox-based system.

Wastewater type	System	Biochar feedstock	Biochar preparation condition	Effects on nitrogen removal rates	Roles of biochar	References
Synthetic wastewater	Anammox	Bamboo	Pyrolyzed at 400 °C for 2.5 h under the pure nitrogen atmosphere	5.6%–18.0% promotion compared to the control group.	Facilitate the anammox biomass to secrete redoxactive EPS. Promote the extracellular electron transfer. Enrich the AnAOB abundance.	(Wang et al., 2022a)
Synthetic wastewater	Anammox	Sewage sludge	Nil	Reach 69.5%–90.9% under the optimum conditions.	Provide attachment sites for the AnAOB.	(Mojiri et al., 2020)

(Continued)

Table 7.1 (Continued)

Wastewater type	System	Biochar feedstock	Biochar preparation condition	Effects on nitrogen removal rates	Roles of biochar	References
Synthetic wastewater	Anammox	Corn stover	Pyrolyzed at 300 °C for 2.5 h under oxygen-free conditions with argon cycling gas protection	Average nitrogen removal efficiencies of 97.9%, 97.5% and 87.9%, respectively	Decrease the effluent nitrate concentration by 40.7–45.2%. Enhance the EPS electroactivity Increase anammox functional genes copies.	(Xu et al., 2021)
Synthetic wastewater	Anammox	Bamboo, peach and coconut	Nil	95.4%–98%	Enrich the AnAOB	(Adams et al., 2021)
Synthetic wastewater	Anammox	Litchi and chinar	Nil	87.39%–88.39%	Promote the maintenance of anammox biomass, bacterial activity and shock resistance capacity.	(Xie et al., 2022b)
Synthetic wastewater	Anammox	Bamboo	Nil	$92.9 \pm 1.1\%$ and $93.7 \pm 1.2\%$	Promote the extracellular electron transfer. Increase the anammox-related genes abundance	(Wang et al., 2022b)
Synthetic wastewater	Anammox	Bamboo	Nil	Over 98%	Improve the anammox start-up	(Chen et al., 2012)
Landfill leachate and domestic sewage	Anammox	Coconut and peach	Nil	Average nitrogen removal rates of 83% and 85%, respectively	Increase the anammox-related functional genes abundance.	(Xie et al., 2022a)

Table 7.1 (Continued)

Wastewater type	System	Biochar feedstock	Biochar preparation condition	Effects on nitrogen removal rates	Roles of biochar	References
Synthetic wastewater	Anammox	Sawdust	Pyrolyzed at 300 °C for 2 h	Increase to 95.4% as compared to 87.3% of the control groups.	Donate the electron to promote the nitrate reduction. Mediate the denitrifying cellular metabolism by inducing protein changes.	(Li et al., 2022b)
Synthetic wastewater	Anammox	Coconut shells	Pyrolyzed at 300, 500 and 700 °C for 2 h under the nitrogen atmosphere and then added into a mixture solution of $FeCl_2$ and $FeCl_3$ to obtain magnetic property	Increase the NH_4^+-N (28.5–38.5%) and NO_2^--N (36.6–60.4%) removal rates as compared to 19.7% and 39.9% of the control groups.	Provide redox active sites for the efficient nitrogen biological removal. Mediate the denitrifying cellular metabolism by inducing protein changes.	(An et al., 2022)

microbial aggregation and even has the potential to facilitate the interaction among different microbial groups (Adams et al., 2021; Qi et al., 2021). The roles of biochar in promoting anammox are discussed in the following subsections.

7.2.1 pH and Inhibitor Buffer

The stability of an anammox system is easily subject to the external condition stress, where a fluctuation in the solution pH or some inhibitors introduced with the wastewater feed might severely hurt microbial functioning (Tomaszewski et al., 2017; Zhang et al., 2020). A narrow pH range of 7.2–7.6 was recommended for AnAOB growth, and the environmental pH would also influence the concentrations of potentially toxic matter (such as free ammonia and nitrite) in the anammox substrate (Tomaszewski et al., 2017). For example, the over-threshold pH would increase the free ammonia concentration, while there was an increase in the nitrite level under low-pH conditions (Tomaszewski et al., 2017). As a versatile buffer, biochar has a wild pH range of 5–12 and stable alkalinity releasing capacity, which indicated its flexible application in different reactors with unstable pH (Kumar et al., 2021). The adsorption ability of biochar on inhibitors has been extensively reported on enhancing the anammox process, due to its diverse structure characteristics, including π-π interaction from the aromatic structures, functional groups, and porous nature (Mojiri et al., 2020; Xie et al., 2022b).

7.2.2 Electron Transfer Promotion

It is reported that some functional groups with redox properties, such as quinone moieties, have been identified to mediate the electron transfer linked to the AnAOB metabolism (Chen et al., 2022a; Kumar et al., 2021). Previous literature reported that the redox active property of biochar was embodied in the electron exchanging capacity, which was closely associated with the surface functional groups of biochar (Xu et al., 2021b). For example, Chen et al. (2018) found that the rice straw biochar prepared under low temperature (i. e. 300 °C) has rich phenol hydroxyl groups, which could facilitate the reduction of nitrate via donating electrons (Chen et al., 2018). A similar promotion of low-temperature biochar on the anammox process was observed in the work of Xu et al., and they guessed that this might be due to an improvement in the NADH/NAD$^+$ ratio, which contributed to the metabolism of AnAOB (Figure 7.3) (Xu et al., 2021b, 2020). Also, the electrons donated from the biochar were likely to participate in the anammox reactions with a relatively slow kinetic rate (Xu et al., 2020). By contrast, the biochar pyrolyzed at 800 °C has superior electron-donating ability due to the disappearance of surface functional groups and the formation of graphite structure, which would compete with ammonium for the electron generating by the denitrification process (Chen et al., 2018). The medium-temperature biochar had a relative equal electron accepting and donating capacity, thus acting as an electron shuttle in the coupling of the anammox and iron reduction processes (Zhou et al., 2016).

Furthermore, biochar could not only mediate the microbial metabolism during the anammox process through direct electron exchanging power but also pose considerable impacts on the electron transport involved in the sludge extracellular polymeric substances (EPS) (Xu et al., 2021a). EPS, including the protein, nucleic acids, lipids, polysaccharides, and other polymeric compounds released by microorganisms, is prevalent in the biochemical process in the natural environment. It is reported that EPS plays a decisive role in the process of anammox sludge aggregation and electron transfer outside microbial cells (Hou et al., 2015). Xu et al. (2021) reported that biochar could contributed to the electron transport involved in the EPS, while the electrochemical results showed that the promoted

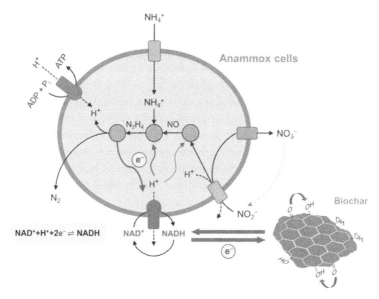

Figure 7.3 The proposed effects of redox active biochar on the anammox process with the permission from ref (Xu et al., 2020 / with permission of Elsevier).

extent of biochar is negatively correlated to its particle size (Xu et al., 2021a). The average nitrogen removal efficiency of 10–30 um corn straw biochar reached 97.9%, which was significantly higher than that of the control group (80.5%). Therefore, the properties of biochar (like surface functional groups and particle size) seem important when using its electron transfer mediating capacities to improve the anammox process.

7.2.3 Microbial Immobilization

Microbial loss has always been a ubiquitous problem that adds to the difficulty of starting anammox process (Li et al., 2022c). Therefore, one of the critical points in stabilizing the anammox process is to maintain high-level microbial interception. The sufficient internal surface and porous structure of biochar could be the favorable attachment sites for vulnerable AnAOB, which is beneficial to its enrichment and further improve the whole system stability (Mojiri et al., 2020). According to the earlier study, there was a significant improvement in the *Planctomycetes* (a typical kind of AnAOB) abundance by a maximum of 54.5% with the jujube biochar addition (Xie et al., 2022b). As a result, the ammonia nitrogen and total nitrogen removal efficiencies increased by 84.11% and 87.92%, respectively, in the jujube biochar-added reactors. Similar promotion ability of biochar on improving the abundance in functional microbial groups was also observed by Xie et al. (2022b).

As a carrier for microbial immobilization, biochar could also enhance its adaptability toward environmental changes and substrate inhibition (Wang et al., 2022a). It is reported that when the COD concentration reached 250 mg/L without biochar addition, the functional microorganisms in the anammox process showed rapid decrease along with the obvious decline of NRR from 89% to 72%, indicating the adverse impact of excessive organic carbon on AnAOB activity (Xie et al., 2022b). But the addition of biochar promoted the AnAOB resistance and maintained the reactor performance at a relatively high level. Furthermore, in the biochar-assisted anammox system, the microbial aggregates consisting of AnAOB and other symbiotic consortia would be more likely to form as biochar provided the suitable colonization zones for microorganisms, which could therefore facilitate the interaction among microbial communities (Adams et al., 2021).

7.3 Future Perspectives

Many studies have reported the great potential and versatile roles of biochar on promoting anammox process (Adams et al., 2021; Chen et al., 2018; Li et al., 2022b). First, biochar functions as a stabilizer to guarantee the suitable conditions of AnAOB, because on the one hand, its abundant porous and conjugated aromatic structure could adsorb the inhibitors in the environment (He et al., 2022). On the other hand, it has diverse surface functional groups and high ash contents, which could facilitate the stabilization of environmental pH value. Besides, biochar plays an enhancer role in the anammox system, which could not only contribute to the microbial interaction by providing favorable habitats but also facilitate the electron transfer involved in the anammox-related metabolism.

However, there still exist technical difficulties for pure culture of AnAOB, so that we can't get explicit information on its growth characteristics and complete metabolism pathways. Therefore, this may also be the reason why there are still many questions and black boxes regarding the effects of biochar on the anammox process. For example, some kinds of biochar, especially those derived from sewage sludge, contains a lot of heavy metals, like Zn^{2+}, Mn^{2+}, and Cd^{2+}, which are likely to be released to disturb the microbial functions (Mojiri et al., 2020). Meanwhile, it has been detected that biochar could introduce some environmentally persistent free radicals (EPFRs), resulting in the generation of reactive oxygen species, which might attack the cells and hinder the anammox process (Liu et al., 2022).

Despite the challenges, biochar-assisted anammox process still promising and deserves more attention. On the one hand, biochar is derived from biomass waste, of which the generation is a value-added process. The application of biochar on improving anammox system could further facilitate circular bioeconomy and exhibit excellent sustainability, which is a hopeful attempt to kill two birds with one stone. Many existing studies have demonstrated that the redox properties and adsorption capacities of biochar could pose considerable effects on the stabilization of anammox process, so more initiatives are needed for the modification of biochar or combination with other materials in the biochar-assisted anammox system.

References

Adams, M., Xie, J., Chang, Y., Kabore, A.W.J., Chen, C., 2021. Start-up of anammox systems with different biochar amendment: Process characteristics and microbial community. *Science of the Total Environment*, 790, 148242.

An, T., Chang, Y., Xie, J., Cao, Q., Liu, Y., Chen, C., 2022. Deciphering physicochemical properties and enhanced microbial electron transfer capacity by magnetic biochar. *Bioresource Technology*, 363, 127894.

Arora, A.S., Nawaz, A., Yun, C.M., Cho, H., Lee, M., 2019. Ecofriendly anaerobic ammonium oxidation system: Optimum operation and inhibition control strategies for enhanced nitrogen removal. *Industrial and Engineering Chemistry Research*, 58(45), 20847–20856.

Cao, Y., van Loosdrecht, M., Daigger, G.T., 2017. Mainstream partial nitritation–anammox in municipal wastewater treatment: Status, bottlenecks, and further studies. *Applied Microbiology and Biotechnology*, 101(4), 1365–1383.

Chen, C., Huang, X., Lei, C., Zhu, W., Chen, Y., Wu, W., 2012. Improving anammox start-up with bamboo charcoal. *Chemosphere*, 89(10), 1224–1229.

Chen, G., Zhang, Z., Zhang, Z., Zhang, R., 2018. Redox-active reactions in denitrification provided by biochars pyrolyzed at different temperatures. *Science of the Total Environment*, 615, 1547–1556.

Chen, Y., Guo, G., Li, Y., 2022a. A review on upgrading of the anammox-based nitrogen removal processes: Performance, stability, and control strategies. *Bioresource Technology*, 364, 127992.

Chen, Y., Guo, Y., Feng, G., Urasaki, K., Guo, G., Qin, Y., Kubota, K., Li, Y., 2023. Key factors improving the stability and the loading capacity of nitrogen removal in a hydroxyapatite (HAP)-enhanced one-stage partial nitrification/anammox process. *Chemical Engineering Journal*, 452, 139589.

Chen, Z., Meng, F., Zhou, C., Wu, X., Jin, C., 2022b. Optimum relative frequency and fluctuating substrate selection in reinforcing anammox-mediated anabolic adaptation. *Water Research*, 228, 119377.

Cui, H., Zhang, L., Zhang, Q., Li, X., Huang, Y., Peng, Y., 2021. Advanced nitrogen removal from low C/N municipal wastewater by combining partial nitrification-anammox and endogenous partial denitrification-anammox (PN/A-EPD/A) process in a single-stage reactor. *Bioresource Technology*, 339, 125501.

Ding, B., Li, Z., Cai, M., Lu, M., Liu, W., 2022. Feammox is more important than anammox in anaerobic ammonium loss in farmland soils around Lake Taihu, China. *Chemosphere*, 305, 135412.

Fernández, I., Dosta, J., Fajardo, C., Campos, J.L., Mosquera-Corral, A., Méndez, R., 2012. Short-and long-term effects of ammonium and nitrite on the Anammox process. *Journal of Environmental Management*, 95, S170–S174.

Gao, X., Zhang, L., Peng, Y., Ding, J., An, Z., 2023. The successful integration of anammox to enhance the operational stability and nitrogen removal efficiency during municipal wastewater treatment. *Chemical Engineering Journal*, 451, 138878.

Han, X., Zhang, S., Yang, S., Zhang, L., Peng, Y., 2020. Full-scale partial nitritation/anammox (PN/A) process for treating sludge dewatering liquor from anaerobic digestion after thermal hydrolysis. *Bioresource Technology*, 297, 122380.

He, M., Xu, Z., Hou, D., Gao, B., Cao, X., Ok, Y.S., Rinklebe, J., Bolan, N.S., Tsang, D.C., 2022. Waste-derived biochar for water pollution control and sustainable development. *Nature Reviews Earth and Environment*, 3(7), 444–460.

Hou, X., Liu, S., Zhang, Z., 2015. Role of extracellular polymeric substance in determining the high aggregation ability of anammox sludge. *Water Research*, 75, 51–62.

Ji, X., Wang, Y., Zhan, X., Wu, Z., Lee, P., 2022. Meta-Omics reveal the metabolic acclimation of freshwater anammox bacteria for saline wastewater treatment. *Journal of Cleaner Production*, 362, 132184.

Jin, R., Yang, G., Yu, J., Zheng, P. 2012. The inhibition of the Anammox process: A review. *Chemical Engineering Journal*, 197, 67–79.

Kadam, P.C. and Boone, D.R., 1996. Influence of pH on ammonia accumulation and toxicity in halophilic, methylotrophic methanogens. *Applied and Environmental Microbiology*, 62(12), 4486–4492.

Kang, D., Lin, Q., Xu, D., Hu, Q., Li, Y., Ding, A., Zhang, M., Zheng, P., 2018. Color characterization of anammox granular sludge: Chromogenic substance, microbial succession and state indication. *Science of the Total Environment*, 642, 1320–1327.

Karthäuser, C., Ahmerkamp, S., Marchant, H.K., Bristow, L.A., Hauss, H., Iversen, M.H., Kiko, R., Maerz, J., Lavik, G., Kuypers, M.M., 2021. Small sinking particles control anammox rates in the Peruvian oxygen minimum zone. *Nature Communications*, 12(1), 1–12.

Kumar, M., Dutta, S., You, S., Luo, G., Zhang, S., Show, P.L., Sawarkar, A.D., Singh, L., Tsang, D.C., 2021. A critical review on biochar for enhancing biogas production from anaerobic digestion of food waste and sludge. *Journal of Cleaner Production*, 305, 127143.

Li, J., Gao, F., Chen, X., Zhang, Y., Dong, H., 2022a. Insights into nitrogen removal from seawater-based wastewater through marine anammox bacteria under ampicillin stress: Microbial community evolution and genetic response. *Journal of Hazardous Materials*, 424, 127597.

Li, Q., Jia, Z., Fu, J., Yang, X., Shi, X., Chen, R., 2022b. Biochar enhances partial denitrification/anammox by sustaining high rates of nitrate to nitrite reduction. *Bioresource Technology*, 349, 126869.

Li, X., Peng, Y., Zhang, J., Du, R., 2022c. Multiple roles of complex organics in polishing THP-AD filtrate with double-line anammox: Inhibitory relief and bacterial selection. *Water Research*, 216, 118373.

Liu, X., Meng, Q., Wu, F., Zhang, C., Tan, X., Wan, C., 2022. Enhanced biogas production in anaerobic digestion of sludge medicated by biochar prepared from excess sludge: role of persistent free radicals and electron mediators. *Bioresource Technology*, 347, 126422.

Lou, T., Peng, Z., Jiang, K., Niu, N., Wang, J., Liu, A., 2022. Nitrogen removal characteristics of biofilms in each area of a full-scale AAO oxidation ditch process. *Chemosphere*, 302, 134871.

Meng, J., Hu, Z., Wang, Z., Hu, S., Liu, Y., Guo, H., Li, J., Yuan, Z., Zheng, M., 2022. Determining factors for nitrite accumulation in an acidic nitrifying system: Influent ammonium concentration, operational pH, and ammonia-oxidizing community. *Environmental Science and Technology*, 56(16), 11578–11588.

Mojiri, A., Ohashi, A., Ozaki, N., Aoi, Y., Kindaichi, T., 2020. Integrated anammox-biochar in synthetic wastewater treatment: Performance and optimization by artificial neural network. *Journal of Cleaner Production*, 243, 118638.

Mulder, A., Van de Graaf, A.A., Robertson, L.A., Kuenen, J.G., 1995. Anaerobic ammonium oxidation discovered in a denitrifying fluidized bed reactor. *Fems Microbiology Ecology*, 16(3), 177–183.

Negi, D., Verma, S., Singh, S., Daverey, A., Lin, J., 2022. Nitrogen removal via Anammox process in constructed wetland-A comprehensive review. *Chemical Engineering Journal*, 437, 135434.

Qi, Q., Sun, C., Zhang, J., He, Y., Tong, Y.W., 2021. Internal enhancement mechanism of biochar with graphene structure in anaerobic digestion: the bioavailability of trace elements and potential direct interspecies electron transfer. *Chemical Engineering Journal*, 406, 126833.

Ren, Z., Wang, H., Zhang, L., Du, X., Huang, B., Jin, R., 2022. A review of anammox-based nitrogen removal technology: From microbial diversity to engineering applications. *Bioresource Technology*, 363, 127896.

Rios-Del Toro, E.E., López-Lozano, N.E., Cervantes, F.J., 2017. Up-flow anaerobic sediment trapped (UAST) reactor as a new configuration for the enrichment of anammox bacteria from marine sediments. *Bioresource Technology*, 238, 528–533.

Rong, C., Luo, Z., Wang, T., Qin, Y., Wu, J., Guo, Y., Hu, Y., Kong, Z., Hanaoka, T., Sakemi, S., 2022. Biomass retention and microbial segregation to offset the impacts of seasonal temperatures for a pilot-scale integrated fixed-film activated sludge partial nitritation-anammox (IFAS-PN/A) treating anaerobically pretreated municipal wastewater. *Water Research*, 225, 119194.

Tomaszewski, M., Cema, G., Ziembińska-Buczyńska, A., 2017. Influence of temperature and pH on the anammox process: A review and meta-analysis. *Chemosphere*, 182, 203–214.

Van der Star, W.R., Abma, W.R., Blommers, D., Mulder, J., Tokutomi, T., Strous, M., Picioreanu, C., van Loosdrecht, M.C., 2007. Startup of reactors for anoxic ammonium oxidation: Experiences from the first full-scale anammox reactor in Rotterdam. *Water Research*, 41(18), 4149–4163.

Wang, W., Liu, Q., Xue, H., Wang, T., Fan, Y., Zhang, Z., Wang, H., Wang, Y., 2022a. The feasibility and mechanism of redox-active biochar for promoting anammox performance. *Science of the Total Environment*, 814, 152813.

Wang, W., Wang, T., Liu, Q., Wang, H., Xue, H., Zhang, Z., Wang, Y., 2022b. Biochar-mediated DNRA pathway of anammox bacteria under varying COD/N ratios. *Water Research*, 212, 118100.

Wang, Z., Gu, Z., Yang, Y., Dai, B., Xia, S., 2022c. Review of biochar as a novel carrier for anammox process: Material, performance and mechanisms. *Journal of Water Process Engineering*, 50, 103277.

Xie, J., Cao, Q., An, T., Mabruk, A., Xie, J., Chang, Y., Guo, M., Chen, C., 2022a. Small biochar addition enhanced anammox granular sludge system for practical wastewater treatment: Performance and microbial community. *Bioresource Technology*, 363, 127749.

Xie, J., Guo, M., Xie, J., Chang, Y., Mabruk, A., Zhang, T.C., Chen, C., 2022b. COD inhibition alleviation and anammox granular sludge stability improvement by biochar addition. *Journal of Cleaner Production*, 345, 131167.

Xu, J., Li, C., Zhu, N., Shen, Y., Yuan, H., 2021a. Particle size-dependent behavior of redox-active biochar to promote anaerobic ammonium oxidation (anammox). *Chemical Engineering Journal*, 410, 127925.

Xu, J., Li, C., Zhu, N., Shen, Y., Yuan, H., 2021b. Alleviating the nitrite stress on anaerobic ammonium oxidation by pyrolytic biochar. *Science of the Total Environment*, 774, 145800.

Xu, J., Wu, X., Zhu, N., Shen, Y., Yuan, H., 2020. Anammox process dosed with biochars for enhanced nitrogen removal: Role of surface functional groups. *Science of the Total Environment*, 748, 141367.

Xu, Z., He, M., Xu, X., Cao, X., Tsang, D.C. 2021. Impacts of different activation processes on the carbon stability of biochar for oxidation resistance. *Bioresource Technology*, 338, 125555.

Yang, Y., Lu, Z., Azari, M., Kartal, B., Du, H., Cai, M., Herbold, C.W., Ding, X., Denecke, M., Li, X., 2022. Discovery of a new genus of anaerobic ammonium oxidizing bacteria with a mechanism for oxygen tolerance. *Water Research*, 226, 119165.

Zhang, G., Zhang, L., Han, X., Zhang, S., Peng, Y., 2021. Start-up of PN-anammox system under low inoculation quantity and its restoration after low-loading rate shock. *Frontiers of Environmental Science & Engineering*, 15(2), 1–11.

Zhang, Q., Cheng, Y., Huang, B., Jin, R., 2022a. A review of heavy metals inhibitory effects in the process of anaerobic ammonium oxidation. *Journal of Hazardous Materials*, 128362.

Zhang, Q., Lin, J., Kong, Z., Zhang, Y., 2022b. A critical review of exogenous additives for improving the anammox process. *Science of the Total Environment*, 833, 155074.

Zhang, T., Wu, X., Li, H., Tsang, D.C., Li, G., Ren, H., 2020. Struvite pyrolysate cycling technology assisted by thermal hydrolysis pretreatment to recover ammonium nitrogen from composting leachate. *Journal of Cleaner Production*, 242, 118442.

Zhou, G., Yang, X., Li, H., Marshall, C.W., Zheng, B., Yan, Y., Su, J., Zhu, Y., 2016. Electron shuttles enhance anaerobic ammonium oxidation coupled to iron (III) reduction. *Environmental Science and Technology*, 50(17), 9298–9307.

8

Application of Biochar for Sludge Dewatering

Dongdong Ge[1], Nanwen Zhu[1], Mingjing He[2], and Daniel C.W. Tsang[3]

[1] School of Environmental Science & Engineering, Shanghai Jiao Tong University, Shanghai, China
[2] Department of Civil and Environmental Engineering, The Hong Kong Polytechnic University, Hung Hom, Kowloon, Hong Kong, China
[3] State Key Laboratory of Clean Energy Utilization, Zhejiang University, China

8.1 Introduction

With the rapid development of the economy and urbanization, the volume of municipal wastewater has surged, and this has resulted in the huge increase in sludge (Wu et al., 2016a). Sludge treatment is paramount; improper treating methods may cause harmful secondary pollution due to its complex constituents, such as pathogenic bacteria, toxic organics, and heavy metals (Ge et al., 2021).

The moisture content of the generated sludge from wastewater treatment plants is as high as 97–99%, seriously affecting the sludge treatment efficiency (Ge et al., 2022). The typical ultimate sludge disposal approaches including incineration, composting, land application, and building materials utilization, which strictly require the sludge moisture below 60% at least. Therefore, sludge dewatering is a crucial unit in the full-chain sludge treatment and disposal process. In fact, sludge dewaterability is very poor due to the numerous hydrophilic extracellular polymeric substances (EPS), which consist of protein, polysaccharide, humic acid, nucleic acid, and so on, maintaining the gel-like network structure of sludge flocks (Xiao et al., 2021). Conventionally, inorganic aluminum salts, ferric salts, and organic polyacrylamide are extensively employed as coagulants and flocculants. Unfortunately, the water content of the dewatered sludge cake after mechanical pressure filtration still ranges from 75% to 85% (Zhen et al., 2018). This is due to the undamaged EPS and high compressibility of the flocculated sludge (Wu et al., 2016a).

Biochar Applications for Wastewater Treatment, First Edition. Edited by Daniel C.W. Tsang and Yuqing Sun.
© 2023 John Wiley & Sons, Inc. Published 2023 by John Wiley & Sons, Inc.

In the recent years, various physical conditioners known as skeleton builders have been popularly exploited to reduce sludge compressibility, construct drainage channels, and enhance mechanical strength and permeability of sludge (Li et al., 2021; Wu et al., 2017). Lime is representatively used to serve as an inorganic backbone supporter to facilitate the structural formation with high porosity, permeability, and rigid lattice, which allows moisture to discharge easily at high filtering pressure. Noticeably, the additive lime with dosage around 20 wt% of dry sludge mass for deep-dewatering will cause a considerable increment in sludge volume, and the collected dewatered sludge cake with very low caloric value is unconducive to subsequent disposal except landfilling. Hence, it is necessary to seek the novel and efficacious skeleton builders for sludge conditioning and dewatering.

Biochar is normally produced from pyrolysis, hydrothermal carbonization, and microwave digestion of various agriculture and municipal solid wastes such as rice straw, corncob, and sewage sludge under the restricted oxygen condition, which is considered an environment-friendly and cost-effective technology (Guo, Z. et al., 2020b). For example, the rice straw with high annual yield of 8×10^8 tons in China can be utilized as cheap feedstock for biochar preparation (Guo et al., 2019), avoiding the secondary pollution in incineration. With the specific pretreatment and modification, the customized biochar may possess large specific surface area and copious functional groups, making it promising in toxic substances adsorption, catalytic oxidation, and sludge conditioning (Chen, Y.-d. et al., 2020). Gu et al. (2012) prepared the Fe_3O_4-loaded magnetic carbon material with superior porous structure and high pore volume of 0.504 mL/g, which render the highly efficient catalytic activity in dyestuff intermediate degradation during Fenton-like treatment (Gu et al., 2012). The biochar prepared by sewage sludge has plenty of soluble salts, active groups, and good porous framework, making it possible to use sludge as a self-produced conditioner (Wu et al., 2019).

Sludge dewaterability is popularly characterized by capillary suction time (CST), specific resistance of filtration (SRF), water content (Wc) of dewatered sludge cake, and the net sludge solids yield (Y_N) (Ge et al., 2019; Wu et al., 2016a). The CST is commonly measured with a CST instrument (Triton, UK). SRF is determined by vacuum filtration method. In brief, the 100 mL sludge sample is loaded into the 9 cm standard Brinell funnel with a premoistened 1.2 μm glass fiber filter paper and placed under a constant 34.5 kPa vacuum pressure for 30 minutes. The filtrate volume (V) and filtration time are recorded in order to calculate the sludge SRF. Then Eq. (8.1) is used to calculate the specific resistance of sludge.

$$SRF = \frac{2PA^2 b}{\mu W} \tag{8.1}$$

where *SRF* (m/kg) denotes specific resistance of sludge, P (N/m^2) denotes filtration pressure, A (m^2) denotes the filter area, μ (N·s/m^2) denotes the filtrate viscosity, W (kg/m^3) denotes the unit mass of filtrate cake solids, and b (s/m^6) is calculated based on the slope of volume and time of filtrate discharge.

Y_N is calculated by the following Eq. (8.2).

$$Y_N = F \cdot \left(\frac{2P\omega}{\mu \cdot t} \cdot \frac{1}{SRF} \right)^{1/2} \tag{8.2}$$

where P denotes the pressure (N/m^2), ω denotes the sludge cake solids weight per volume of filtrate, which is assumed as constant (kg/m^3), μ denotes the filtrate viscosity (N·s/m^2), t denotes the filtration time for the filtrate volume reaching to half of sludge volume (s), and F denotes a correction factor shown by the Eq. (8.3).

$$F = \frac{SS_{original}}{SS_{original} + SS_{conditioned}} \tag{8.3}$$

where $SS_{original}$ and $SS_{conditioned}$ denote the masses of original sludge solids and conditioned solids per unit sludge volume, respectively (g/L).

8.2 Preparation of Biochar-Based Sludge Conditioner

Tao et al. (2019) used the dewatered sludge, which was first conditioned by Fenton reagent and red mud, as feedstock to prepare pyrolysis biochar at the temperature ranging from 200 to 900 °C (Tao et al., 2019). Sometimes, the feedstock was soaked in acid solution to remove carbonates (Guo et al., 2019). The calcinated biochar was further washed using acid solution to remove the surface impurities (Wu et al., 2016b), but the pickling could result in the breakage of pore walls, thereby increasing the mesopores (Yang, Y. et al., 2022). Moreover, the $ZnCl_2$ was frequently selected as an activator to be impregnated into the feedstock before pyrolysis and work as a pore-forming reagent for mass transfer of gases (Kong et al., 2013), and also a skeleton for the deposition of carbon polymers during carbonization (Yang, Y. et al., 2022). The three sludge-based biochar (SBB) materials commonly have distinctive surface features. The raw SBB surface is smooth and blocky, whereas the $ZnCl_2$-activated and post-pickling SBBs showed numerous debris and their surface microstructures were obviously rough and porous (Figure 8.1), potentially providing more active sites (Yang, Y. et al., 2022).

Reportedly, the adsorption isotherms of the three various SBBs are class B hysteresis regression lines, suggesting the mesoporous structure with parallel wall slit pores (Figure 8.2a). The pore size of the three variously prepared SBBs was

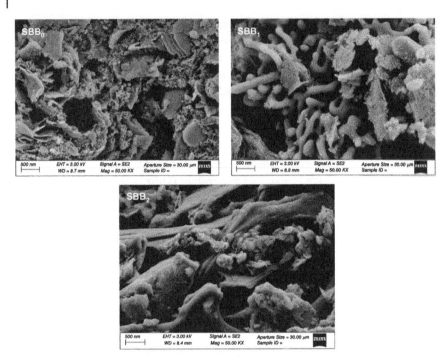

Figure 8.1 Scanning electron microscopes images of the direct carbonization sludge-based biochar (SBB_0), $ZnCl_2$-activated sludge-based biochar (SBB_1), and HCl pickling sludge-based biochar (SBB_2) (Yahong Yang et al., 2022 / with permission from Elsevier).

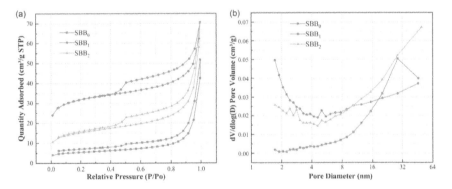

Figure 8.2 N_2 adsorption isotherm (a) and pore size distribution (b) of the direct carbonization sludge-based biochar (SBB_0), $ZnCl_2$-activated sludge-based biochar (SBB_1), and HCl pickling sludge-based biochar (Yang et al., 2022 / with permission of Elsevier).

mesoporous dominantly, and there were minor microporous and macroporous (Figure 8.2b). Especially, the specific surface area of the $ZnCl_2$-activated SBB reached 116.00 m^2/g, and the total pore volume was 0.077 cm^3/g.

The higher temperature of 1000 °C would result in the compact network conformation due to the sintering procedure making the inorganic materials deformed, thereby covering the metals (Tay et al., 2001). Heating temperature exhibited a crucial impact on the pore size distribution, surface area, and the adsorption capacity of the prepared biochar (Tay et al., 2001). The bulky micronsized particles were gradually formed with higher pyrolytic temperature, and 700 °C was the optimal for the maximum specific surface area (Guo, Z. et al., 2020b).

The rich husk biochar was further subjected to $FeCl_3$ solution soaking with ultrasonic treatment to increase the positive charges on the biochar surface (Wu et al., 2016b). The various surface morphology of the rich husk biochar before/after acid soaking and Fe loading are shown in Figure 8.3. Similarly, rice straw biochar was treated with $AlCl_3$ solution under ultrasonic condition to improve the adsorption of Al^{3+} on biochar surface (Guo et al., 2019). The biochar doped with multiple transition metals was also synthesized by pyrolysis of the mixed sludge with Fe_2O_3 and Al_2O_3, and the $FeAl_2O_4$ coated biochar was obtained due to the reactions between aluminum and iron compounds at 800 °C (Tao et al., 2020). Similarly, Kou et al. (2021) used the Fe-rich sludge conditioned with $KMnO_4$ as feedstock to prepare pyrolysis biochar for catalyzing peroxymonosulfate to degrade organic phenolic contaminants, and found that the Mn-Fe bimetallic oxides boost the direct electron transfer of the two elements, thereby enhancing the catalytic oxidation effect (Kou et al., 2021).

The corn-core powder was also used to jointly prepare the sludge-based biochar (SBB). In brief, the corn-core powder was firstly treated by alkalization and etherification treatment in 60 °C water bath for 3 h. Then, the modified corn-core powder (MCCP) was introduced into sludge and the vacuum filtration was performed

Figure 8.3 The various morphological structures of rice husk biochar with different treatments (Yan Wu et al., 2016b / with permission from Elsevier).

for collecting the MCCP-sludge cake. Finally, the sludge cake was sintered at 200–900 °C nitrogen atmosphere to gain the SBB (Guo, Z. et al., 2020b). Corn straw was employed to produce biochar in a 500 °C muffle for 2 h with N_2 protection for persulfate activation to improve sludge dewatering (Guo, J. et al., 2020).

8.3 Efficacy of Biochar Conditioning on Enhanced Sludge Dewaterability

SBB could increase the shear resistance of sludge flocks and reduce compressibility (Guo, Z. et al., 2020b). Some scholars compared three synthesized sludge-based biochars by the direct carbonization, $ZnCl_2$ pretreatment and carbonization, and carbonization combined with HCl treatment for recycling in sludge conditioning and dewatering, and the SBB with $ZnCl_2$ preactivation displayed the superior conditioning performance for sludge dewatering than others (Yang, Y. et al., 2022). With the Al^{3+}-doped rice straw biochar dosage of 0.3 g/g DS, the SRF and Wc declined to 1.2×10^{12} m/kg and 38 s, respectively, and Y_N increased by 23-fold to 19.4 kg/(m^2·h) (Guo et al., 2019). Although the individual persulfate (5 mM) exhibited light improvement of sludge dewaterability, the coupled persulfate with 2.1g/L biochar could reduce CST by 2.46 times and Wc declined to 66.3% (Guo, J. et al., 2020). Pyrolysis processing could facilitate the release of heavy metals bound to organic matters and the co-precipitation with the prepared biochar, which is conducive to the immobilization of heavy metals (Chen, C. et al., 2020; Li et al., 2018).

The optimal pyrolysis temperature for preparing rice husk biochar ranges from 400 to 500 °C (Wu et al., 2016a, 2016b). With the increased pyrolysis temperature, the organic functional groups, soluble salt ions would be diminished and the porous biochar structure would be subjected to damage. Therefore, the sludge filterability deteriorated using the sludge biochar sintered at higher temperature (> 500 °C) (Wu et al., 2019). Nevertheless, the pyrolysis temperature is very crucial for the composition and transformation of minerals in biochar. The reduction of Fe_2O_3 to Fe_3O_4 appeared at 400–600 °C; $Fe_{0.95}C_{0.05}$ and monoplasmatic Fe were presented at 800–900 °C (Gu et al., 2012; Tao et al., 2019). The customized sludge-based biochar at pyrolysis temperature of 800 °C might accelerate Fe^{2+} dissolution from iron and $Fe_{0.95}C_{0.05}$ in aqueous phase than that at low temperature. Thus, it can enhance the H_2O_2 decomposition into ·OH for EPS decomposition and the generated Fe^{3+} worked as coagulant for improved sludge filterability (Tao et al., 2019).

The Fe-doped rice husk biochar presented superior conditioning effect for sludge dewatering than the ordinary rich husk biochar. At the optimal dosage of 60% DS, sludge SRF declined by 97.9% and the Wc of dewatered sludge cake decreased to 77.97% and Y_N rose by 28-fold (Wu et al., 2016b). Interestingly, with the successive addition of 20 g/kg DS $KMnO_4$, 138.09 g/kg DS $FeCl_3$, and 70% DS sludge-based biochar prepared by the thermal processing of dried sludge cake at 400 °C for 2 h, the conditioned sludge SRF was dropped by 99.03%, and Y_N was elevated by 24.6 times and Wc declined to 60.63% (Wu et al., 2017). Similar report presented that Fe^{2+}-activated peroxymonosulfate combined with 120 mg/g VS rice straw biochar conditioning process harvested the very low Wc of the dewatered sludge cake of 38.5%.

Overdosed SBB has the adverse effects on the amelioration of sludge dewaterability due to the alkaline characteristic of biochar, inducing the increased pH and electrostatic repulsion (Guo, Z. et al., 2020a, 2020b). Moreover, excess biochar would block the drainage channels and hinder the waster lease.

Hydrothermal carbonization occurs in an aqueous solution with various waste biomass or sludge at the temperature of 150–250 °C and pressure of 1–1.5 bar. Conventionally, the sludge is first mixed with transition metals under alkaline condition for the specific time. Then, the mixture is transferred into the hydrothermal reactor under high temperature and pressure. Finally, the resultant product is washed by large amount of deionized water and organic solution till the pH became neutral and the particles is further dried and stored (Li, S. et al., 2022). Microwave digestion is another practicable method to synthesize biochar with less energy consumption and little yield of hazardous gases. At first, the feedstock is conditioned by alkaline reagent and dried. Subsequently, acid treatment is conducted and the treated substrate is subjected to microwave carbonization (Gu et al., 2012). However, there is little literature reporting the application of hydrothermal biochar and microwave digested biochar for sludge dewatering.

The artificially supplemented transition metals elements can facilitate the charge neutralization effect and catalytic capacity. Noteworthily, excessively loaded biochar with metal components showed undesirable results, which was ascribed to the agglomerated metals blocking the pore channels and weakening the biochar's activity (Mian and Liu, 2019). Furthermore, the SBB was employed as conductor, catalyst, and skeleton builder in electrolysis sludge conditioning, and with addition of the prepared carbon electrodes from sludge-based particle sintered at 800 °C in three-dimensional electrochemical reactor, the conditioned sludge dewaterability was enhanced with the CST declined by 2.39-fold and the increased Y_N of 6.88 kg·m^3/h and the low SRF of 6.20×10^{12} m·kg (Yu et al., 2022).

8.4 Variations of Sludge Physicochemical Characteristics via Biochar Conditioning

Cationic ions loaded biochar has a certain of charge neutralization (Guo et al., 2019), and thereby, the positively charged biochar might electronically neutralize the negative charges from organic particles, boosting the destabilization of a sludge gel-like system, congregation of fine bio-colloids, and solid–liquid separation.

Biochar-activated persulfate might increase the zeta potential of the conditioned sludge, and with the addition of 2.1 g/L corn straw–based biochar and 7.5 mM persulfate, the zeta potential of sludge rose to −4.6 mV (Guo, J. et al., 2020). The biochar catalytic ozonation could attack the biological macromolecules with negative charges and facilitate the sulfhydryl groups into disulfide bonds or sulfur-containing intermediates (Xiao et al., 2021).

The average sludge particle size apparently increased after biochar application, which was attributed to bridging, sweeping, and agglomerating effects (Guo, Z. et al., 2020b). Sometimes, the oxidative pretreatment caused smaller sludge particle size due to the breakage of bio-flocs and cells, whereas the SBB enabled the dispersed colloidal fragments to aggregate and enlarge the particle size, which was favorable for the water discharge (Xiao et al., 2021).

Biochar showed special removal effect on extracellular protein (Guo et al., 2019; Yang, Y. et al., 2022), which has been popularly perceived to play a detrimental role in sludge dewatering. On the contrary, polysaccharide hardly influences sludge dewaterability. The SBB with poriferous structure was capable of adsorbing tiny organics, enhancing particle agglomeration, and improving flocs precipitation (Guo, Z. et al., 2020b).

Also, after the SBB conditioning, the relative hydrophobicity of sludge increased, which would facilitate the release of bound water and solid–liquid separation (Ge et al., 2020; Yang, Y. et al., 2022).

SBB showed little influence on extracellular organic substances dissolution and decomposition, since it exerts the main role of skeleton supporting through the porous and rigid reticular structure (Guo, Z. et al., 2020a). However, it was reported that the biochar with the larger specific surface area has more active sites for adsorbing tiny colloidal particles, facilitating the generation of lump flocs and the improvement of sedimentation, which is favorable for sludge dewatering (Guo, Z. et al., 2020b). After biochar conditioning, sludge microstructure distinctly changed with the appeared large cracks, as depicted in Figure 8.4 (Guo et al., 2019).

8.5 Technical Mechanism and Implementation Prospects

Biochar contributed to constructing convenient drainage channels for the outflow of water-entrapped sludge interior, and thereby, the biochar-conditioned sludge cake deformed and collapsed in mechanical pressure filtration, as shown in Figure 8.5 (Wu et al., 2019).

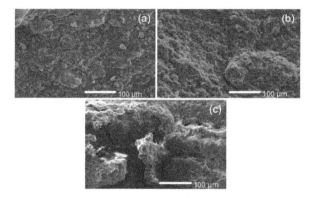

Figure 8.4 Various microstructures of (a) raw sludge, (b) rice straw biochar conditioned sludge, and (c) sludge conditioned by Al^{3+} modified rice straw biochar (Junyuan Guo et al., 2019 / with permission from Elsevier).

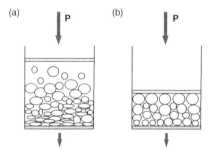

Figure 8.5 Schematic diagram of raw sludge dewatering and biochar-conditioned sludge dewatering mechanism (Wu et al., 2017 / with permission of Elsevier).

The Fe-modified biochar could provide positive charges, neutralizing the negatively charged functional groups and elevate the zeta potential of sludge (Wu et al., 2016b). Compared with the compressibility of 1.50 of raw sludge cake, the Fe-doped biochar reduce the compressibility coefficient of sludge cake to 0.86, indicating the increased rigid strength and the permeability with the biochar as a skeleton builder (Wu et al., 2016b). Wang et al. investigated the application of rice husk flour, rice husk biochar, and sludge-based biochar prepared by rice husk and ferric ions for improved sludge dewaterability comparatively, and found that Fe content affected sludge compressibility noticeably, and zeta potential was considered the predominant factor governing sludge dewaterability (Wang et al., 2020). The abundant functional groups on biochar surface may react with sludge flocks to improve sludge agglomeration and settlement performance (Wu et al., 2019). Moreover, the successive treatment method showed the distinctive roles of oxidant, coagulant, and biochar in sludge property modification. Wu et al. (2017) pointed out that the micro-disintegration induced by oxidant such as $KMnO_4$ caused the leakage of EPS and the wrapped water, and the successive ferric ions flocculated the destroyed fragments to form compact flocs by electrostatic neutralization, and sludge biochar constructed a permeable and incompressible sludge cake for enhancing dewatering performance (Figure 8.6). Wu et al. (2019) stated the mesoporous in biochar increased water outflow. The macro pores (>0.5 μm) were reported to be available for water discharge, whereas smaller pores had little influence (Thapa et al., 2009).

Noteworthily, the persistent free radicals (PFRs) were observed after the corn straw was prepared into biochar, and PFRs could boost $S_2O_8^{2-}$ decomposition to produce $SO_4^{\cdot-}$ for EPS disruption (Fang et al., 2015; Guo, J. et al., 2020). Especially, $SO_4^{\cdot-}$ might impair hydrophobic organics of EPS into small molecules to discharge the wrapped water. It was verified that the hydrophobic organic matters are positively associated with sludge dewaterability (Guo, J. et al., 2020).

The concentrations of heavy metals in sludge cake and filtrate decreased after sludge was conditioned with rice husk biochar, which was probably attributed to the strong adsorption capacity (Wang et al., 2020). Also, the "dilution effect" of additive biochar would cause the decrement of the concentration of heavy metals

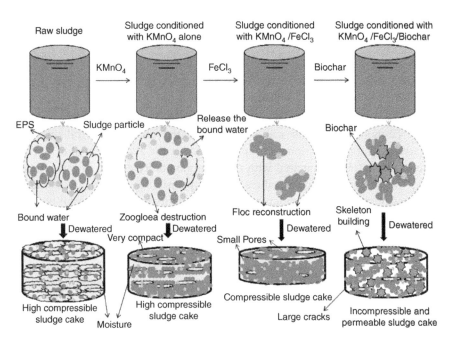

Figure 8.6 Variations of the sludge cakes after the successive $KMnO_4$/$FeCl_3$/Biochar treatments (Wu et al., 2017 / with permission of Elsevier).

(Guo, Z. et al., 2020a). However, the biochar altered species distribution of heavy metals by adsorption and immobilization capacity (Xiong et al., 2018).

If the transition metal-loaded biochar was coupled with oxidant for sludge conditioning, the superior sludge dewaterability was harvested due to the multiple effects involving homogeneous and heterogeneous catalytic oxidation (Li, H. et al., 2022; Li et al., 2021), cationic electroneutralization, and biochar skeleton supporting (Tao et al., 2019). Xiao et al. stated that the recycling of valence states of Mn and Fe loaded on the prepared $MnFe_2O_4$-doped SBB surface advanced the decomposition of O_3 into highly reactive radicals (Xiao et al., 2021). Yang, X. et al. (2022) also synthesized the $MnFe_2O_4$-doped tea-based biochar in peroxymonosulfate activation for ameliorating sludge dewatering and simultaneously confirmed that the added tannic acid could promote the valence transformation of Mn and Fe (Figure 8.7), which not only enhanced the oxidative performance of process but also facilitated the compressibility by precipitating proteins (Yang, X. et al., 2022). Moreover, the introduced $FeAl_2O_4$ phase existing in biochar could accelerate the Fenton reaction by the synergy of Fe and Al to weaken O–O bond of H_2O_2 and improve sludge conditioning effect (Tao et al., 2020).

In the biochar-coupled electrolysis process, the sludge-based carbon electrode with the strong anodic current response has powerful electrochemically oxidative capacity on the hydrophilic proteinaceous compounds, and thus, the treated sludge protein structure became looser, and more hydrophobic groups were exposed. The multiple effects of sludge-based carbon electrode are presented in Figure 8.8.

Tao et al. conceived a closed-loop sludge recycling system for sludge dewatering and sludge-based biochar reuse (Figure 8.9). The cost of rice husk-sludge biochar application for sludge conditioning was estimated as 0.82 USD/kg DS, which saved 88.4% in cost (Wang et al., 2020).

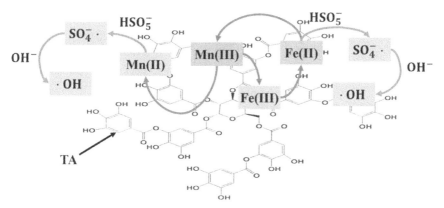

Figure 8.7 The valence transformation of Mn and Fe loaded on biochar surface in the presence of tannic acid (Yang et al., 2022 / with permission of Elsevier).

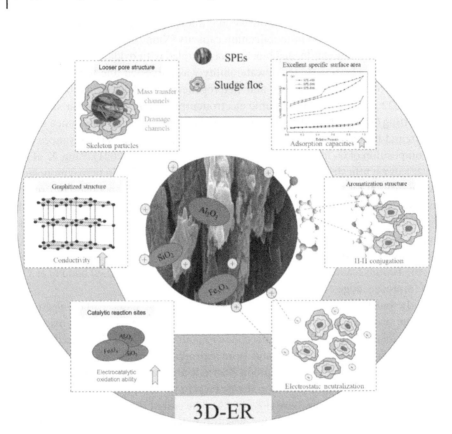

Figure 8.8 The proposed multiple effects of sludge-based carbon electrode on enhanced dewaterability in three-dimensional electrochemical reactor (Haixiang Yu et al., 2022 / with permission from Elsevier).

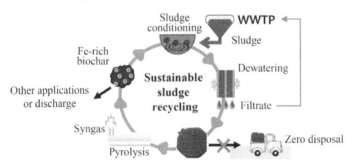

Figure 8.9 The closed-loop sludge recycling system (Tao et al., 2019 / with permission of Elsevier).

Various kinds of biomass resources still need to be investigated to explore the surface characteristics of the generated biochar, and the optimized processes to prepare biochar for the conditioning of sludge with various physicochemical properties should be estimated comprehensively. The environmental risks connected to converting hazardous substances such as heavy-metal toxic organics should be clearly ascertained. Furthermore, large-scale pilot study is necessary to assess the practical results such as biochar properties and effects, sludge and filtrate characteristics, and economic benefits in engineering implementation.

References

Chen, C., Liu, G., An, Q., Lin, L., Shang, Y., Wan, C., 2020. From wasted sludge to valuable biochar by low temperature hydrothermal carbonization treatment: Insight into the surface characteristics. *Journal of Cleaner Production*, 263, 121600.

Chen, Y.-d., Wang, R., Duan, X., Wang, S., Ren, N.-q., Ho, S.-H., 2020. Production, properties, and catalytic applications of sludge derived biochar for environmental remediation. *Water Research*, 187, 116390.

Fang, G., Liu, C., Gao, J., Dionysiou, D.D., Zhou, D., 2015. Manipulation of persistent free radicals in biochar to activate persulfate for contaminant degradation. *Environmental Science & Technology*, 49(9), 5645–5653.

Ge, D., Dong, Y., Zhang, W., Yuan, H., Zhu, N., 2020. A novel Fe^{2+}/persulfate/tannic acid process with strengthened efficacy on enhancing waste activated sludge dewaterability and mechanism insight. *Science of the Total Environment*, 733, 139146.

Ge, D., Wu, W., Li, G., Wang, Y., Li, G., Dong, Y., Yuan, H., Zhu, N., 2022. Application of CaO_2-enhanced peroxone process to adjust waste activated sludge characteristics for dewaterability amelioration: Molecular transformation of dissolved organic matters and realized mechanism of deep-dewatering. *Chemical Engineering Journal*, 437, 135306.

Ge, D., Zhang, W., Yuan, H., Zhu, N. 2019. Enhanced waste activated sludge dewaterability by tannic acid conditioning: Efficacy, process parameters, role and mechanism studies. *Journal of Cleaner Production*, 241, 118287.

Ge, D., Zhu, Y., Li, G., Yuan, H., Zhu, N., 2021. Identifying the key sludge properties characteristics in Fe^{2+}-activated persulfate conditioning for dewaterability amelioration and engineering implementation. *Journal of Environmental Management*, 296, 113204.

Gu, L., Zhu, N., Zhou, P., 2012. Preparation of sludge derived magnetic porous carbon and their application in Fenton-like degradation of 1-diazo-2-naphthol-4-sulfonic acid. *Bioresource Technology*, 118, 638–642.

Guo, J., Jia, X., Gao, Q., 2020. Insight into the improvement of dewatering performance of waste activated sludge and the corresponding mechanism by biochar-activated persulfate oxidation. *Science of the Total Environment*, 744, 140912.

Guo, J., Jiang, S., Pang, Y., 2019. Rice straw biochar modified by aluminum chloride enhances the dewatering of the sludge from municipal sewage treatment plant. *Science of the Total Environment*, 654, 338–344.

Guo, Z., Ma, L., Dai, Q., Ao, R., Liu, H., Wei, Y., Mu, L., 2020a. Role of extracellular polymeric substances in sludge dewatering under modified corn-core powder and sludge-based biochar pretreatments. *Ecotoxicology and Environmental Safety*, 202, 110882.

Guo, Z., Ma, L., Dai, Q., Ao, R., Liu, H., Yang, J., 2020b. Combined application of modified corn-core powder and sludge-based biochar for sewage sludge pretreatment: Dewatering performance and dissipative particle dynamics simulation. *Environmental Pollution*, 265, 115095.

Kong, L., Xiong, Y., Tian, S., Luo, R., He, C., Huang, H., 2013. Preparation and characterization of a hierarchical porous char from sewage sludge with superior adsorption capacity for toluene by a new two-step pore-fabricating process. *Bioresource Technology*, 146, 457–462.

Kou, L., Wang, J., Zhao, L., Jiang, K., Xu, X., 2021. Coupling of $KMnO_4$-assisted sludge dewatering and pyrolysis to prepare Mn, Fe-codoped biochar catalysts for peroxymonosulfate-induced elimination of phenolic pollutants. *Chemical Engineering Journal*, 411, 128459.

Li, C., Wang, X., Zhang, G., Li, J., Li, Z., Yu, G., Wang, Y., 2018. A process combining hydrothermal pretreatment, anaerobic digestion and pyrolysis for sewage sludge dewatering and co-production of biogas and biochar: Pilot-scale verification. *Bioresource Technology*, 254, 187–193.

Li, H., Chen, J., Zhang, J., Dai, T., Yi, H., Chen, F., Zhou, M., Hou, H., 2022a. Multiple environmental risk assessments of heavy metals and optimization of sludge dewatering: Red mud–reed straw biochar combined with Fe^{2+} activated H_2O_2. *Journal of Environmental Management*, 316, 115210.

Li, H., Dai, T., Chen, J., Chen, L., Li, Y., Kan, X., Hou, H., Han, Y., 2021. Enhanced sludge dewaterability by Fe-rich biochar activating hydrogen peroxide: Co-hydrothermal red mud and reed straw. *Journal of Environmental Management*, 296, 113239.

Li, S., Huang, D., Cheng, M., Wei, Z., Du, L., Wang, G., Chen, S., Lei, L., Chen, Y., Li, R. 2022. Application of sludge biochar nanomaterials in Fenton-like processes: Degradation of organic pollutants, sediment remediation, sludge dewatering. *Chemosphere*, 307, 135873.

Mian, M.M., Liu, G., 2019. Sewage sludge-derived $TiO_2/Fe/Fe_3C$-biochar composite as an efficient heterogeneous catalyst for degradation of methylene blue. *Chemosphere*, 215, 101–114.

Tao, S., Liang, S., Chen, Y., Yu, W., Hou, H., Qiu, J., Zhu, Y., Xiao, K., Hu, J., Liu, B., 2020. Enhanced sludge dewaterability with sludge-derived biochar activating hydrogen peroxide: Synergism of Fe and Al elements in biochar. *Water Research*, 182, 115927.

Tao, S., Yang, J., Hou, H., Liang, S., Xiao, K., Qiu, J., Hu, J., Liu, B., Yu, W., Deng, H., 2019. Enhanced sludge dewatering via homogeneous and heterogeneous Fenton reactions initiated by Fe-rich biochar derived from sludge. *Chemical Engineering Journal*, 372, 966–977.

Tay, J., Chen, X., Jeyaseelan, S., Graham, N., 2001. Optimising the preparation of activated carbon from digested sewage sludge and coconut husk. *Chemosphere*, 44(1), 45–51.

Thapa, K.B., Qi, Y., Clayton, S., Hoadley, A.F.A., 2009. Lignite aided dewatering of digested sewage sludge. *Water Research*, 43(3), 623–634.

Wang, M., Wu, Y., Yang, B., Deng, P., Zhong, Y., Fu, C., Lu, Z., Zhang, P., Wang, J., Qu, Y., 2020. Comparative study of the effect of rice husk-based powders used as physical conditioners on sludge dewatering. *Scientific Reports*, 10(1), 1–9.

Wu, J., Lu, T., Bi, J., Yuan, H., Chen, Y., 2019. A novel sewage sludge biochar and ferrate synergetic conditioning for enhancing sludge dewaterability. *Chemosphere*, 237, 124339.

Wu, Y., Zhang, P., Zeng, G., Liu, J., Ye, J., Zhang, H., Fang, W., Li, Y., Fang, Y., 2017. Combined sludge conditioning of micro-disintegration, floc reconstruction and skeleton building ($KMnO_4$/$FeCl_3$/Biochar) for enhancement of waste activated sludge dewaterability. *Journal of the Taiwan Institute of Chemical Engineers*, 74, 121–128.

Wu, Y., Zhang, P., Zeng, G., Ye, J., Zhang, H., Fang, W., Liu, J., 2016a. Enhancing sewage sludge dewaterability by a skeleton builder: Biochar produced from sludge cake conditioned with rice husk flour and $FeCl_3$. *ACS Sustainable Chemistry & Engineering*, 4(10), 5711–5717.

Wu, Y., Zhang, P., Zhang, H., Zeng, G., Liu, J., Ye, J., Fang, W., Gou, X., 2016b. Possibility of sludge conditioning and dewatering with rice husk biochar modified by ferric chloride. *Bioresource Technology*, 205, 258–263.

Xiao, T., Dai, X., Wang, X., Chen, S., Dong, B., 2021. Enhanced sludge dewaterability via ozonation catalyzed by sludge derived biochar loaded with $MnFe_2O_4$: Performance and mechanism investigation. *Journal of Cleaner Production*, 323, 129182.

Xiong, Q., Zhou, M., Liu, M., Jiang, S., Hou, H., 2018. The transformation behaviors of heavy metals and dewaterability of sewage sludge during the dual conditioning with Fe^{2+}-sodium persulfate oxidation and rice husk. *Chemosphere*, 208, 93–100.

Yang, X., Zeng, L., Huang, J., Mo, Z., Guan, Z., Sun, S., Liang, J., Huang, S., 2022. Enhanced sludge dewaterability by a novel $MnFe_2O_4$-Biochar activated peroxymonosulfate process combined with Tannic acid. *Chemical Engineering Journal*, 429, 132280.

Yang, Y., Yang, X., Wang, X., Yang, Q., Xu, W., Li, Y., 2022. Explore the closed-loop disposal route of surplus sludge: Sludge self-circulation preparation of sludge-based biochar (SBB) to enhance sludge dewaterability. *Colloids and Surfaces A: Physicochemical and Engineering Aspects*, 638, 128304.

Yu, H., Zhang, D., Gu, L., Wen, H., Zhu, N., 2022. Coupling sludge-based biochar and electrolysis for conditioning and dewatering of sewage sludge: Effect of char properties. *Environmental Research*, 214, 113974.

Zhen, G., Lu, X., Su, L., Kobayashi, T., Kumar, G., Zhou, T., Xu, K., Li, Y.Y., Zhu, X., Zhao, Y., 2018. Unraveling the catalyzing behaviors of different iron species (Fe^{2+} vs. Fe^0) in activating persulfate-based oxidation process with implications to waste activated sludge dewaterability. *Water Research*, 134, 101–114.

9

Effects of Biochar on Sludge Composting

Dong Li[1], Dongdong Ge[3], Yuqing Sun[4], and Daniel C.W. Tsang[2]

[1] Teleader Solid Waste Disposal (Shandong) Co., Ltd., Jinan China
[2] State Key Laboratory of Clean Energy Utilization, Zhejiang University, China
[3] School of Environmental Science & Engineering, Shanghai Jiao Tong University, Shanghai, China
[4] School of Agriculture, Sun Yat-Sen University, Guangzhou, Guangdong, China

Sewage sludge contains a large number of organic pollutants, toxic heavy metals, and pathogenic microorganisms, which will pose a serious threat to the environment and human health. Although sludge composting process can effectively transform sludge into high-value fertilizer, it still has the drawbacks of organic and heavy metal pollution, greenhouse gas emission, and nutrient loss. Many studies suggest that biochar addition in sludge composting can effectively improve the composting environment and overcome these drawbacks due to its high carbon content, developed pore structure, rich surface functional groups, and adjustable performance. This chapter examines the effects of biochar addition on sludge composting parameters, heavy metals, organic pollutants, greenhouse gas emissions, and compost quality. The addition of biochar has the following advantages: improving compost parameters; passivating heavy metals and immobilizing organic pollutant to avoid soil pollution; enhancing microbial activity to promote organic decomposition; reducing greenhouse gas and NH_3 emission; reducing nutrient loss; and improving composting quality. We suggest future researches to promote practical application of biochar and sludge co-composting and explore the mechanism of co-composting fertilizer on soil in long-term application.

9.1 Introduction

Water treatment is an important technology to provide clean water resources for cities. Sewage sludge, as the main by-product of wastewater treatment, generally contains high number of organic pollutants, heavy metals, and active microbes [1]. With the increase of urban construction and population, sludge production and treatment cost increase every year [2]. For instance, more than 39 million tons of sludge were produced in China in 2019 [3]. The sludge will naturally undergo anaerobic digestion due to its own microbes, releasing unpleasant-smelling volatile organic and inorganic compounds, such as NH_3 and H_2S [4]. Moreover, the organic pollutants and heavy metals in sludge will be released into the soil and groundwater, causing serious environmental pollution [5]. Therefore, effective and economical treatment of sewage sludge is particularly important. At present, the practical treatment technologies for sewage sludge include landfill, constructed wetlands, incineration, pyrolysis, gasification, hydrothermal, anaerobic digestion and composting [6–9]. Among these technologies, composting can utilize organic matter in sludge to produce value-added fertilizer and is a cost-effective and socially acceptable sludge treatment technology [10, 11].

Composting is a biochemical process with three distinct stages (mesophilic, thermophilic, and mature stage), carried out by different species of microorganisms and nematodes that dominate at different stages [12]. During sludge composting, the unstable organic matter in sludge is transformed into stable organic fertilizer by spontaneous biological decomposition, and the pathogenic bacteria in sludge can be inactivated at high temperature condition (more than 45 °C) [13]. However, traditional sludge composting has the disadvantages of low conversion efficiency, organic matter and heavy metal pollution, time-consuming, nitrogen loss via ammonia volatilization, and greenhouse gases emission [14]. The produced organic fertilizer also has the risk of leaching heavy metals and organic pollutants into the soil and groundwater, which greatly reduces the practical application value of fertilizers [10, 15].

The parameters of sludge composting process include sludge source, C/N ratio, biochemical composition, moisture, temperature, aeration condition, and pH. These affect microbial activity and gas emission in composting process, and organic content and pollution risk of composting products [12]. For instance, microplastics in sludge cannot be completely biodegradable in composting, especially synthetic plastics such as polyethylene and polyvinyl chloride [16]; the high C/N of the sludge results in a longer composting cycle, while the low C/N ratio results in a nitrogen loss (the optimal C/N ratio ranged from 25 to 30) [17]; the pH value of sludge compost has influence on the microbial activity and ammonia volatilization [18]; the moisture in sludge compost also affects nutrient and gas exchange [19].

Researchers have thus focused on overcoming these defects of traditional sludge composting process by changing the raw material formula, adding improvers (inorganic, organic, and microbial), adjusting environmental conditions, and other strategies to provide the best conditions for its application [4, 20, 21]. In particular, biochar addition in co-composting to impact the composting process and further improve the composting product has attracted considerable attention [22–24].

Biochar is a carbon-rich solid formed from the thermochemical decomposition of biomass or other organic materials under anoxic or anaerobic atmosphere [25]. It shows the characteristics of high carbon content, large specific surface area, developed porous structure, abundant surface functional groups and adjustable properties, and is considered to be suitable for improving the composting process and quality [13]. At present, using agricultural and forestry wastes (e.g., straw waste and livestock manure) and municipal solid wastes (e.g., sludge and food waste) as feedstocks to prepare biochar has practical application potential due to its wide source and low cost [26]. The preparation method of biochar mainly includes pyrolysis, gasification, and hydrothermal carbonization, among which pyrolysis is a mature method applied in commercial production. Pyrolysis to produce biochar is proceeded under anoxic or anaerobic atmosphere at temperature of 300–900 °C [27]. Feedstock, pyrolysis temperature, residence time, and heating rate all affect the properties of biochar, which have different effects on sludge composting. Table 9.1 summarized the feedstock and production of biochar for

Table 9.1 Summary of the feedstock and production of biochar and effects on sludge composting.

Biochar feedstock	Biochar production	Dosage	Effect on sludge composting	References
Bamboo	Pyrolysis (450–500 °C)	50%	Reduce NH_3 emission and N loss	[62]
Corn straw	–	~10%	Reduce the release of N_2O, CH_4 gas Improve the value of compost products	[29]
Pig carcasses/	Pyrolysis (700–800 °C)	8%	Promote the accumulation of bioavailable organic N	[28]
Coconut shell			Reduce N loss by 24.40% and 35.50%	
Wheat straw	Pyrolysis (550 °C)	5%	Passivate heavy metals Promote transformation of ammonium to nitrite nitrogen	[31]
Coal	Pyrolysis (600–800 °C)	20%	Increase the abundance and activity of microbials	[63]
Willow	Pyrolysis (350 °C)	10%	Improve compost products for soil water retention	[64]

(Continued)

Table 9.1 (Continued)

Biochar feedstock	Biochar production	Dosage	Effect on sludge composting	References
Rice hull	Pyrolysis (600 °C)	2, 4, 8%	Increase the pH and nutrient content Alleviate the bioavailability heavy metals	[61]
Rice straw	Pyrolysis (500 °C)	5, 10, 20%	Increase the diversity and activity of bacteria	[2]
Willow	Pyrolysis (350 °C)	8%	Reduce the mobility of heavy metals in compost products	[32]
Municipal sludge	Pyrolysis (550 °C)	8%	Reduce the toxic of heavy metals	[65]
Wheat straw	Pyrolysis (350–550 °C)	1, 3, 5, 7%	Passivate heavy metal effectively 5% biochar effectively improve composting quality	[30]
Tree leaves	Pyrolysis (550 °C)	2, 5%	Reduce antibiotic resistance genes abundances	[66]
Litchi wood	Pyrolysis (400–500 °C)	10%	Enhance activities of functional enzymes Reduce peak temperature Inhibit degradation of organic matters	[67]
Willow	Pyrolysis (600 °C)	10%	Reduce the toxicity of composts	[68]
Corncob Beech	Pyrolysis (500 °C)	10%	Increase the composting temperature Reduce the activity of sulfur-oxidizing bacteria	[37]

co-composting and the effects of biochar addition on sludge composting. In practice, straw, bamboo, rice husk, and other biomass raw materials are widely used in biochar production based on the production cost and carbon yield. Based on the characteristics of biochar, adding it to co-compost can increase microbial activity, accelerate the humification process, improve the performance of compost products, and reduce greenhouse gas emissions [10]. More importantly, biochar can effectively fix toxic metals and difficult biodegradable organic pollutants in sludge to reduce the pollution going into soil and groundwater [13, 22].

Therefore, this chapter looks at the effects of biochar addition on sludge composting process and quality, with potential prospects and conclusions, and suggests future areas of research. This chapter provides recent findings on the effects of biochar addition on sludge composting, promotes the practical application of biochar and sludge co-composting in the future, and expands on the high-value application direction of biochar.

9.2 Effects of Biochar Addition on Sludge Composting

9.2.1 Effects on Compost Parameters Effect on C/N

The C/N ratio is one of the important factors affecting the composting process, which can be used to evaluate the compost maturity. The different properties and the amount of biochar addition can regulate the initial C/N ratio of the compost. The suitable initial C/N ratio of organic aerobic compost is about 25:1, and the C/N ratio of nitrogen-rich sludge can be improved by carbon-rich biochar [10]. In addition, biochar can facilitate the composting process by promoting C mineralization and N preservation, thus reducing the post-composting C/N ratio [28].

Effect on temperature. Temperature is one of the most important parameters of composting. It not only affects the efficiency of composting but also is a factor to eliminate potential pathogenic microorganisms in sludge [23]. In the thermophilic stage of sludge compost, the microbial activity and organic matter conversion efficiency are the highest, and temperature plays an important role in this stage. Biochar can make the compost obtain a higher temperature in thermophilic stage. For example, Zhou et al. found that after the addition of biochar, the maximum temperature of sludge compost in the thermophilic period reached 53.5 °C, which was higher than 49.9 °C in the control group [29]. This may be attributed to the biochar improving the pore structure of the compost pile, thereby reducing heat loss. In addition, biochar increases the abundance and activity of microorganisms, which facilitates the conversion of organic matter and generates additional heat.

Effect on pH. The pH in sludge compost has an important effect on the fluidity of heavy metals, and low pH will lead to heavy metal leaching. Biochar is rich in oxygen-containing functional groups, and the alkaline groups in it can improve the pH of the compost. However, biochar enhances microbial activity, which will lead to the decomposition of organic matter to produce organic and inorganic acids to reduce pH and also promote the degradation of organic nitrogen to produce ammonia to increase pH [22].

Effect on moisture content. The appropriate moisture content for composting should be kept at 50–60%, as high moisture content inhibits the flow of oxygen while low moisture content reduces microbial activity. For sludge, the moisture content is generally higher than 85%, but the addition of biochar can make the water content of a sludge compost pile reach an appropriate level. Although biochar causes the compost to heat up, its high-water holding capacity prevents water loss and thus reduces the loss of nutrients [23].

9.2.2 Effects on Heavy Metals

The co-composting of biochar and sludge can affect the conversion of heavy metals. The heavy metals in biochar mainly come from the feedstocks used to produce biochar, such as the enrichment of soil-heavy metals by plant feedstocks. The heavy metals in sludge are mainly the residues of all kinds of wastewater treatment. Biochar can passivate heavy metals during sludge composting and reduce the migration and bioavailability of heavy metals [30, 31]. The mechanism of biochar stabilization of heavy metals includes adsorption, complexation, precipitation, ion exchange, electrostatic interaction, and reduction [24]. The rich specific surface area of biochar can adsorb and fix some heavy metals (such as Pb). There are many oxygen-containing functional groups (especially carboxyl groups) on the surface of biochar, which can bind to the surface of heavy metals. Biochar also immobilizes heavy metals by co-precipitation of humus produced by microorganisms with metal oxides [10]. Heavy metals passivated by biochar will still exist in compost, but their fluidity will be significantly reduced when eluted or applied to soil with compost products. Therefore, the passivation ability of biochar to heavy metals in sludge compost greatly improves the quality of compost products and reduces the risk of compost products to soil and groundwater environment.

Qin et al. found that biochar passivated Zn in sludge compost through electrostatic interaction, replacement of exchanged cations and oxygen-containing functional groups in biochar [31]. Liu et al. found that biochar had different passivation effects on heavy metals by principal component analysis, and had the best passivation effect on Pb, followed by Cr and As [30]. However, Gondek et al. found that biochar reduced the leaching of Zn and Pb from heavy metals in compost, but increased the leaching of Cd and Cu [32]. It can be concluded that biochar has passivation effect on heavy metals in compost, but this largely depends on the properties of biochar and the types of heavy metals contained in sludge compost. In order to study the passivation of specific heavy metals in sludge compost, biochar with specific properties should be considered.

9.2.3 Effects on Organic Matters

The organic pollutants in sludge compost are polycyclic aromatic hydrocarbons (PAHs), polychlorinated dibenzo-p-dioxins (PCDDs), polychlorinated dibenzofurans (PCDFs), polychlorinated biphenyls (PCBs), phthalic acid esters (PAEs), and other organic compounds [33]. The physical and chemical properties of sludge, composting methods, and composting conditions all affect the dissipation of organic matters [34]. In general, organic matters in sludge compost can be dissipated through biodegradation, mineralization, volatilization, and other ways. Some organic components in sludge can also be adsorbed to sequester organic matters. The pore structure of biochar can remove organic pollutants from sludge compost by improving microbial activity and absorbing organic matter. Biochar increases the porosity of compost piles and can increase air flow in the compost, thus promoting microbial growth. Increased microbial activity leads to accelerated degradation of organic matter in the compost to simpler organic compounds (amino acids, small carbohydrates, and phenolic compounds) [24]. Therefore, adding biochar to compost can improve the degradation of dissolved organic carbon (DOC). The high specific surface area of biochar contributes to the adsorption of NH_3, H_2S, and other organic matter generated by microbial activities, thus reducing the leaching of organic matter [22]. In addition, biochar can enhance the aromatics of organic matter in compost, thus improving the stability of humic substances [23].

The most reported organic pollutants in sludge compost are PAHs, which are volatile organic compounds with serious environmental hazards and health risks. The addition of biochar can effectively reduce PAHs in compost by improving microbial activity and adsorption [35]. In addition, Pang et al. found that biochar extended the thermophilic stage and promoted the decomposition of organic matter. The degradation of organophosphate esters is mainly caused by the oxidation reduction and nucleophilic substitution of quinone groups and other oxygen-containing groups on the surface of biochar [36]. Ouyang et al. also proposed that biochar on the one hand reduces the production of H_2S by improving the pore structure, and on the other hand reduces the release of H_2S through adsorption [37].

9.2.4 Effects on Gaseous Emissions

Greenhouse gas (including CO_2, CH_4, N_2O, CO) and NH_3 emissions are disadvantages in the composting process. On the one hand, the loss of carbon and nitrogen will reduce the composting quality; on the other hand, the release of these gases will contribute to global warming. Many researchers have explored ways to reduce gas emissions during composting, and adding biochar for co-composting is considered an effective strategy [11, 29, 38–41].

Reducing greenhouse gas (GHG) emissions. The availability of O_2 was found to be a key factor in GHGs production during composting [42]. The high specific surface area and well-developed pore structure of biochar can affect the density and particle size of the compost pile, thus promoting oxygen supply and reducing anaerobic zone formation [29]. In the early stage of compost, the pH neutral and anaerobic environment are conducive to the emission of CH_4, but most of the generated CH_4 will be oxidized to CO_2 by methanotrophic bacteria [43]. Biochar addition can decrease the activity of methanogenic bacteria and increase the activity of methanogenic bacteria, resulting in the decrease emission of CH_4 [44]. Zhou et al. explored the effects of biochar with different pore structures on CH_4 emission and found that biochar with small pore size improved the condition of compost pile, which was conducive to the growth of aerobic microorganisms and resulted in less CH_4 emission [45]. Dume et al. co-composted wheat straw with sludge and found that CO_2 and CH_4 emissions were reduced by 70% and 80%, respectively [46].

The release of N_2O during composting will not only increase the greenhouse effect but also cause the loss of N. During composting, two opposite processes, denitrification and incomplete nitrification, result in the production of N_2O [47]. However, due to the coexistence of aerobic and anaerobic environments, these two processes can occur simultaneously in composting process, and their intermediates can affect each other's reaction rates [48]. Some studies suggest that N_2O is mainly produced by denitrification in anoxic microenvironment [49]. Therefore, using biochar to reduce anaerobic zone in composting can effectively reduce the production of N_2O. Meng et al. used spent mushroom substrate (SMS) as an additive to co-compost with sludge and found that the release of N_2O was reduced by 86.2%. They suggested that the addition of SMS improved oxygen diffusion in sludge composting and provided an aerobic environment for microorganisms [50]. Zhou et al. added corn straw biochar in sludge composting and found greatly reduced N_2O emission (reduced by 31.3%) and total N loss (reduced by 38.2%), and this change was caused by the increase in compost temperature and pH due to the addition of biochar [29].

Reducing NH_3 emission. During the composting process, NH_3 gas produced by the decomposition of nitrogenous substances will be released in the thermophilic stage, resulting in the loss of N [43]. Furthermore, ammonia odor pollution caused by NH_3 gas is one of the major disadvantages of sludge composting. Generally, the changes of composting conditions such as temperature, pH, and aeration rate all affect NH_3 release [47, 51]. Many studies have shown that the addition of biochar can effectively fix NH_3 to increase the fertilizer value of compost and reduce NH_3 emission to solve odor pollution [28, 31, 52, 53]. Biochar with high specific surface area and surface functional groups can adsorb NH_3 to reduce N loss during composting process [54]. In addition, it can also provide a suitable growth environment for microorganisms, especially improve the activity of nitrifying bacteria, and realize the efficient conversion of NH_3 to NO_3^- to achieve NH_3 fixation [55].

Jiang et al. found that biochar reduced 67.4% NH_3 emissions in sludge composting and significantly enhanced the total N and the end compost, which caused the adsorption of NH_4-N and NO_3-N and the improvement of enzyme and N-fixing bacteria activities [52]. Liu et al. found that plant-derived biochar promoted the accumulation of bioavailable organic nitrogen in sludge composting to shorten the NH_3 emission cycle [28]. Qin et al. found that the adsorption of urea, uric acid, and NH_3 and the increase of compost pile pH by biochar addition reduced NH_3 emission and improved N retention capacity [31].

9.2.5 Effects on Microbial Community and Activities

The pore structure of biochar is similar to the size of microorganisms, which can be used as the breeding site. This porous structure can retain nutrients such as DOC and nitrogen compounds for microbial growth, has breathability and water retention, has pH buffer capacity, and can fix and inhibit substances harmful to microorganisms such as NH_3 and heavy metals, thus inhibiting toxic compounds, pathogen competition, and adverse compost environment [22]. The studies showed that adding biochar could effectively increase the abundance of bacteria, fungi, and actinomycetes in sludge compost.

Du et al. found that the addition of more than 10% biochar inhibited protease but promoted cellulase and peroxidase, thus increasing microbial diversity during sludge composting, especially the activity of *bacillus* in the later stage of composting [2]. Jiang et al. found that the biochar addition increased total N, cellulase, and fluorescein diacetate hydrolase hydrolase, thus increasing the abundance of nitrogen-fixing bacteria at genus level, especially Proteobacteria [52]. Malinowski et al. found that adding 5% biochar increased the number of thermophilic bacteria in the first 14 days of the composting (mean abundance ratio 192% compared to control) and reduced the abundance of potentially pathogenic microorganisms [56].

Biochar affects the biological properties of compost, and the change of the biological properties will also affect the change of the physical and chemical properties of compost. The enhancement of microbial activity promoted the degradation of organic matter, nitrification, and denitrification, thus improving the quality of compost.

9.2.6 Effects on Quality of Sludge Compost

Compost converts organic matter into humic substances (HS) through humification and can be used as soil fertilizer or added to improve soil fertility and quality. As functional organic matters, HS are mainly composed of fulvic acid (FA), humic acid (HA), and humins (HU) [57]. The increase of HS content can reduce the compost toxicity, improve carbon sequestration capacity, and promote crop growth, which is an important evaluation parameter of compost quality [58].

Several studies have confirmed that biochar addition can enhance humification and promote the formation of FA and HA in sludge compost to produce high-quality compost fertilizer. Feng et al. found that the addition of biochar and lactic acid increased the decomposition rate of organic matter by 157%, and increased the production and retention of nutrients in compost [59]. Zhou et al. proposed that the synergistic effect of biochar and manganese ore promoted the conversion of humus precursors and FA into stable HA, which was conducive to reducing plant toxicity and improving carbon fixation ability of sludge composts [60].

The mineralization and retention of nitrogen in the sludge-composting process is also the key to improving compost quality. N loss in the composting process will not only reduce the composting efficiency but also cause secondary pollution to the environment. Similar to the above description of reducing NH_3 emission by biochar, adding biochar to sludge compost can absorb NH_3 and N_2O and provide suitable growth environment for nitrifying bacteria to convert NH_3 into NO_3^-, thus achieving the purpose of reducing N loss. Jiang et al. proposed that the addition of biochar enhanced the activity of nitrogen-fixing bacteria and increased the NO_3-N content at the end of compost by 105.8% [52]. Qin et al. also found that the addition of biochar promoted the conversion of NH_4-N into NO_3-N, thus reducing the N loss in the composting process, and the NO_3-N content reached 3.11g/kg after 21 days composting [31].

In addition to N, biochar has a positive effect on the retention of organic carbon, K, Ca, Mg, P, and other nutrients [47]. These elements may be present in biochar itself due to the wide range of biochar feedstock. The surface of biochar is negatively charged, and K, Ca, and Mg can be fixed by electrostatic adsorption to prevent leaching from the compost. Wang et al. found that 4% biochar can increase TN, TP, and TK in sludge composts, and this increase is caused by microbial mineralization of organic carbon and the release of K from biochar [61]. Feng et al. proposed that the improvement of microbial activity and adsorption of organic molecules by biochar promoted the decomposition and mineralization of organic matter by microorganisms, which significantly increased the content of available P and available K in compost [59].

In summary, the improvement of microbial activity and the fixation of nutrient elements by biochar addition resulted in more nutrients retaining in sludge compost, thus improving the quality of compost.

9.3 Future Perspectives

The addition of biochar can improve the conditions of sludge composting, improve the microbial activity, promote the decomposition of organic matter, reduce the leaching of pollutants and gas emissions, and improve the quality of composting. The fertilizer produced by compost has positive effects on soil improvement and crop health. However, the large-scale application of biochar and sludge co-compost fertilizer in soil still needs further investigation. Here, we present some ideas that need further study:

1) *Biochar has important effect on sludge composting.* Standardized quantitative methods are needed to evaluate the effects of different feedstocks, production methods, and biochar properties (particle size, surface morphology and structure, surface chemistry) on the composting process and compost products.
2) *Heavy metal contamination may limit the use of compost in soil.* Although biochar can passivate and immobilize heavy metals, the total amount of heavy metals in compost is constant. Studies have shown that the presence of biochar can reduce the uptake of heavy metals by crops when compost is added to the soil, but in order to realize the practical application of compost products containing heavy metals, the mechanism and influencing factors of biochar fixation of heavy metals must be investigated in detail.
3) *The stability of sludge compost products used as soil fertilizer must be evaluated after medium- and long-term application.* Many studies have confirmed that biochar-added sludge compost products have positive effects on soil. However, long-term follow-up studies have not been reported, and the mechanism and changes of this fertilizer in soil over a long period of time still need to be studied. For example, the continuous application of sludge compost products as fertilizers can increase the N and P elements in soil, but will these elements cause eutrophication in water in the long-term application? Will heavy metals in compost leach into the soil due to other factors during long-term application?
4) *Life cycle evaluation and carbon balance calculation is needed.* This applies to the whole process of biochar and sludge co-composting fertilizer production and soil improvement so that its effect on the environment and carbon neutrality can be evaluated.

9.4 Summary

Biochar can effectively improve the composting condition and accelerate the composting efficiency in the sludge-composting process. The well-developed porous structure and abundant surface functional groups of biochar can reduce the leaching of heavy metals and organic matter, reduce greenhouse gas and NH_3 emissions, and prevent loss of nutrients. It is an economical and feasible carbon sequestration strategy to convert biomass into biochar, which is not easily degraded, and use it in sludge composting to produce valuable fertilizer. Biochar and sludge co-composted fertilizers have positive effects on soil improvement and crop health. Detailed mechanism exploration and long-term risk assessment are needed to determine the environmental impact of this strategy.

References

1 Liew, C.S., Kiatkittipong, W., Lim, J.W., Lam, M.K., Ho, Y.C., Ho, C.D., Ntwampe, S.K.O., Mohamad, M., Usman, A., 2021. Stabilization of heavy metals loaded sewage sludge: Reviewing conventional to state-of-the-art thermal treatments in achieving energy sustainability. *Chemosphere*, 277.

2 Du, J.J., Zhang, Y.Y., Qu, M.X., Yin, Y.T., Fan, K., Hu, B., Zhang, H.Z., Wei, M.B., Ma, C., 2019. Effects of biochar on the microbial activity and community structure during sewage sludge composting. *Bioresource Technology*, 272, 171–179.

3 Zhang, W.J., Wei, L.L., Zhou, H., Wang, D.S., 2022. Environmental impacts and optimizing strategies of municipal sludge treatment and disposal routes in China based on life cycle analysis. *Environment International*, 166.

4 Liew, C.S., Yunus, N.M., Chidi, B.S., Lam, M.K., Goh, P.S., Mohamad, M., Sin, J.C., Lam, S.M., Lim, J.W., Lam, S.S., 2022. A review on recent disposal of hazardous sewage sludge via anaerobic digestion and novel composting. *Journal of Hazardous Materials*, 423.

5 Wei, L.L., Zhu, F.Y., Li, Q.Y., Xue, C.H., Xia, X.H., Yu, H., Zhao, Q.L., Jiang, J.Q., Bai, S.W., 2020. Development, current state and future trends of sludge management in China: Based on exploratory data and CO_2-equivaient emissions analysis. *Environment International*, 144.

6 Chen, Z., Zheng, Z.J., He, C.L., Liu, J.M., Zhang, R., Chen, Q., 2022. Oily sludge treatment in subcritical and supercritical water: A review. *Journal of Hazardous Materials*, 433.

7 Jain, M., Upadhyay, M., Gupta, A.K., Ghosal, P.S., 2022. A review on the treatment of septage and faecal sludge management: A special emphasis on constructed wetlands. *Journal of Environmental Management*, 315.

8 Quan, L.M., Kamyab, H., Yuzir, A., Ashokkumar, V., Hosseini, S.E., Balasubramanian, B., Kirpichnikova, I., 2022. Review of the application of gasification and combustion technology and waste-to-energy technologies in sewage sludge treatment. *Fuel*, 316.

9 Ezzariai, A., Hafidi, M., Khadra, A., Aemig, Q., El Felsa, L., Barret, M., Merlina, G., Patureau, D., Pinelli, E., 2018. Human and veterinary antibiotics during composting of sludge or manure: Global perspectives on persistence, degradation, and resistance genes. *Journal of Hazardous Materials*, 359, 465–481.

10 Guo, X.X., Liu, H.T., Zhang, J., 2020. The role of biochar in organic waste composting and soil improvement: A review. *Waste Manage*, 102, 884–899.

11 Pan, J.T., Cai, H.Z., Zhang, Z.Q., Liu, H.B., Li, R.H., Mao, H., Awasthi, M.K., Wang, Q., Zhai, L.M., 2018. Comparative evaluation of the use of acidic additives on sewage sludge composting quality improvement, nitrogen conservation, and greenhouse gas reduction. *Bioresource Technology*, 270, 467–475.

12. Onwosi, C.O., Igbokwe, V.C., Odimba, J.N., Eke, I.E., Nwankwoala, M., Iroh, I.N., Ezeogu, L.I., 2017. Composting technology in waste stabilization: On the methods, challenges and future prospects. *Journal of Environmental Management*, 190, 140–157.
13. Xiao, R., Awasthi, M.K., Li, R.H., Park, J., Pensky, S.M., Wang, Q., Wang, J.J., Zhang, Z.Q., 2017. Recent developments in biochar utilization as an additive in organic solid waste composting: A review. *Bioresource Technology*, 246, 203–213.
14. Ayilara, M.S., Olanrewaju, O.S., Babalola, O.O., Odeyemi, O., 2020. Waste management through composting: Challenges and potentials. *Sustainability-Basel*, 12.
15. Chen, Z., Li, Y.Z., Peng, Y.Y., Mironov, V., Chen, J.X., Jin, H.X., Zhang, S.H., 2022. Feasibility of sewage sludge and food waste aerobic co-composting: Physicochemical properties, microbial community structures, and contradiction between microbial metabolic activity and safety risks. *Science of the Total Environment*, 825.
16. El Hayany, B., Rumpel, C., Hafidi, M., El Fels, L., 2022. Occurrence, analysis of microplastics in sewage sludge and their fate during composting: A literature review. *Journal of Environmental Management*, 317.
17. Osman, A.I., Fawzy, S., Farghali, M., El-Azazy, M., Elgarahy, A.M., Fahim, R.A., Maksoud, M., Ajlan, A.A., Yousry, M., Saleem, Y., Rooney, D.W., 2022. Biochar for agronomy, animal farming, anaerobic digestion, composting, water treatment, soil remediation, construction, energy storage, and carbon sequestration: A review. *Environmental Chemistry Letters*, 20, 2385–2485.
18. Zhang, L., Sun, X.Y., 2017. Addition of fish pond sediment and rock phosphate enhances the composting of green waste. *Bioresource Technology*, 233, 116–126.
19. Juarez, M.F.D., Prahauser, B., Walter, A., Insam, H., Franke-Whittle, I.H., 2015. Co-composting of biowaste and wood ash, influence on a microbially driven-process. *Waste Manage*, 46, 155–164.
20. Huang, D.L., Gao, L., Cheng, M., Yan, M., Zhang, G.X., Chen, S., Du, L., Wang, G.F., Li, R.J., Tao, J.X., Zhou, W., Yin, L.S., 2022. Carbon and N conservation during composting: A review. *Science of the Total Environment*, 840.
21. Thomson, A., Price, G.W., Arnold, P., Dixon, M., Graham, T., 2022. Review of the potential for recycling CO_2 from organic waste composting into plant production under controlled environment agriculture. *Journal of Cleaner Production*, 333.
22. Antonangelo, J.A., Sun, X., Zhang, H.L., 2021. The roles of co-composted biochar (COMBI) in improving soil quality, crop productivity, and toxic metal amelioration. *Journal of Environmental Management*, 277.
23. Godlewska, P., Schmidt, H.P., Ok, Y.S., Oleszczuk, P., 2017. Biochar for composting improvement and contaminants reduction: A review. *Bioresource Technology*, 246, 193–202.

24 Sanchez-Monedero, M.A., Cayuela, M.L., Roig, A., Jindo, K., Mondini, C., Bolan, N., 2018. Role of biochar as an additive in organic waste composting. *Bioresource Technology*, 247, 1155–1164.

25 Khan, N., Chowdhary, P., Gnansounou, E., Chaturvedi, P., 2021. Biochar and environmental sustainability: Emerging trends and techno-economic perspectives. *Bioresource Technology*, 332.

26 Wang, J.L., Wang, S.Z., 2019. Preparation, modification and environmental application of biochar: A review. *Journal of Cleaner Production*, 227, 1002–1022.

27 Pan, X.Q., Gu, Z.P., Chen, W.M., Li, Q.B., 2021. Preparation of biochar and biochar composites and their application in a Fenton-like process for wastewater decontamination: A review. *Science of the Total Environment*, 754.

28 Liu, X.M., Wang, Y.Q., Zhou, S.Q., Cui, P., Wang, W.W., Huang, W.F., Yu, Z., Zhou, S.G., 2022. Differentiated strategies of animal-derived and plant-derived biochar to reduce nitrogen loss during paper mill sludge composting. *Bioresource Technology*, 360, 10.

29 Zhou, S.X., Li, Y., Jia, P.Y., Wang, X., Kong, F.L., Jiang, Z.X., 2022. The co-addition of biochar and manganese ore promotes nitrous oxide reduction but favors methane emission in sewage sludge composting. *Journal of Cleaner Production*, 339, 10.

30 Liu, W., Huo, R., Xu, J.X., Liang, S.X., Li, J.J., Zhao, T.K., Wang, S.T., 2017. Effects of biochar on nitrogen transformation and heavy metals in sludge composting. *Bioresource Technology*, 235, 43–49.

31 Qin, X., Wu, X.S., Teng, Z.N., Lou, X.Y., Han, X.B., Li, Z.X., Han, Y.H., Zhang, F., Li, G., 2022. Effects of adding biochar on the preservation of nitrogen and passivation of heavy metal during hyperthermophilic composting of sewage sludge. *Journal of the Air and Waste Management*, 73, 15–24.

32 Gondek, K., Mierzwa-Hersztek, M., Kopec, M., 2018. Mobility of heavy metals in sandy soil after application of composts produced from maize straw, sewage sludge and biochar. *Journal of Environmental Management*, 210, 87–95.

33 Lu, H., Chen, X.H., Mo, C.H., Huang, Y.H., He, M.Y., Li, Y.W., Feng, N.X., Katsoyiannis, A., Cai, Q.Y., 2021. Occurrence and dissipation mechanism of organic pollutants during the composting of sewage sludge: A critical review. *Bioresource Technology*, 328, 124847.

34 Aemig, Q., Doussiet, N., Danel, A., Delgenes, N., Jimenez, J., Houot, S., Patureau, D., 2019. Organic micropollutants' distribution within sludge organic matter fractions explains their dynamic during sewage sludge anaerobic digestion followed by composting. *Environmental Science and Pollution Research*, 26, 5820–5830.

35 Stefaniuk, M., Oleszczuk, P., 2016. Addition of biochar to sewage sludge decreases freely dissolved PAHs content and toxicity of sewage sludge-amended soil. *Environmental Pollution*, 218, 242–251.

36 Pang, L., Huang, Z., Yang, P., Wu, M., Zhang, Y., Pang, R., Jin, B., Zhang, R., 2023. Effects of biochar on the degradation of organophosphate esters in sewage sludge aerobic composting. *Journal of Hazardous Materials*, 442, 130047.

37 Ouyang, X., Lin, H.Y., Hu, Z.B., Zheng, Y.K., Li, P.Y., Huang, W.B., 2022. Effect of biochar structure on H_2S emissions during sludge aerobic composting: Insights into microscale characterization and microbial mechanism. *Biomass Conversion and Biorefinery*.

38 Xue, S.D., Zhou, L.A., Zhong, M.Z., Awasthi, M.K., Mao, H., 2021. Bacterial agents affected bacterial community structure to mitigate greenhouse gas emissions during sewage sludge composting. *Bioresource Technology*, 337, 8.

39 Zukowska, G., Mazurkiewicz, J., Myszura, M., Czekala, W., 2019. Heat energy and gas emissions during composting of sewage sludge. *Energies*, 12, 13.

40 Wang, M.J., Awasthi, M.K., Wang, Q., Chen, H.Y., Ren, X.N., Zhao, J.C., Li, R.H., Zhang, Z.Q., 2017. Comparison of additives amendment for mitigation of greenhouse gases and ammonia emission during sewage sludge co-composting based on correlation analysis. *Bioresource Technology*, 243, 520–527.

41 Awasthi, M.K., Wang, M.J., Chen, H.Y., Wang, Q., Zhao, J.C., Ren, X.N., Li, D.S., Awasthi, S.K., Shen, F., Li, R.H., Zhang, Z.Q., 2017. Heterogeneity of biochar amendment to improve the carbon and nitrogen sequestration through reduce the greenhouse gases emissions during sewage sludge composting. *Bioresource Technology*, 224, 428–438.

42 Ge, J.Y., Huang, G.Q., Li, J.B., Sun, X.X., Han, L.J., 2018. Multivariate and multiscale approaches for interpreting the mechanisms of nitrous oxide emission during pig manure-wheat straw aerobic composting. *Environmental Science & Technology*, 52, 8408–8418.

43 Awasthi, M.K., Zhang, Z.Q., Wang, Q., Shen, F., Li, R.H., Li, D.S., Ren, X.N., Wang, M.J., Chen, H.Y., Zhao, J.C., 2017. New insight with the effects of biochar amendment on bacterial diversity as indicators of biomarkers support the thermophilic phase during sewage sludge composting. *Bioresource Technology*, 238, 589–601.

44 Yin, Y.A., Yang, C., Li, M.T., Zheng, Y.C., Ge, C.J., Gu, J., Li, H.C., Duan, M.L., Wang, X.C., Chen, R., 2021. Research progress and prospects for using biochar to mitigate greenhouse gas emissions during composting: A review. *Science of the Total Environment*, 798.

45 Zhou, Q., Liu, G., Hu, Z., Zheng, Y., Lin, Z., Li, P., 2022. Impact of different structures of biochar on decreasing methane emissions from sewage sludge composting. *Waste Management Research*, 41, 723–732.

46 Dume, B., Hanc, A., Svehla, P., Michal, P., Chane, A.D., Nigussie, A., 2021. Carbon dioxide and methane emissions during the composting and vermicomposting of sewage sludge under the effect of different proportions of straw pellets. *Atmosphere-Basel*, 12, 13.

47 Agyarko-Mintah, E., Cowie, A., Van Zwieten, L., Singh, B.P., Smillie, R., Harden, S., Fornasier, F., 2017. Biochar lowers ammonia emission and improves nitrogen retention in poultry litter composting. *Waste Manage*, 61, 129–137.

48 He, S., Ding, L.L., Pan, Y., Hu, H.D., Ye, L., Ren, H.Q., 2018. Nitrogen loading effects on nitrification and denitrification with functional gene quantity/transcription analysis in biochar packed reactors at 5 degrees C. *Scientific Reports-UK*, 8.

49 Yang, Y.J., Awasthi, M.K., Du, W., Ren, X.N., Lei, T., Lv, J.L., 2020. Compost supplementation with nitrogen loss and greenhouse gas emissions during pig manure composting. *Bioresource Technology*, 297.

50 Meng, L.Q., Zhang, S.M., Gong, H.N., Zhang, X.C., Wu, C.D., Li, W.G., 2018. Improving sewage sludge composting by addition of spent mushroom substrate and sucrose. *Bioresource Technology*, 253, 197–203.

51 Ma, T., 2020. Effect of processing conditions on nitrogen loss of sewage sludge composting. *Compost Science and Utilization*, 28, 117–128.

52 Jiang, J.S., Wang, Y., Yu, D., Zhu, G.F., Cao, Z.G., Yan, G.X., Li, Y.B., 2021. Comparative evaluation of biochar, pelelith, and garbage enzyme on nitrogenase and nitrogen-fixing bacteria during the composting of sewage sludge. *Bioresource Technology*, 333, 10.

53 Ding, Y., Xiong, J.S., Zhou, B.W., Wei, J.J., Qian, A.A., Zhang, H.J., Zhu, W.Q., Zhu, J., 2019. Odor removal by and microbial community in the enhanced landfill cover materials containing biochar-added sludge compost under different operating parameters. *Waste Manage*, 87, 679–690.

54 Awasthi, M.K., Duan, Y.M., Awasthi, S.K., Liu, T., Zhang, Z.Q., 2020. Influence of bamboo biochar on mitigating greenhouse gas emissions and nitrogen loss during poultry manure composting. *Bioresource Technology*, 303.

55 Zhou, G.X., Qiu, X.W., Wu, X.Y., Lu, S.B., 2021. Horizontal gene transfer is a key determinant of antibiotic resistance genes profiles during chicken manure composting with the addition of biochar and zeolite. *Journal of Hazardous Materials*, 408.

56 Malinowski, M., Wolny-Koladka, K., Vaverkova, M.D., 2019. Effect of biochar addition on the OFMSW composting process under real conditions. *Waste Manage*, 84, 364–372.

57 Guo, X.X., Liu, H.T., Wu, S.B., 2019. Humic substances developed during organic waste composting: Formation mechanisms, structural properties, and agronomic functions. *Science of the Total Environment*, 662, 501–510.

58 Zhou, S.X., Kong, F.L., Lu, L., Wang, P., Jiang, Z.X., 2022. Biochar – An effective additive for improving quality and reducing ecological risk of compost: A global meta-analysis. *Science of the Total Environment*, 806.

59 Feng, X.Q. and Zhang, L., 2022. Combined addition of biochar, lactic acid, and pond sediment improves green waste composting. *Science of the Total Environment*, 852.

60 Zhou, S.X., Zhang, C.Y., Xu, H.X., Jiang, Z.X., 2022. Co-applying biochar and manganese ore can improve the formation and stability of humic acid during co-composting of sewage sludge and corn straw. *Bioresource Technology*, 358.
61 Wang, X., Chu, Z., Fan, T., Liang, S., Li, G., Zhang, J., Zhen, Q., 2022. Application of rice husk biochar and earthworm on concentration and speciation of heavy metals in industrial sludge treatment. *International Journal of Environmental Research and Public Health*, 19, 14.
62 Li, Y.B., Jin, P.F., Liu, T.T., Lv, J.H., Jiang, J.S., 2019. A novel method for sewage sludge composting using bamboo charcoal as a separating material. *Environmental Science and Pollution Research*, 26, 33870–33881.
63 Jiang, J., Wang, Y., Yu, D., Hou, R., Ma, X., Liu, J., Cao, Z., Cheng, K., Yan, G., Zhang, C., Li, Y., 2022. Combined addition of biochar and garbage enzyme improving the humification and succession of fungal community during sewage sludge composting. *Bioresource Technology*, 346, 126344.
64 Glab, T., Zabinski, A., Sadowska, U., Gondek, K., Kopec, M., Mierzwa-Hersztek, M., Tabor, S., Stanek-Tarkowska, J., 2020. Fertilization effects of compost produced from maize, sewage sludge and biochar on soil water retention and chemical properties. *Soil and Tillage Research*, 197, 10.
65 Malinska, K., Golanska, M., Caceres, R., Rorat, A., Weisser, P., Slezak, E., 2017. Biochar amendment for integrated composting and vermicomposting of sewage sludge – The effect of biochar on the activity of *Eisenia fetida* and the obtained vermicompost. *Bioresource Technology*, 225, 206–214.
66 Fu, Y., Zhang, A., Guo, T., Zhu, Y., Shao, Y., 2021. Biochar and hyperthermophiles as additives accelerate the removal of antibiotic resistance genes and mobile genetic elements during composting. *Materials (Basel)*, 14, 13.
67 Du, J.J., Zhang, Y.Y., Hu, B., Qv, M.X., Ma, C., Wei, M.B., Zhang, H.Z., 2019. Insight into the potentiality of big biochar particle as an amendment in aerobic composting of sewage sludge. *Bioresource Technology*, 288, 9.
68 Kopec, M., Baran, A., Mierzwa-Hersztek, M., Gondek, K., Chmiel, M.J., 2018. Effect of the addition of biochar and coffee grounds on the biological properties and ecotoxicity of composts. *Waste and Biomass Valorization*, 9, 1389–1398.

10

Sludge Utilization as Biochar for Nutrient Recovery

Deng Pan[1], Dongdong Ge[2], and Daniel C.W. Tsang[3]

[1] EIT Institute for Advanced Study, Ningbo, China
[2] School of Environmental Science & Engineering, Shanghai Jiao Tong University, Shanghai, China
[3] State Key Laboratory of Clean Energy Utilization, Zhejiang University, China

10.1 Sewage Sludge (SS) Management

Various forms of sludge are produced at the distinctive processing stages in a wastewater treatment facility. Primary sludge is first generated when heavy particles, grease, and oil are removed during the primary treatment, and it contains 2%–9% solids and approximately 90% moisture (Gherghel et al., 2019). Secondary sludge, also known as activated sludge, derives from wastewater biological treatment, where microorganisms break down the biodegradable organic components for metabolism. Generally, the activated sludge consists of 50%–55% carbon (C), 25%–30% oxygen (O), 10%–15% nitrogen (N), 6%–10% hydrogen (H), 1%–3% phosphorous (P), and 0.5%–1.5% sulfur (S). The total solids concentration ranges from 0.8% to 3.3%, depending on the differential biological treatment procedures (Gherghel et al., 2019). To enhance the quality of the effluent in the environment, the tertiary sludge is produced in the advanced wastewater treatment stage for the further removal of nutrients, such as N and P (Mustapha et al., 2017).

There is a potential issue concerning the reuse and disposal of sewage sludge. The overall amount of sludge produced globally has greatly grown, and this trend is anticipated to continue (Yuan et al., 2015). The production, storage, usage, and disposal of sewage sludge are linked to significant environmental issues (Li et al., 2018; Tarrago et al., 2017; Villar et al., 2016), due to its propensity for fermentation

Biochar Applications for Wastewater Treatment, First Edition. Edited by Daniel C.W. Tsang and Yuqing Sun.
© 2023 John Wiley & Sons, Inc. Published 2023 by John Wiley & Sons, Inc.

and the presence of hazardous organic and inorganic compounds, such as pathogenic organisms and heavy metals (Villar et al., 2016). Sewage sludge has an immediate impact on leaching production and CO_2 released into the atmosphere when it is stored in landfills (Kacprzak et al., 2017). Also, vast amounts of waste are produced in settling, causing the challenging management and disposal in sewage treatment facilities (Villar et al., 2016). Environmental issues like odors or an increase in waste volume may be avoided with the use of suitable sludge treatment and disposal techniques, and the nutrient in the sludge can be recovered finally (Cao et al., 2016).

10.2 Importance of Sludge as a Feedstock for Biochar

It is necessary to cleanse sewage sludge before using it as a soil addition or fertilizer for lowering its water content, unpleasant smell, and harmful organism content (Turunen et al., 2018). The quality and safety of applying sewage sludge on the soil surface are essential. Therefore, proper sewage sludge treatment should be given significant consideration in the majority of industrialized nations (Fijalkowski et al., 2017). For instance, Ahmad et al. (2022) examined the effects of using biochar with soil and provided evidence that sludge-derived biochar (SDBC) was valuable because it can increase plant biomass, microbial diversity, and soil nutrients.

The amount of organic matter, N, P, potassium (K), and pollutants, as well as their presence, impact the effectiveness of fertilizers made from sewage sludge. Noticeably, the manufacture of organic fertilizers or organic minerals from sewage sludge is currently popular (Grobelak et al., 2019).

From an economic and ecological point of view, pyrolysis is regarded as a suitable technology to transform sewage sludge into sludge-derived biochar (SDBC) (Manara and Zabaniotou, 2012). Pyrolysis can reduce the volume of sludge to only 20% of what it was before. Pathogens and dangerous compounds are removed and the metal is fixed in the solid residue and will not exude (Li et al., 2018; Wang et al., 2017). The biochar produced by pyrolyzing sewage sludge is rich in nutrients and without pathogenic organisms (Song et al., 2014).

10.3 Factors Affecting the Properties of SDBC

The properties of SDBC will be affected by the raw material, temperature, heating rates, and retention time. Table 10.1 shows the different process parameters for SDBC.

Table 10.1 Process parameters for SDBC.

Raw materials	Type of process	Temp. (°C)	Heating rate (°C/min)	Retention Time (h)	Volatile content (%)	Fixed carbon content (%)	Ash content (%)	Pore volume	BET surface area (m2/g)	References
Electroplating industry sludge	Carbonization	550	–	1	22.87	–	–	–	19.6	Bhatnagar et al. (2008)
Electroplating industry sludge	Pyrolysis	500	–	1	–	21.6	–	–	19.6	Bhatnagar and Minocha (2009)
Paper industry sludge	Carbonization	650	3	2	–	–	81	42.51 cm^3/g	87.4	Méndez et al. (2009)
Paper industry sludge	Pyrolysis	650	3	2	–	–	–	42.51 nm	88.4	Méndez et al. (2010)
Sewage sludge	carbonization	600	–	2	–	–	–	–	51	Bouzid et al. (2008)
Sewage sludge	Pyrolysis	500	15	3	–	–	–	10 cm^3/g	–	Fan and Zhang (2008)
Sewage sludge	Pyrolysis	950	10	1	–	–	–	0.049 cm^3/g	116	Pietrzak and Bandosz (2008)
Sewage sludge	Pyrolysis	950	10	1	–	–	–	0.05 cm^3/g	105	Seredych et al. (2008)
Sewage sludge	Pyrolysis	950	10	1	–	–	–	0.352 cm^3/g	181	Gutiérrez-Segura et al. (2009)

(Continued)

Table 10.1 (Continued)

Raw materials	Type of process	Temp. (°C)	Heating rate (°C/min)	Retention Time (h)	Volatile content (%)	Fixed carbon content (%)	Ash content (%)	Pore volume	BET surface area (m2/g)	References
Sewage sludge	Pyrolysis	500	–	1	–	–	–	0.84 nm	100	Gutiérrez-Segura et al. (2009)
Sewage sludge	Pyrolysis	800	10	–	46.04	5.66	42.58	–	–	Xiaohua and Jiancheng (2012)
Sewage sludge	Carbonization	550	10	2	–	–	–	–	24.73	Lu et al. (2012)
Sewage sludge	Pyrolysis	300	–	1	73.7	0.4	25.9	–	51	Agrafioti et al. (2013)
Sewage sludge	Pyrolysis	600	10	20	16.7	4.7	78.53	0.007 cm^3/g	37.18	Méndez et al. (2013)
Tannery sludge	Carbonization	600	–	–	23.85	55.9	18.75	–	167.4	Geethakarthi and Phanikumar (2011)

Adapted from Singh et al. (2020).

10.3.1 Raw Material

The qualities of the raw material used to manufacture biochar and the technique used to extract it affect the biochar's basic composition, together with some harmful substances, including heavy metals and polycyclic aromatic hydrocarbons (Cha et al., 2016). Compared with biochar made from wood, which has a low nutritional level, sewage sludge has a high nutrient concentration. While the characteristics of biochar created at extremely high temperatures differ from the employed raw material, those produced at moderate temperatures share a chemical composition with the biomass used for pyrolysis (Dahlawi et al., 2018).

10.3.2 Temperature

Temperature is a crucial thermodynamic factor that affects the structural and physicochemical properties of SDBC (Zhao et al., 2017). Generally speaking, low pyrolysis temperatures result in lesser biochar adsorption capacity, whereas biochar produced at high temperatures (500–1000 °C) covers a larger area and has more microporosity (Qambrani et al., 2017). In detail, temperature increases are accompanied by an increase in the concentration of components, including calcium (Ca), K, and P, as well as the ratio of C : N and C : O in biochar (Dahlawi et al., 2018). The increased temperature during pyrolysis causes increased biochar pH, larger surface area, and improved hydrophobicity, whereas low pyrolysis temperatures (500 °C) promote partial carbonization and produce biochar with smaller pores, less surface area, and higher O content (Oliveira et al., 2017). Additionally, it was noted that high moisture content in the sludge results in steam-enriched conditions at high temperatures, which causes biochar to partially gasify and accelerate the formation of micropores (Rajapaksha et al., 2016).

10.3.3 Heating Rates

The time it takes to achieve the temperature for the biomass reaction is called the heating rate (Hadi et al., 2015). The structure of biochar and the generation of bio-oil and biogas are significantly impacted by heating rates (Gul et al., 2015). Processes with heating rates in the range of 1 °C/min to 100 °C/min are called slow pyrolysis processes, and processes with heating rates higher than 1000 °C/min are called fast pyrolysis processes (Montoya et al., 2015). Slow heating rates are more conducive to the creation of biochar, whereas faster heating rates have a good effect on the production of bio-oil, resulting in high yields of bio-oil from biomass with noncondensable gasses and solid biochar (Qambrani et al., 2017).

10.3.4 Retention Time

According to Hadi et al. (2015), when pyrolyzing sewage sludge, a shorter retention time is needed at a high carbonization temperature. According to Méndez et al. (2009), the ideal retention time is 4 h at 500 °C and 2 h at 650 °C. It is also noted that at 650 °C, mesopores and BET surface area decrease while micropore production is consistent with an increase in residence time (Smith et al., 2009). In addition, increasing stay time reduces the micropore volume of biochar at the same carbonization temperature (Mohan et al., 2014).

10.4 Nutrients in SDBC

10.4.1 Nitrogen (N)

Due to its high demand of crops and the potential for losses through leaching, runoff, and volatilization, N is one of the elements in soils that is most limiting for plant development and production (Nguyen et al., 2017). For many agricultural soils to continue producing during cropping seasons, N must be continuously applied in all of its forms (Fageria and Baligar, 2005).

Plants may be able to obtain N from biochar. In addition to organic forms of N (such as water-soluble, nonhydrolyzable, and hydrolysable forms), biochar also contains N in inorganic forms (Liu et al., 2019). Despite the fact that most biomasses have low levels of N, during pyrolysis, the biomass loses mass (mostly moisture), which increases the N concentration. Due to the element's gaseous emissions, there may be some losses of N during the pyrolysis of biomass. Because of this, not all of the N species present in the feedstock can be found in biochar. For instance, during biomass pyrolysis, some amino acids, like arginine, which possesses amide groups, are mostly transformed into ammonia or other gaseous forms of N (Leng et al., 2020). Moreover, the conversion of N-containing compounds and the quantity and types of N species in final biochar products may be affected by the presence of metal elements in the feedstock (Xiao et al., 2018).

The N content of the SDBC dropped as the pyrolysis temperature rose (Yuan et al., 2016). The elimination of N by volatilization was thought to be the cause of its loss in the higher-temperature SDBC (Cao and Harris, 2010). Furthermore, part of the nitrogen in the sewage sludge was converted into tar-N, and some of it was converted into gas-N, including N_2, and NH_3 (Liu et al., 2015). N is coupled to numerous organic molecules and begins to volatilize at a very low temperature (i.e., 200 °C) since it is mostly present as protein-N in sewage sludge.

Tian et al. (2014) investigated the N speciation in the SDBC made at temperatures between 150 °C and 800 °C. Raw sludge's pyrrolic and pyridinic N were converted during pyrolysis into biochar that was mostly rich in heterocyclic-N. In addition, amine, nitrile, and other organic N compounds were included in the sludge biochar. However, when these N species gradually changed, SDBC began to generate new N compounds, primarily including amino, pyridine, pyrrole, quaternary N, NH_4^+, NO_2^-, and NO_3^-. As a result of its ability to prevent the creation of volatile species such as NO, NH_3, HCN compounds, and other compounds, Tian et al. (2017) indicated that retention of N is considerably higher in the CO_2 atmosphere than in the N_2. Thus, the manufacture of N-enriched biochar was often favored by the CO_2 atmosphere and lower pyrolysis temperature (preferred to be around 300 °C).

10.4.2 Phosphorus (P)

The P concentration is positively associated with the pyrolytic temperature (Hossain et al., 2020). The concentration effect brought on by the lower biochar production with increased temperature can be blamed for the rise in P content in biochar with higher pyrolytic temperatures (Bridle and Pritchard, 2004).

According to Xu et al. (2017), in comparison to woody-based biochar, agricultural residue-based biochar, and animal dung-based biochar, SDBC showed higher P concentrations. Additionally, they found that the quantity of accessible P (also known as Olsen-P) in biochar increased from 280 mg/kg to 676 mg/kg when the pyrolytic temperature went from 300 °C to 600 °C. Yuan et al. (2016) mentioned that with the increase of pyrolysis temperature from 300 °C to 700 °C, the PO_4^{3-} content increased from 280 mg/kg to 676 mg/kg. However, the SDBC can still function as a useful slow-release P resource due to the low proportion of water-soluble PO_4^{3-} ranging from 0.72% to 1.37%.

10.4.3 Potassium (K)

The K concentration in biochar also fluctuates depending on the pyrolysis temperature and kind of feedstock. Similar to P, the concentration effect may be used to explain how the K content of biochar rises with rising pyrolytic temperature (Hossain et al., 2020). A rise in the K content in biochar produced from giant reeds at temperatures between 300 °C and 600 °C was also noted by Zheng et al. (2013).

According to Yuan et al. (2016), as the pyrolysis temperature rose, the water-soluble K content in the SDBC fell from 1236 mg/kg at 300 °C to 316 mg/kg at 700 °C. As a result, the fraction of water-soluble K decreased from 16.55% to 1.90%. The majority of the lost K was in the form of water-soluble state.

10.5 SDBC for Soil Amendment and Nutrient Utilization

In order to lessen the amount of sludge and remediate the soil, sludge can be converted to biochar and used as a material for soil amendment. In agronomic application, the presence of micronutrients, macronutrients, and some organic matter is crucial, demonstrating the benefits of employing sludge as a soil supplement.

Figure 10.1 shows the properties of SDBC for soil amendment and nutrient utilization. As a result of its use in agriculture, the pH of the soil drops while it increases N fixation, which boosts crop output. By doing this, farmers become less reliant on suppliers and the use of fertilizers. According to research by Hossain et al. (2010), the soil cation exchange capacity (CEC) increases by 40% with the application of SDBC. This improved plant availability of nutrients and carbon sequestration make crops more drought-resistant. SDBC also benefits soil microbial biomass, such as arbuscular mycorrhizal fungi and earthworms (Atkinson et al., 2010).

The quality of nutrient-poor soils, such as acidic, dry, humid, and tropical soil, is improved; adding SDBC thus leads to increased crop yields. According to Callegari and Capodaglio (2018), biochar can boost soil retention of N and P fertilizers and decrease the migration of various organic and inorganic contaminants, such as heavy metals, in a mobile form.

Biochar can improve plant absorption of N, reduce soil N_2O emissions, and reduce soil N leaching (Liu et al., 2018). Figure 10.2 shows the impact of biochar on soil nitrogen. Chu et al. (2020) researched the SDBC's N-fertilizer retention

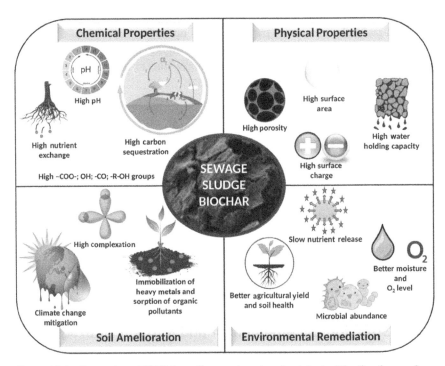

Figure 10.1 Properties of SDBC for soil amendment and nutrient utilization from ref (Simranjeet Singh et al.2020 / with permission from Elsevier).

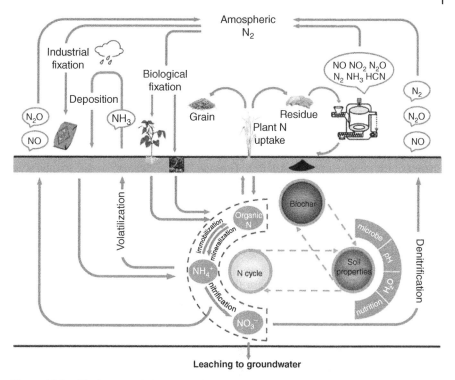

Figure 10.2 The impact of biochar on the soil nitrogen from ref (Liu et al., 2021 / from Springer Nature).

capability to help keep the soil fertile. They discovered that SDBC could reduce ammonia volatilization, thereby boosting soil N retention and rice fertilizer uptake. P-fertilizer is crucial for soil as well. Frišták et al. (2018) created P-fertilizer by subjecting sludge to pyrolysis, and the resultant sludge biochar possessed stable and valuable compounds, and the P content could be enhanced by two to three times.

10.6 Current Challenges for SDBC

Since sewage sludge may contain heavy metals and organic compounds harmful to the soil and water, an adequate ecotoxicological analysis should be done prior to applying biosolids to the soil (Groth et al., 2016). However, chemical analysis is insufficient for estimating pollutant levels following the use of sludge biochar in the soil as a fertilizer. Given that biochar is mostly utilized in agriculture, it is crucial to determine how hazardous it is to various organisms or groups of organisms (Zielińska and Oleszczuk, 2015). A consequence of biochar can be its direct effect on organisms because of its composition of organic or inorganic contaminants. Biochar can have both beneficial and negative effects on plants, animals, and microbes. In order to achieve minimal toxicity levels, it is advised to employ suitable pyrolysis parameters; nevertheless, not all forms of biochar are appropriate for use in agriculture (Wang et al., 2018).

10.7 Conclusions

Numerous sewage sludge recycling and reuse techniques have emerged as a result of the slowly growing amount of sewage sludge. Sewage sludge pyrolysis is recognized as an appropriate technology for the practical reuse of sewage sludge to produce SDBC from an economic and ecological perspective. The characteristics of SDBC are influenced by various compositions of raw material and process conditions. Biochar produced at high temperatures (500–1000 °C) covers a wider area and has higher microporosity than the biochar prepared at low pyrolysis temperatures, which has a lower adsorption capacity. Slow pyrolysis was demonstrated to result in a large increase in surface area, while quick pyrolysis was proven to have little effect on the ratio of C : O in biochar. Slower heating rates are better for producing biochar, while faster heating rates work well for producing bio-oil. It is crucial to ascertain whether SDBC has a detrimental influence on the soil and how it affects plant growth. The form and content of the main nutrients of SDBC were discussed. As the pyrolysis temperature increases, the N and water-soluble N content of the SDBC decrease while the P and water-soluble P and K content increase. This chapter discussed how SDBC can be used to alter soil and use nutrients. For its agricultural application, it could decrease the pH of the soil and increase N fixation, thereby improving crop yield. Moreover, SDBC also enhances the quality of nutrient-poor soils, including acidic, dry, humid, and tropical soil.

References

Agrafioti, E., Bouras, G., Kalderis, D., Diamadopoulos, E., 2013. Biochar production by sewage sludge pyrolysis. *Journal of Analytical and Applied Pyrolysis*, 101, 72–78.

Ahmad, A., Chowdhary, P., Khan, N., Chaurasia, D., Varjani, S., Pandey, A., Chaturvedi, P., 2022. Effect of sewage sludge biochar on the soil nutrient, microbial abundance, and plant biomass: A sustainable approach towards mitigation of solid waste. *Chemosphere*, 287, 132112.

Atkinson, C.J., Fitzgerald, J.D., Hipps, N.A., 2010. Potential mechanisms for achieving agricultural benefits from biochar application to temperate soils: A review. *Plant and Soil*, 337(1), 1–18.

Bhatnagar, A. and Minocha, A.K., 2009. Utilization of industrial waste for cadmium removal from water and immobilization in cement. *Chemical Engineering Journal*, 150(1), 145–151.

Bhatnagar, A., Minocha, A.K., Pudasainee, D., Chung, H.-K., Kim, S.-H., Kim, H.-S., Lee, G., Min, B., Jeon, B.-H., 2008. Vanadium removal from water by waste metal sludge and cement immobilization. *Chemical Engineering Journal*, 144(2), 197–204.

Bouzid, J., Elouear, Z., Ksibi, M., Feki, M., Montiel, A., 2008. A study on removal characteristics of copper from aqueous solution by sewage sludge and pomace ashes. *Journal of Hazardous Materials*, 152(2), 838–845.

Bridle, T. and Pritchard, D., 2004. Energy and nutrient recovery from sewage sludge via pyrolysis. *Water Science and Technology*, 50(9), 169–175.

Callegari, A. and Capodaglio, A.G., 2018. Properties and beneficial uses of (bio) chars, with special attention to products from sewage sludge pyrolysis. *Resources*, 7(1), 20.

Cao, X. and Harris, W., 2010. Properties of dairy-manure-derived biochar pertinent to its potential use in remediation. *Bioresource Technology*, 101(14), 5222–5228.

Cao, X., Jiang, Z., Cui, W., Wang, Y., Yang, P., 2016. Rheological properties of municipal sewage sludge: Dependency on solid concentration and temperature. *Procedia Environmental Sciences*, 31, 113–121.

Cha, J.S., Park, S.H., Jung, S.-C., Ryu, C., Jeon, J.-K., Shin, M.-C., Park, Y.-K., 2016. Production and utilization of biochar: A review. *Journal of Industrial and Engineering Chemistry*, 40, 1–15.

Chu, Q., Xue, L., Singh, B.P., Yu, S., Müller, K., Wang, H., Feng, Y., Pan, G., Zheng, X., Yang, L., 2020. Sewage sludge-derived hydrochar that inhibits ammonia volatilization, improves soil nitrogen retention and rice nitrogen utilization. *Chemosphere*, 245, 125558.

Dahlawi, S., Naeem, A., Rengel, Z., Naidu, R., 2018. Biochar application for the remediation of salt-affected soils: Challenges and opportunities. *Science of the Total Environment*, 625, 320–335.

Fageria, N.K. and Baligar, V.C., 2005. Enhancing Nitrogen Use Efficiency in Crop Plants. in: D. L. Spark *Advances in Agronomy*. Academic Press, Vol. 88, pp. 97–185.

Fan, X. and Zhang, X., 2008. Adsorption properties of activated carbon from sewage sludge to alkaline-black. *Materials Letters*, 62(10–11), 1704–1706.

Fijalkowski, K., Rorat, A., Grobelak, A., Kacprzak, M.J., 2017. The presence of contaminations in sewage sludge–The current situation. *Journal of Environmental Management*, 203, 1126–1136.

Frišták, V., Pipíška, M., Soja, G., 2018. Pyrolysis treatment of sewage sludge: A promising way to produce phosphorus fertilizer. *Journal of Cleaner Production*, 172, 1772–1778.

Geethakarthi, A., Phanikumar, B., 2011. Adsorption of reactive dyes from aqueous solutions by tannery sludge developed activated carbon: Kinetic and equilibrium studies. *International Journal of Environmental Science & Technology*, 8(3), 561–570.

Gherghel, A., Teodosiu, C., De Gisi, S., 2019. A review on wastewater sludge valorisation and its challenges in the context of circular economy. *Journal of Cleaner Production*, 228, 244–263.

Grobelak, A., Grosser, A., Kacprzak, M., Kamizela, T., 2019. Sewage sludge processing and management in small and medium-sized municipal wastewater treatment plant-new technical solution. *Journal of Environmental Management*, 234, 90–96.

Groth, V.A., Carvalho-Pereira, T., da Silva, E.M., Niemeyer, J.C., 2016. Ecotoxicological assessment of biosolids by microcosms. *Chemosphere*, 161, 342–348.

Gul, S., Whalen, J.K., Thomas, B.W., Sachdeva, V., Deng, H., 2015. Physico-chemical properties and microbial responses in biochar-amended soils: Mechanisms and future directions. *Agriculture, Ecosystems & Environment*, 206, 46–59.

Gutiérrez-Segura, E., Colín-Cruz, A., Fall, C., Solache-Ríos, M., Balderas-Hernández, P., 2009. Comparison of Cd–Pb adsorption on commercial activated carbon and carbonaceous material from pyrolysed sewage sludge in column system. *Environmental Technology*, 30(5), 455–461.

Hadi, P., Xu, M., Ning, C., Lin, C.S.K., McKay, G., 2015. A critical review on preparation, characterization and utilization of sludge-derived activated carbons for wastewater treatment. *Chemical Engineering Journal*, 260, 895–906.

Hossain, M.K., Strezov, V., Chan, K.Y., Nelson, P.F., 2010. Agronomic properties of wastewater sludge biochar and bioavailability of metals in production of cherry tomato (Lycopersicon esculentum). *Chemosphere*, 78(9), 1167–1171.

Hossain, M.Z., Bahar, M.M., Sarkar, B., Donne, S.W., Ok, Y.S., Palansooriya, K.N., Kirkham, M.B., Chowdhury, S., Bolan, N., 2020. Biochar and its importance on nutrient dynamics in soil and plant. *Biochar*, 2(4), 379–420.

Kacprzak, M., Neczaj, E., Fijałkowski, K., Grobelak, A., Grosser, A., Worwag, M., Rorat, A., Brattebo, H., Almås, Å., Singh, B.R., 2017. Sewage sludge disposal strategies for sustainable development. *Environmental Research*, 156, 39–46.

Leng, L., Xu, S., Liu, R., Yu, T., Zhuo, X., Leng, S., Xiong, Q., Huang, H., 2020. Nitrogen containing functional groups of biochar: An overview. *Bioresource Technology*, 298, 122286.

Li, M., Tang, Y., Ren, N., Zhang, Z., Cao, Y., 2018. Effect of mineral constituents on temperature-dependent structural characterization of carbon fractions in sewage sludge-derived biochar. *Journal of Cleaner Production*, 172, 3342–3350.

Liu, H., Zhang, Q., Hu, H., Liu, P., Hu, X., Li, A., Yao, H. (2015). Catalytic role of conditioner CaO in nitrogen transformation during sewage sludge pyrolysis. *Proceedings of the Combustion Institute*, 35(3), 2759–2766.

Liu, L., Tan, Z., Gong, H., Huang, Q., 2019. Migration and transformation mechanisms of nutrient elements (N, P, K) within biochar in Straw–Biochar–Soil–Plant systems: A review. *ACS Sustainable Chemistry & Engineering*, 7(1), 22–32.

Liu, Q., Zhang, Y., Liu, B., Amonette, J.E., Lin, Z., Liu, G., Ambus, P., Xie, Z., 2018. How does biochar influence soil N cycle? A meta-analysis. *Plant and Soil*, 426(1), 211–225.

Lu, H., Zhang, W., Yang, Y., Huang, X., Wang, S., Qiu, R., 2012. Relative distribution of Pb2+ sorption mechanisms by sludge-derived biochar. *Water Research*, 46(3), 854–862.

Manara, P. and Zabaniotou, A., 2012. Towards sewage sludge based biofuels via thermochemical conversion – A review. *Renewable and Sustainable Energy Reviews*, 16(5), 2566–2582.

Méndez, A., Barriga, S., Fidalgo, J., Gascó, G., 2009. Adsorbent materials from paper industry waste materials and their use in Cu (II) removal from water. *Journal of Hazardous Materials*, 165(1–3), 736–743.

Méndez, A., Barriga, S., Saa, A., Gascó, G., 2010. Removal of malachite green by adsorbents from paper industry waste materials: Thermal analysis. *Journal of Thermal Analysis and Calorimetry*, 99(3), 993–998.

Méndez, A., Terradillos, M., Gascó, G., 2013. Physicochemical and agronomic properties of biochar from sewage sludge pyrolysed at different temperatures. *Journal of Analytical and Applied Pyrolysis*, 102, 124–130.

Mohan, D., Sarswat, A., Ok, Y.S., Pittman, C.U., Jr, 2014. Organic and inorganic contaminants removal from water with biochar, a renewable, low cost and sustainable adsorbent–a critical review. *Bioresource Technology*, 160, 191–202.

Montoya, J.I., Chejne-Janna, F., Garcia-Pérez, M., 2015. Fast pyrolysis of biomass: A review of relevant aspects. Part I: Parametric study. *Dyna*, 82(192), 239–248.

Mustapha, M.A., Manan, Z.A., Alwi, S.R.W., 2017. A new quantitative overall environmental performance indicator for a wastewater treatment plant. *Journal of Cleaner Production*, 167, 815–823.

Nguyen, T.T.N., Xu, C.-Y., Tahmasbian, I., Che, R., Xu, Z., Zhou, X., Wallace, H.M., Bai, S.H., 2017. Effects of biochar on soil available inorganic nitrogen: A review and meta-analysis. *Geoderma*, 288, 79–96.

Oliveira, F.R., Patel, A.K., Jaisi, D.P., Adhikari, S., Lu, H., Khanal, S.K., 2017. Environmental application of biochar: Current status and perspectives. *Bioresource Technology*, 246, 110–122.

Pietrzak, R. and Bandosz, T.J., 2008. Interactions of NO2 with sewage sludge based composite adsorbents. *Journal of Hazardous Materials*, 154(1–3), 946–953.

Qambrani, N.A., Rahman, M.M., Won, S., Shim, S., Ra, C., 2017. Biochar properties and eco-friendly applications for climate change mitigation, waste management, and wastewater treatment: A review. *Renewable and Sustainable Energy Reviews*, 79, 255–273.

Rajapaksha, A.U., Vithanage, M., Lee, S.S., Seo, D.-C., Tsang, D.C., Ok, Y.S., 2016. Steam activation of biochars facilitates kinetics and pH-resilience of sulfamethazine sorption. *Journal of Soils and Sediments*, 16(3), 889–895.

Seredych, M., Strydom, C., Bandosz, T.J., 2008. Effect of fly ash addition on the removal of hydrogen sulfide from biogas and air on sewage sludge-based composite adsorbents. *Waste Management*, 28(10), 1983–1992.

Singh, S., Kumar, V., Dhanjal, D.S., Datta, S., Bhatia, D., Dhiman, J., Samuel, J., Prasad, R., Singh, J., 2020. A sustainable paradigm of sewage sludge biochar: Valorization, opportunities, challenges and future prospects. *Journal of Cleaner Production*, 269, 122259.

Smith, K., Fowler, G., Pullket, S., Graham, N.J.D., 2009. Sewage sludge-based adsorbents: A review of their production, properties and use in water treatment applications. *Water Research*, 43(10), 2569–2594.

Song, X., Xue, X., Chen, D., He, P., Dai, X., 2014. Application of biochar from sewage sludge to plant cultivation: Influence of pyrolysis temperature and biochar-to-soil ratio on yield and heavy metal accumulation. *Chemosphere*, 109, 213–220.

Tarrago, M., Garcia-Valles, M., Aly, M., Martínez, S., 2017. Valorization of sludge from a wastewater treatment plant by glass-ceramic production. *Ceramics International*, 43(1), 930–937.

Tian, K., Liu, W.-J., Qian, -T.-T., Jiang, H., Yu, H.-Q., 2014. Investigation on the evolution of N-containing organic compounds during pyrolysis of sewage sludge. *Environmental Science & Technology*, 48(18), 10888–10896.

Tian, S., Tan, Z., Kasiulienė, A., Ai, P., 2017. Transformation mechanism of nutrient elements in the process of biochar preparation for returning biochar to soil. *Chinese Journal of Chemical Engineering*, 25(4), 477–486.

Turunen, V., Sorvari, J., Mikola, A., 2018. A decision support tool for selecting the optimal sewage sludge treatment. *Chemosphere*, 193, 521–529.

Villar, I., Alves, D., Pérez-Díaz, D., Mato, S., 2016. Changes in microbial dynamics during vermicomposting of fresh and composted sewage sludge. *Waste Management*, 48, 409–417.

Wang, H., Feng, M., Zhou, F., Huang, X., Tsang, D.C., Zhang, W., 2017. Effects of atmospheric ageing under different temperatures on surface properties of sludge-derived biochar and metal/metalloid stabilization. *Chemosphere*, 184, 176–184.

Wang, M., Zhu, Y., Cheng, L., Andserson, B., Zhao, X., Wang, D., Ding, A., 2018. Review on utilization of biochar for metal-contaminated soil and sediment remediation. *Journal of Environmental Sciences*, 63, 156–173.

Xiao, R., Wang, J.J., Gaston, L.A., Zhou, B., Park, J.-H., Li, R., Dodla, S.K., Zhang, Z., 2018. Biochar produced from mineral salt-impregnated chicken manure: Fertility properties and potential for carbon sequestration. *Waste Management*, 78, 802–810.

Xiaohua, W. and Jiancheng, J., 2012. Effect of heating rate on the municipal sewage sludge pyrolysis character. *Energy Procedia*, 14, 1648–1652.

Xu, X., Zhao, Y., Sima, J., Zhao, L., Mašek, O., Cao, X., 2017. Indispensable role of biochar-inherent mineral constituents in its environmental applications: A review. *Bioresource Technology*, 241, 887–899.

Yuan, H., Lu, T., Huang, H., Zhao, D., Kobayashi, N., Chen, Y., 2015. Influence of pyrolysis temperature on physical and chemical properties of biochar made from sewage sludge. *Journal of Analytical and Applied Pyrolysis*, 112, 284–289.

Yuan, H., Lu, T., Wang, Y., Chen, Y., Lei, T., 2016. Sewage sludge biochar: Nutrient composition and its effect on the leaching of soil nutrients. *Geoderma*, 267, 17-.

Zhao, S.-X., Ta, N., Wang, X.-D., 2017. Effect of temperature on the structural and physicochemical properties of biochar with apple tree branches as feedstock material. *Energies*, 10(9), 1293.

Zheng, H., Wang, Z., Deng, X., Zhao, J., Luo, Y., Novak, J., Herbert, S., Xing, B., 2013. Characteristics and nutrient values of biochars produced from giant reed at different temperatures. *Bioresource Technology*, 130, 463–471.

Zielińska, A. and Oleszczuk, P., 2015. The conversion of sewage sludge into biochar reduces polycyclic aromatic hydrocarbon content and ecotoxicity but increases trace metal content. *Biomass and Bioenergy*, 75, 235–244.

11

Biochar for Electrochemical Treatment of Wastewater

Dong Li[1], Yang Zheng[3], Yuqing Sun[4], and Daniel C.W. Tsang[2]

[1] Teleader Solid Waste Disposal (Shandong) Co., Ltd.,Jinan, China
[2] State Key Laboratory of Clean Energy Utilization, Zhejiang University, China
[3] School of Materials Science and Engineering, Ocean University of China, Qingdao, China
[4] School of Agriculture, Sun Yat-Sen University, Guangzhou, Guangdong, China

Electrochemical water treatment has been widely studied as a green and efficient water treatment technology. However, due to the high cost and unsustainability of electrode materials, there is an urgent need to develop cost-effective and green low-carbon alternative materials. Biochar has good electrochemical performance and is a potential substitute for carbon electrode in electrochemical water treatment. Due to its advantages of environmental protection, economy and renewability, the research of biochar electrode materials has attracted more and more attention. This chapter reviews the electrochemical behavior, preparation and application of biochar in electrochemical water treatment. In the future, it is necessary to further study biochar with high electrochemical performance, develop new biochar preparation strategies, and comprehensively evaluate the action mechanism of biochar in electrochemical water treatment, so as to promote the application of biochar in actual electrochemical water treatment

11.1 Introduction

To solve the problem of water pollution and meet the increasing requirements for water resources, efficient and cost-effective wastewater treatment technologies are essential [1]. Traditional wastewater treatment strategies include physical, chemical, and biological strategies, but they have some drawbacks such as incomplete removal of pollutants, residual sludge, and inability to deal with refractory organic pollutants [2]. In comparison, electrochemical technologies applied to water and wastewater treatment are receiving extensive attention due to their environmental compatibility, high removal efficiency, and potential cost effectiveness [3]. Electrochemical techniques include electrosorption, electrocoagulation, electrooxidation, electrofenton, electrodisinfection, and electrodeposition [4–6]. These techniques show potential for the removal of various contaminants and pathogenic microorganisms. Many researchers have explored the electrochemical properties and degradation mechanisms of various electrode materials [7–9]. Generally, the ideal electrode material should have high specific surface area, hierarchical pore structure, high conductivity, good wettability, and stable electrochemical performance. Based on these properties, carbon materials are considered to have potential for electrode material applications, and researchers have developed various carbon materials as electrode materials, such as graphene [10], carbon nanotube [11], and biochar [12–14]. Among these electrode materials, biochar has attracted the attention of researchers because of its adjustable structure, low cost, and wide sources.

Biochar is produced by thermochemical decomposition of biomass under anoxic or anaerobic atmosphere. The properties of biochar are determined by feedstocks and process conditions such as thermal treatment, temperature, time, and catalyst [15, 16]. In addition, various modification methods can improve the physical and electrochemical properties of biochar, so as to obtain high-performance electrode materials [17, 18]. The prepared biochar typically shows high specific surface area, hierarchical pore structure, and good electrical conductivity [19]. The natural presence of N, S, and Fe-based minerals in biochar allows it to be used as a catalyst for redox reactions. The role of biochar in the electrochemical treatment of wastewater can be divided into three categories: the oxygen-containing functional groups and persistent free radicals of biochar affect the electron exchange [20]; biochar acts as electron donor or acceptor to participate in redox reaction [21]; the electrosorption capacity associated with the pore structure of biochar for capacitive deionization [22]. Based on these studies, biochar has been widely used in wastewater treatment, such as organic removal, heavy metal removal, deionization (ion removal and seawater desalination), and disinfection [23].

Herein, we reviewed recent research on the electrochemical treatment of wastewater by biochar, discussed the different electrochemical behavior of biochar, summarized the preparation of biochar electrode, and introduced the application of biochar electrode in wastewater treatment. According to the knowledge gap and challenges in electrochemical treatment of wastewater by biochar, the issues needing further study of biochar electrode in the future were proposed, which laid the foundation for the design of high efficiency and low-cost biochar electrode.

11.2 Different Electrochemical Behavior of Biochar

11.2.1 Electron Exchange

The surface functional groups and carbon matrix structure of biochar affect the electron exchange capacity. In general, surface functional groups affect electron transfer from electron donor to electron acceptor on the biochar surface, while the carbon matrix structure affects the electrical conductivity of biochar to affects electron transfer within the biochar. These two electronic exchange mechanisms may occur independently or simultaneously [24].

The electron exchange on biochar surface is related to surface functional groups and persistent free radicals. In general, biochar prepared at low temperatures (300–400 °C) has more surface functional groups than biochar prepared at high temperatures (600–800 °C). The C=O and OH groups on the biochar surface can act as intermediaries between electron-producing microorganisms and electron-receiving contaminants [25]. In addition, the biochar surface electrochemical groups (carbon atom in pyridine N) react with the sulfide to form reactive sulfur intermediate, which also demonstrate the transfer of electrons from the reductant to contaminants [26]. Furthermore, biochar can also improve microbial activity via electron exchange. Wu et al. found that biochar with more OH groups improved the electron exchange capacity and promoted microbial denitrification [27]. Similarly, hydrothermal carbon with abundant C=O group promotes the activity of methanogens during the anaerobic digestion of glucose [28]. Surface O-centered radicals, such as semiquinone-type radicals, can accept electrons from lactate and transfer them to Cr(VI) [29].

The transfer of electrons within the biochar is called bulk conductivity. Biochar has an adsorption effect on pollutants, and the high bulk conductivity is favorable to the transfer of electrons within the biochar to the surface-adsorbed pollutants [30]. Wang et al. found that the introduction of N atoms into the carbon lattice was conducive to donating electrons to the peroxydisulfate, which promoted the acceptance of electrons by adsorbed organic pollutants [31]. In general, the graphitization degree of biochar increased with the increase of thermal treatment temperature. The biochar prepared at 700 °C showed a graphite-like structure and a higher bulk conductivity [32]. Chen et al. verified by electrochemical measurements that the biochar with higher graphitization degree showed higher bulk conductivity [33]. Therefore, the graphitic structure in biochar should be the critical structure of the bulk conductivity. In addition, porous biochar with conjugated aromatic structure also showed high bulk conductivity [34].

11.2.2 Electron Donor or Acceptor

As electron donor of acceptor, biochar has been used in wastewater treatment through electrochemical free radical or non-free radical pathways. As an electron donor, biochar provides electrons to oxidative system to conduct or facilitate the reaction. Conversely, as an electron acceptor, biochar accepts electrons from reductive system to conduct or facilitate the reaction [21].

The functional groups and chemical composition of biochar determine whether it is a donor or acceptor. For electron donor, the groups related to the electron donating capacity are OH, C–O, O–C=O, C=O, and aromatic C=C. For electron acceptor, O–C=O, C=O, aromatic C=C and high oxidation-state Fe and Mn are related to electron accepting capacity [23]. In addition, COOH groups of biochar can be involved in both Cr(VI)reduction and As(III) oxidation [35]. Chemical composition is a direct factor affecting biochar electrochemical behaviors. As raw materials of biochar, biomass is rich in K, Ca, Na, Fe, Si, etc., but different species of biomass contain different substances and contents [36]. In general, sewage sludge and animal waste have higher mineral content than plant biomass, and these mineral components may affect the surface functional groups of biochar [37]. Moreover, biochar with well-developed pore structure is beneficial to its electron donating or accepting capacity. Li et al. proposed that the surface area of biochar is positively correlated with its electron donating or accepting capacity [38]. The pore structure and surface functional groups of biochar are affected by heat treatment temperature, residence time, heating rate, and carbonization methods. Among them, heat treatment temperature is a crucial factor affecting the structure and electrochemical behavior of biochar (Figure 11.1). For example, Xu et al. found that the biochar prepared above 600 °C resulted in the formation of aromatic-C–O, which could donate electrons for Cr(VI) reduction [30]. In addition, surface modification and pretreatment can also affect the electron donating or accepting capacity of biochar. HNO_3, O_3 and H_2O_2 treatment are common surface modification methods, which are conducive to the formation of surface oxygen-containing groups of biochar [39].

11.2.3 Electrosorption Capacity

Capacitive deionization is a new electrochemical water treatment technology based on electrosorption capacity of electrode, which has the advantages of high energy efficiency, environmental friendliness, and low energy consumption [40].

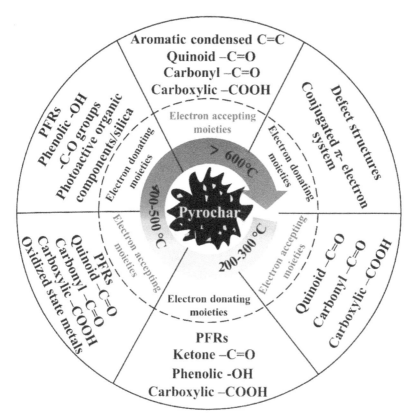

Figure 11.1 The potential electron accepting or donating moieties of biochar at different heat treatment temperature range [23].

By applying a constant DC voltage or current to the electrodes, an electrostatic field between the electrodes is created, and anions and cations are driven to the anode and cathode, respectively, under the action of an electric field. After the electrodes adsorption ions achieve saturation, the electrodes are regenerated by zero voltage desorption (ZVD), reverse current desorption, and reverse voltage desorption (RVD) modes [41]. In the process of electrode adsorption of anions and cations, two storage mechanisms are mainly used to treat saltwater, namely electric double layer capacitive ion storage and Faraday ion storage [42]. As shown in Figure 11.2, there are three architectures of capacitive deionization system: classical capacitive deionization (CDI), membrane-capacitive deionization (MCDI), and flow electrode-capacitive deionization (FCDI). Traditional CDI consists of two electrodes separated by an insulator, which is easy to set but inefficient [43]. MCDI is based on the form of traditional CDI, and the corresponding ion exchange

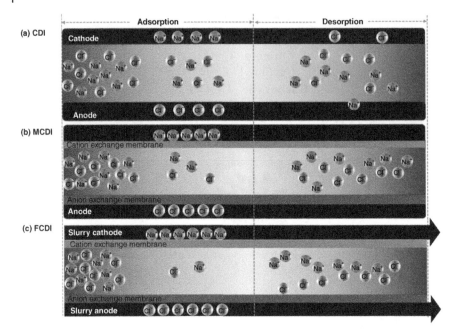

Figure 11.2 Architectures of capacitive deionization system: (a) CDI, (b) MCDI, and (c) FCDI [22].

membrane is introduced on the surface of two electrodes, which has better desalination efficiency and anti-fouling performance [44]. However, these two systems adopt static electrode structure, and the total adsorption capacity of each operation is lower than the maximum adsorption capacity of the electrode. In addition, regeneration process is required after electrons are saturated, causing the incontinuous operation of the system [45]. Therefore, researchers have proposed the concept of FCDI to overcome these drawbacks. FCDI replaces the static electrode with a slurry-based electrode that can be recirculated during operation, enabling the continuous operation of system [46, 47].

The virtual factor to the electrochemical treatment of water in capacitive deionization system is the electrosorption capacity of electrode materials. Materials with high specific surface area, appropriate pore size distribution, high conductivity, and high electrochemical stability have better electrosorption capacity [22]. Biochar can improve its physical and electrochemical properties through various forms of modification, so it is developed for use in CDI electrode materials. Biochar generally has hierarchical pore structure and high specific surface area, which can provide a large interface with the solution for ion storage. Moreover, the macropores in the hierarchical pore structure can form an ion buffer reservoir, which is beneficial to shorten the ionic diffusion distance and promote ion transport [48]. The mesopore allows for rapid mass transfer kinetics and low mass transfer resistance in the process of ion transport to promote the rapid transfer of ions to the micropore. The large specific surface area contributed by micropores provides abundant active sites, which are conducive to ion separation [49]. Therefore, the electrosorption capacity of biochar can be improved by structural regulation.

11.3 Preparation of Biochar Electrode Materials

The preparation of biochar involves thermochemical transformation, which usually uses high temperature to carbonize biomass into stable carbon-rich solid under oxygen-free or oxygen-limited conditions. The commonly used thermochemical techniques are pyrolysis (include rapid and slow pyrolysis), hydrothermal carbonization, and microwave cracking, which are different in heating methods, temperature, and residence time. The key step to prepare biochar electrode materials is the carbonization of biomass to obtain stable carbon materials. However, direct carbonization will lead to poor electrochemical performance of biochar, which cannot meet the needs of practical applications. Therefore, researchers have developed activation methods, template methods, and complex methods to improve the physical and chemical properties of biochar, resulting in electrode materials with high electrochemical performance.

11.3.1 Carbonization

Carbonization is the decomposition of biomolecules such as lignin and cellulose in biomass into small molecules, thus forming biochar [50]. The carbonization conditions such as heating rate, thermal treatment temperature, and residence time have important effects on the electrochemical properties of biochar. According to different process conditions, the production of biochar by carbonization can be divided into pyrolysis, hydrothermal carbonization, and microwave heating.

Pyrolysis is to carbonize biomass by thermal decomposition under anoxic or anaerobic atmosphere, according to different heating rates divided into fast pyrolysis and slow pyrolysis [51]. Pyrolysis includes three stages: dehydration, covalent bond breaking, and polycyclic condensation [52]. Fast pyrolysis is advantageous to the biochar with developed pores, while slow pyrolysis is advantageous to the biochar with abundant surface functional groups [53]. Pyrolysis temperature and time affect the decomposition degree of lignin and cellulose in biomass. Higher temperature and longer time can promote the complete decomposition of biomass and improve the degree of polymerization and carbonization of biochar [54, 55].

Hydrothermal carbonization is the conversion of biomass into biochar at relatively low temperatures (120–260 °C) by relying on the autogenic pressure of water in a closed autoclave. Hydrolysis, dehydration, decarboxylation, polycondensation, and aromatization occur in this process [56]. Biochar prepared by this method shows high H/C and O/C ratios, abundant surface functional groups (such as hydroxyl, carboxyl, and carbonyl groups), and good hydrophilicity, but low specific surface area and pore structure [57]. Hydrothermal temperature, residence time, and liquid-solid ratio are important process parameters. Higher temperature and time will lead to the reduction of carbonyl and carboxyl groups in biochar and the increase of aromaticity [58]. Because water molecules are related to the hydrolysis of biomass organic matter in hydrothermal carbonization, reducing the liquid-solid ratio will also reduce oxygen-containing functional groups of biochar [59].

Microwave heating generally refers to the use of microwave-assisted biomass pyrolysis or hydrothermal carbonization, which has the characteristics of fast heating rate, heating uniformity, and low energy consumption [60]. The surface structure and chemical composition of biochar are affected by microwave power, thermal treatment temperature, and residence time [61]. With the increase of microwave power, the specific surface area and aromatic structure of biochar can be improved, and the H/C ratio and O/C ratio can be decreased. Biochar with well-developed pore structure can be obtained by increasing heat treatment temperature and residence time, but the surface functional groups will be reduced.

11.3.2 Activation

Activation is generally divided into physical activation and chemical activation, which is used to improve the pore structure and increase the specific surface area of biochar. Biochar with abundant pores and high specific surface area has more adsorption sites, which can accelerate the rapid transport of ions and the adsorption of electrolyte, so as to improve the electrosorption capacity of biochar electrodes.

Physical activation is a thermal treatment process (600–1000 °C) of biomass under specific atmospheres (such as H_2O, CO_2, and NH_3) to increase the porosity of the biochar [62]. In the process of physical activation, H_2O or CO_2 reacts with the more active carbon atoms in biochar to produce gas (CO and H_2), resulting in high specific surface area and developed pore structure [63]. In general, the increased activation time will increase the specific surface area of biochar, which is conducive to the rapid transport of ions in the electrosorption process. However, too-long activation time will lead to the collapse of pore structure and reduce the specific surface area and pore volume of biochar. In addition, activation yield is an important factor to be considered in the activation process. The conventional carbonization yield of lignocellulosic materials is 20–30 wt.%, but the total yield after physical activation is about 10 wt.% [64]. Under microwave-assisted heating, the yield of CO_2-activated biochar is higher than that of conventional heating, but has no effect on the yield of H_2O-activated biochar [65]. However, under microwave-assisted heating, the specific surface area and pore volume of H_2O-activated biochar doubled compared with that under traditional heating conditions [66].

Chemical activation is a method to obtain porous carbon by mixing activator with raw material and pyrolyzing mixture at high temperature under inert atmosphere. Commonly used activators include, for example, KOH, $KHCO_3$, K_2CO_3, $ZnCl_2$, and H_3PO_4 [67]. Among them, KOH is the chemical activator commonly used in biochar. For example, Li et al. mixed biomass tar with KOH and activated mixture at 800 °C to obtain porous carbon foam with a specific surface area of 1667 m^2/g [68]. Du et al. found that the electrosorption capacity of biochar increased by about 40% in CDI system after KOH activation [43]. H_3PO_4 is also used to activate biomass to produce biochar with high capacitive properties [69]. The activation temperature and the ratio of activator to raw material are the key factors affecting the properties of activated biochar. Low activation temperature and ratio are not conducive to complete activation and pore connectivity, while high activation temperature and ratio will lead to pore structure collapse. Therefore, through the experiment to determine the appropriate temperature and ratio of biochar can make the pore structure of rich and high specific surface area, so as to obtain the electrode material with excellent electrochemical performance [70]. Compared with physical activation, chemical activation can significantly improve the specific surface area and pore volume of biochar.

11.3.3 Template

Template is a method to prepare biochar with required pore structure, which is usually carried out simultaneously with carbonization. The biomass is mixed with the template to fill the pores of it, and then the template is carbonized and removed to obtain the template biochar. The obtained biochar has uniform and orderly pore structure, high specific surface area, and high pore volume. It is often used for capacitive deionization treatment of wastewater. According to the different types of templates used, template methods are divided into the hard template and soft template method.

In the hard template method, rigid or semi-rigid materials (such as mesoporous silica, metal-organic frameworks, and zeolites) with stable properties and specific structural morphology are used as templates to prepare porous carbon materials through the reverse replication effect [71]:

1) A rigid material with nanoporous structure is designed as a template mold.
2) Biochar precursor is adsorbed on the surface of the template or filled in its pores to form composite precursor.

3) The composite precursor is carbonized and the hard template is removed by physical and chemical methods to obtain a porous material that replicates in reverse [22].

Cuong et al. prepared hierarchal porous carbon with a specific surface area of 1839 m^2/g by using SiO_2 as a natural hard template combined with chemical activation, which had an electrosorption capacity of 8.11 mg/g in a single-pass CDI system [14]. However, the hard template method requires strong corrosive chemicals to remove the template, and the biochar may hinder template removal, resulting in no or less-porous carbon. The surfactants, amphiphilic molecules, and triblock copolymers are used as templating agents in the soft template method [72]. These templates interact with biomass precursors to form composite nanoparticles in aqueous or organic phases through hydrogen bonding, and then they are decomposed and volatilized at high temperature to form hierarchical porous biochar [73].

11.3.4 Composite Materials

By combining biochar with other materials, the properties of biochar and other materials can be effectively utilized to improve the properties of the composite materials. Biochar is used for support loading or composite active substance due to its environmental protection, low cost, good conductivity, and porous characteristics. It has been reported that the combination of biochar with MnO_2, which has poor conductivity, has produced supercapacitor electrode materials that are cost-effective and environmentally friendly, with high specific capacitance (1370 F/g) [74]. Adorna et al. co-precipitated coconut-shell biochar with MnO_2 to prepare composite CDI electrode material, which has good Na electrosorption capacity (34–68 mg/g) [75]. Yao et al. prepared the cathode electrode for the electrocatalytic removal of bromate by carbonation-impregnation method [76]. The biochar had the structure of macroporous three-dimensional network and was combined with Pd; the removal efficiency reached 96.7% at 10 mA current intensity. In addition, Mansoori et al. synthesized bimetallic composite of biochar and iron-copper oxide for electro-Fenton degradation of ciprofloxacin [77]. The porous structure of biochar provided good accessibility of active sites for active metals.

Although many studies have developed new composite electrode materials and shown excellent performance in electrochemical water treatment, most of them are limited to the laboratory scale, and more attention should be paid to the practical application of electrode materials with low cost and easy large-scale preparation.

11.4 Application in Electrochemical Wastewater Treatment

11.4.1 Electrochemical Oxidation

Electrochemical oxidation has been widely used in electrochemical water treatment because of its environmental friendliness, energy efficiency, and simple operation. Hydroxyl radicals ($^\cdot$OH) can be produced by oxygen reduction (ORR) in cathodic or water oxidation in anodic, which has strong oxidation capacity to decompose contaminants from water. Biochar is a good choice for electrode materials, especially cathode materials, due to its electrical conductivity, high selectivity toward the two-electrons ORR, and chemical stability [78, 79]. For example, Zhang et al. prepared biochar electrode materials from pyrolysis sludge for electrochemical oxidation of azo dyes, and the removal rate of methyl orange reached 94.5% [80]. In addition, there are many studies on the use of biochar as electrode material for wastewater treatment by electric Fenton process, and this process utilize $^\cdot$OH produced by Fe^{2+} and electro-generated H_2O_2 at cathode to decompose contaminant in wastewater [81]. Traditional electric Fenton process has problems with pH range and iron mud precipitation, but biochar electrodes can effectively overcome these drawbacks. Biochar electrode can concentrate organic pollutants on its surface through adsorption to minimize the diffusion loss of reactive oxygen species, and effectively degrade pollutants to achieve self-cleaning effect [82]. Deng et al. used waste wood-derived biochar as cathode for electro-Fenton treatment of sulfathiazole, and found that this system could operate under neutral conditions in the presence of pyrophosphate ligands [83]. They also used reed-derived biochar composite nickel foam to prepare cathodes for electro-Fenton efficient degradation of sulfamerazine [84].

The above electrochemical oxidation process is generally called two-dimensional electrochemical oxidation, which has the disadvantages of short electrode life, limited mass transfer, high process temperature, and low current efficiency [85]. Therefore, researchers proposed to overcome these drawbacks by loading adsorbent particles (particle electrodes) between the anode and cathode of the electrochemical system through a three-dimensional electrochemical process. Soares et al. used waste-derived biochar from vineyards as particle electrodes for the treatment of drugs in wastewater by a three-dimensional electrochemical process. They proposed that adsorption of biochar particles made a limited contribution to pollutant removal, thus allowing the use of biochar with a low specific surface area as an economically sustainable particle alternative [86]. Zhao et al. processed the almond shell biochar to prepare dispersed graphene nanosheets, which were used as particle electrodes in the three-dimensional electrolysis system to efficiently mineralize Cu-ethylene diamine tetraacetic acid, and the Cu(II) removal efficient reached 96.8% [87]. Wang et al. prepared particle electrode of activated biochar for electrochemical oxidation of Cu-1-hydroxyethylidene-1,1-diphosphonic acid in a three-dimensional electrolysis system, and the maximum Cu(II) removal (90.7%) was achieved after 180 min of reaction at 40 mA [88].

11.4.2 Electrochemical Deposition

Electrochemical deposition using biochar as electrode has been widely studied for its low cost and high removal efficiency to treat heavy metal pollution in industrial wastewater. Wang et al. prepared biochar electrodes by high temperature carbonization of rice straw to remove Cd from simulated wastewater, and the removal rate reached 76.6%. They regenerated biochar electrodes by acid washing and simultaneously recovered Cd due to the stability of biochar [8]. In addition, using biochar with porous structure as carrier of microorganisms can improve the air-water interface to obtain oxygen as terminal electron acceptor on the one hand, and provide deposition sites for inorganic salts and surface organic matter on the other hand [89].

11.4.3 Electro-adsorption

Electro-adsorption has the advantages of stable operation, good adaptability, high energy efficiency, high removal efficiency, low consumption, easy regeneration, and environmental friendliness [40]. It is through the application of electrostatic field to drive electrolyte ions to carry the opposite charge electrode and accumulate in the electrode, so as to realize the desalination and purification of wastewater. After the adsorbed ions of the electrode reach saturation, the electrode can be regenerated by applying zero voltage or reverse voltage to release the ions back into the solution [41]. This technology is also called capacitive deionization, and is an effective electrochemical technology for seawater desalination and wastewater purification. Stephanie et al. modified activated biochar with sulfonate and amine functional groups to develop cation and anion selectivity [90]. The prepared biochar electrodes were used in the asymmetric capacitive deionization cell, showing high NaCl adsorption capacity (9.25 mg/g), which can effectively desalinate high-salt seawater. Sun et al. prepared N-doped porous biochar, and achieved an electrical adsorption capacity of 17.67 mg/g in 500 mg/L NaCl solution under 1.2 V through the synergistic effect of pseudocapacitance mechanism and double electric layer ion storage [91].

Ion removal is another important application of electro-adsorption technology. The selective removal of NO_3^-, Ca^{2+}, Mg^{2+}, etc. can be achieved by adjusting the specific surface area and surface functional groups and adding chemical charge in biochar. Chen et al. prepared biochar with interconnected hierarchical pore structure using eggshell membrane, and this biochar showed strong selectivity for NO_3^- over SO_4^- and

Cl⁻ [12]. Wang et al. used quaternary ammonium nitrogen functionalized mesoporous biochar as electrode for perchlorate electro-adsorption removal [92]. The results showed that mesoporous biochar promoted electron transfer and reduced ion diffusion resistance, and the electro-adsorption capacity of ClO_4^- reached 28.31 mg/g.

11.4.4 Electrochemical Disinfection

Electrochemical disinfection technology is environmentally friendly, easy to operate, and highly efficient. Its mechanism includes irreversible damage to microbe by strong electric field, electron transfer between anode and microbe, and indirect oxidation of active substances [93]. Biochar has the advantages of stable electrochemical performance and low cost, and it has been applied in electrochemical disinfection. Sun et al. used orange peel biochar as the cathode in electrochemical disinfection, and found that the continuous porous structure of biochar provided adsorption sites to reduce mass transfer resistance and thus produced more ˙Cl and ˙O_2^- for high-efficiency electrochemical disinfection of *P. putida* [94]. However, the research on using biochar as an electrode for electrochemical disinfection has just started; it is still necessary to develop new biochar materials for electrochemical disinfection in the future.

11.5 Future Perspectives

Researchers are exploring the use of biochar derived from a variety of feedstocks. For electrochemical water treatment, the search for low-cost, high-performance, mass-produced biochar feedstocks has been a key concern. In addition, it is also necessary to consider green and environmentally friendly preparation strategies for biochar preparation. For example, conventional carbonization generally requires high temperature, so in the future, microwave-assisted carbonization or hydrothermal carbonization with lower energy consumption could be a way to reduce the cost of biochar preparation. KOH and other chemical agents are used for the activation of biochar to form hierarchical pore structures, but these chemical agents increase the preparation cost on the one hand, and harm the environment on the other hand. Similarly, hard and soft templates also lead to cost improvement, but biomass-containing natural templates or those rich in heteroatoms can be directly prepared into biochar with certain electrochemical performance. Combining biomass with wastes containing electrochemically active substances also reduces biochar preparation cost.

For electrochemical water treatment technology, it is necessary to continue to explore the feasibility of the actual implementation of electrochemical water treatment. Most current studies are carried out in small installations on the laboratory scale. Some pilot experiments are still needed to comprehensively evaluate the economic cost, energy consumption, operation efficiency, and lifespan of electrochemical water treatment technology. Different electrochemical technologies have their own limitations. For example, electrochemical oxidation technology has poor selectivity, and electrosorption technology can only deal with specific ions. In some complex water conditions, it is important to improve the selectivity of target pollutants to prevent interference by background factors. Therefore, various electrochemical techniques can be combined to achieve complex processing objectives.

To date, there have been few studies on the application of biochar electrodes in electrochemical deposition and electrochemical disinfection. In the future, the application and mechanism of biochar in these technologies can be further studied. Some studies have also mentioned the presence of defect sites (edges, vacancies, etc.) in biochar, but have not investigated their electrochemical behavior in depth. Therefore, it is necessary to explore the relationship between electrochemical behavior and defect sites in biochar.

11.6 Summary

Electrochemical technology is regarded as a promising technology in the field of wastewater treatment, and biochar is regarded as an emerging alternative to traditional carbon electrode materials. Application of biochar in electrochemical treatment of water is being gradually developed. The pore structure, surface properties, and chemical composition of biochar all affect its electrochemical performance. The preparation of biochar electrode includes carbonization, activation, template method, and compound method. At present, the application of biochar in the field of electrochemical water treatment includes heavy metal removal, organic removal, desalination purification, disinfection, and other fields. In the future, it will be necessary to develop new biochar preparation strategies, comprehensively evaluate the mechanism of action and cost-effectiveness of biochar in electrochemical water treatment, and further promote the application of low-cost biochar in actual electrochemical water treatment.

References

1 Zheng, T.L., Wang, J., Wang, Q.H., Meng, H.M., Wang, L.H., 2017. Research trends in electrochemical technology for water and wastewater treatment. *Applied Water Science*, 7, 13–30.
2 Li, X.H., Chen, S., Angelidaki, I., Zhang, Y.F., 2018. Bio-electro-Fenton processes for wastewater treatment: Advances and prospects. *Chemical Engineering Journal*, 354, 492–506.

3. Feng, Y.J., Yang, L.S., Liu, J.F., Logan, B.E., 2016. Electrochemical technologies for wastewater treatment and resource reclamation. *Environmental Science: Water Research*, 2, 800–831.
4. Fitch, A., Balderas-Hernández, P., Ibanez, J.G., 2022. Electrochemical technologies combined with physical, biological, and chemical processes for the treatment of pollutants and wastes: A review. *Journal of Environmental Chemical Engineering*, 10.
5. Srimuk, P., Su, X., Yoon, J., Aurbach, D., Presser, V., 2020. Charge-transfer materials for electrochemical water desalination, ion separation and the recovery of elements. *Nature Reviews. Materials*, 5, 517–538.
6. Moussa, D.T., El-Naas, M.H., Nasser, M., Al-Marri, M.J., 2017. A comprehensive review of electrocoagulation for water treatment: Potentials and challenges. *Journal of Environmental Management*, 186, 24–41.
7. Du, X.D., Oturan, M.A., Zhou, M.H., Belkessa, N., Su, P., Cai, J.J., Trellu, P.E., Mousset, E., 2021. Nanostructured electrodes for electrocatalytic advanced oxidation processes: From materials preparation to mechanisms understanding and wastewater treatment applications. *Applied Catalysis B: Environmental*, 296.
8. Wang, Z.W., Tan, Z.X., Yuan, S.N., Li, H., Zhang, Y., Dong, Y.F., 2022. Direct current electrochemical method for removal and recovery of heavy metals from water using straw biochar electrode. *Journal of Cleaner Production*, 339.
9. Malpass, G.R.P., Motheo, A.D., 2021. Recent advances on the use of active anodes in environmental electrochemistry. *Current Opinion in Electrochemistry*, 27.
10. Li, J., Ji, B.X., Jiang, R., Zhang, P.P., Chen, N., Zhang, G.F., Qu, L.T., 2018. Hierarchical hole-enhanced 3D graphene assembly for highly efficient capacitive deionization. *Carbon*, 129, 95–103.
11. Shao, D., Li, W.J., Wang, Z.K., Yang, C.A., Xu, H., Yan, W., Yang, L., Wang, G.B., Yang, J., Feng, L., Wang, S.Z., Li, Y., Jia, X.H., Song, H.J., 2022. Variable activity and selectivity for electrochemical oxidation wastewater treatment using a magnetically assembled electrode based on Ti/PbO$_2$ and carbon nanotubes. *Separation and Purification Technology*, 301.
12. Chen, J., Zuo, K.C., Li, Y.L., Huang, X.C., Hu, J.H., Yang, Y., Wang, W.P., Chen, L., Jain, A., Verduzco, R., Li, X.Y., Li, Q.L., 2022. Eggshell membrane derived nitrogen rich porous carbon for selective electrosorption of nitrate from water. *Water Research*, 216.
13. Ji, X.Y., Liu, X., Yang, W.L., Xu, T., Wang, X., Zhang, X.Q., Wang, L.M., Mao, X.H., Wang, X., 2022. Sustainable phosphorus recovery from wastewater and fertilizer production in microbial electrolysis cells using the biochar-based cathode. *Science of the Total Environment*, 807.
14. Cuong, D.V., Wu, P.C., Liu, N.L., Hou, C.H., 2020. Hierarchical porous carbon derived from activated biochar as an eco-friendly electrode for the electrosorption of inorganic ions. *Separation and Purification Technology*, 242.

15 Foong, S.Y., Chan, Y.H., Chin, B.L.F., Lock, S.S.M., Yee, C.Y., Yiin, C.L., Peng, W.X., Lam, S.S., 2022. Production of biochar from rice straw and its application for wastewater remediation: An overview. *Bioresource Technology*, 360.

16 Tomczyk, A., Sokolowska, Z., Boguta, P., 2020. Biochar physicochemical properties: Pyrolysis temperature and feedstock kind effects. *Reviews in Environmental Science and Bio/Technology*, 19, 191–215.

17 Singh, A., Sharma, R., Pant, D., Malaviya, P., 2021. Engineered algal biochar for contaminant remediation and electrochemical applications. *Science of the Total Environment*, 774.

18 de Almeida, L.S., Oreste, E.Q., Maciel, J.V., Heinemann, M.G., Dias, D., 2020. Electrochemical devices obtained from biochar: Advances in renewable and environmentally-friendly technologies applied to analytical chemistry. *Trends in Environmental Analytical*, 26.

19 Gopinath, A., Divyapriya, G., Srivastava, V., Laiju, A.R., Nidheesh, P.V., Kumar, M.S., 2021. Conversion of sewage sludge into biochar: A potential resource in water and wastewater treatment. *Environmental Research*, 194.

20 Ruan, X.X., Sun, Y.Q., Du, W.M., Tang, Y.Y., Liu, Q., Zhang, Z.Y., Doherty, W., Frost, R.L., Qian, G.R., Tsang, D.C.W., 2019. Formation, characteristics, and applications of environmentally persistent free radicals in biochars: A review. *Bioresource Technology*, 281, 457–468.

21 Prevoteau, A., Ronsse, F., Cid, I., Boeckx, P., Rabaey, K., 2016. The electron donating capacity of biochar is dramatically underestimated. *Scientific Reports UK*, 6.

22 Chu, M.L., Tian, W.J., Zhao, J., Zou, M.Y., Lu, Z.Y., Zhang, D.T., Jiang, J.F., 2022. A comprehensive review of capacitive deionization technology with biochar-based electrodes: Biochar-based electrode preparation, deionization mechanism and applications. *Chemosphere*, 307.

23 Tian, R., Dong, H.R., Chen, J., Li, R., Xie, Q.Q., Li, L., Li, Y.J., Jin, Z.L., Xiao, S.J., Xiao, J.Y., 2021. Electrochemical behaviors of biochar materials during pollutant removal in wastewater: A review. *Chemical Engineering Journal*, 425.

24 Lu, Y., Xie, Q.Q., Tang, L., Yu, J.F., Wang, J.J., Yang, Z.H., Fan, C.Z., Zhang, S.J., 2021. The reduction of nitrobenzene by extracellular electron transfer facilitated by Fe-bearing biochar derived from sewage sludge. *Journal of Hazardous Materials*, 403.

25 Sathishkumar, K., Li, Y., Sanganyado, E., 2020. Electrochemical behavior of biochar and its effects on microbial nitrate reduction: Role of extracellular polymeric substances in extracellular electron transfer. *Chemical Engineering Journal*, 395.

26 Ding, L.Z., Zhang, P.P., Luo, H., Hu, Y.F., Banis, M.N., Yuan, X.L., Liu, N., 2018. Nitrogen-doped carbon materials as metal-free catalyst for the dechlorination of trichloroethylene by sulfide. *Environmental Science & Technology*, 52, 14286–14293.

27 Wu, Z.S., Xu, F., Yang, C., Su, X.X., Guo, F.C., Xu, Q.Y., Peng, G.L., He, Q., Chen, Y., 2019. Highly efficient nitrate removal in a heterotrophic denitrification system amended with redox-active biochar: A molecular and electrochemical mechanism. *Bioresource Technology*, 275, 297–306.

28 Ren, S., Usman, M., Tsang, D.C.W., O-Thong, S., Angelidaki, I., Zhu, X.D., Zhang, S.C., Luo, G., 2020. Hydrochar-facilitated anaerobic digestion: evidence for direct interspecies electron transfer mediated through surface oxygen-containing functional groups. *Environmental Science & Technology*, 54, 5755–5766.

29 Xu, X.Y., Huang, H., Zhang, Y., Xu, Z.B., Cao, X.D., 2019. Biochar as both electron donor and electron shuttle for the reduction transformation of Cr(VI) during its sorption. *Environmental Pollution*, 244, 423–430.

30 Xu, Z.B., Xu, X.Y., Zhang, Y., Yu, Y.L., Cao, X.D., 2020. Pyrolysis-temperature depended electron donating and mediating mechanisms of biochar for Cr(VI) reduction. *Journal of Hazardous Materials*, 388.

31 Wang, H.Z., Guo, W.Q., Liu, B.H., Wu, Q.L., Luo, H.C., Zhao, Q., Si, Q.S., Sseguya, F., Ren, N.Q., 2019. Edge-nitrogenated biochar for efficient peroxydisulfate activation: An electron transfer mechanism. *Water Research*, 160, 405–414.

32 Xu, Z.B., Xu, X.Y., Tao, X.Y., Yao, C.B., Tsang, D.C.W., Cao, X.D., 2019. Interaction with low molecular weight organic acids affects the electron shuttling of biochar for Cr(VI) reduction. *Journal of Hazardous Materials*, 378.

33 Chen, Y.D., Duan, X.G., Zhang, C.F., Wang, S.B., Ren, N.Q., Ho, S.H., 2020. Graphitic biochar catalysts from anaerobic digestion sludge for nonradical degradation of micropollutants and disinfection. *Chemical Engineering Journal*, 384.

34 Oh, S.Y., Son, J.G., Lim, O.T., Chiu, P.C., 2012. The role of black carbon as a catalyst for environmental redox transformation. *Environmental Geochemistry and Health*, 34, 105–113.

35 Dong, X.L., Ma, L.Q., Gress, J., Harris, W., Li, Y.C., 2014. Enhanced Cr(VI) reduction and As(III) oxidation in ice phase: Important role of dissolved organic matter from biochar. *Journal of Hazardous Materials*, 267, 62–70.

36 Xu, X.Y., Zhao, Y.H., Sima, J.K., Zhao, L., Masek, O., Cao, X.D., 2017. Indispensable role of biochar-inherent mineral constituents in its environmental applications: A review. *Bioresource Technology*, 241, 887–899.

37 Sun, K., Kang, M.J., Zhang, Z.Y., Jin, J., Wang, Z.Y., Pan, Z.Z., Xu, D.Y., Wu, F.C., Xing, B.S., 2013. Impact of deashing treatment on biochar structural properties and potential sorption mechanisms of phenanthrene. *Environmental Science & Technology*, 47, 11473–11481.

38 Li, S.S., Shao, L.M., Zhang, H., He, P.J., Lu, F., 2020. Quantifying the contributions of surface area and redox-active moieties to electron exchange capacities of biochar. *Journal of Hazardous Materials*, 394.

39 Wu, S., Fang, G.D., Wang, Y.J., Zheng, Y., Wang, C., Zhao, F., Jaisi, D.P., Zhou, D.M., 2017. Redox-active oxygen-containing functional groups in activated

carbon facilitate microbial reduction of ferrihydrite. *Environmental Science & Technology*, 51, 9709–9717.
40 Tang, W.W., He, D., Zhang, C.Y., Waite, T.D., 2017. Optimization of sulfate removal from brackish water by membrane capacitive deionization (MCDI). *Water Research*, 121, 302–310.
41 Qu, Y.T., Campbell, P.G., Gu, L., Knipe, J.M., Dzenitis, E., Santiago, J.G., Stadermann, M., 2016. Energy consumption analysis of constant voltage and constant current operations in capacitive deionization. *Desalination*, 400, 18–24.
42 Li, G.X., Hou, P.X., Zhao, S.Y., Liu, C., Cheng, H.M., 2016. A flexible cotton-derived carbon sponge for high-performance capacitive deionization. *Carbon*, 101, 1–8.
43 Du, Z.Y., Tian, W.J., Qiao, K.L., Zhao, J., Wang, L., Xie, W.L., Chu, M.L., Song, T.T., 2020. Improved chlorine and chromium ion removal from leather processing wastewater by biocharcoal-based capacitive deionization. *Separation and Purification Technology*, 233.
44 Jeon, S.I., Park, H.R., Yeo, J.G., Yang, S., Cho, C.H., Han, M.H., Kim, D.K., 2013. Desalination via a new membrane capacitive deionization process utilizing flow-electrodes. *Energy and Environmental Sciences*, 6, 1471–1475.
45 Choi, J., Dorji, P., Shon, H.K., Hong, S., 2019. Applications of capacitive deionization: Desalination, softening, selective removal, and energy efficiency. *Desalination*, 449, 118–130.
46 Yang, F., He, Y.F., Rosentsvit, L., Suss, M.E., Zhang, X.R., Gao, T., Liang, P., 2021. Flow-electrode capacitive deionization: A review and new perspectives. *Water Research*, 200.
47 Ma, J.X., He, C., He, D., Zhang, C.Y., Waite, T.D., 2018. Analysis of capacitive and electrodialytic contributions to water desalination by flow-electrode CDI. *Water Research*, 144, 296–303.
48 Shi, Q.Q., Zhang, X., Shen, B.X., Ren, K., Wang, Y.T., Luo, J.Z., 2021. Enhanced elemental mercury removal via chlorine-based hierarchically porous biochar with CaCO3 as template. *Chemical Engineering Journal*, 406.
49 Liu, S.H., Tang, Y.H., 2020. Hierarchically porous biocarbons prepared by microwave-aided carbonization and activation for capacitive deionization. *Journal of Electroanalytical Chemistry*, 878.
50 Pan, X.Q., Gu, Z.P., Chen, W.M., Li, Q.B., 2021. Preparation of biochar and biochar composites and their application in a Fenton-like process for wastewater decontamination: A review. *Science of the Total Environment*, 754.
51 Chen, Y.D., Wang, R.P., Duan, X.G., Wang, S.B., Ren, N.Q., Ho, S.H., 2020. Production, properties, and catalytic applications of sludge derived biochar for environmental remediation. *Water Research*, 187.
52 Wan, J., Liu, L., Ayub, K.S., Zhang, W., Shen, G.X., Hu, S.Q., Qian, X.Y., 2020. Characterization and adsorption performance of biochars derived from three key biomass constituents. *Fuel*, 269.

53 Yuan, T., He, W.J., Yin, G.J., Xu, S., 2020. Comparison of bio-chars formation derived from fast and slow pyrolysis of walnut shell. *Fuel*, 261.
54 Zhang, Y., Xu, X.Y., Cao, L.Z., Ok, Y.S., Cao, X.D., 2018. Characterization and quantification of electron donating capacity and its structure dependence in biochar derived from three waste biomasses. *Chemosphere*, 211, 1073–1081.
55 Cimo, G., Kucerik, J., Berns, A.E., Schaumann, G.E., Alonzo, G., Conte, P., 2014. Effect of heating time and temperature on the chemical characteristics of biochar from poultry manure. *Journal of Agricultural and Food Chemistry*, 62, 1912–1918.
56 Sharma, H.B., Sarmah, A.K., Dubey, B., 2020. Hydrothermal carbonization of renewable waste biomass for solid biofuel production: A discussion on process mechanism, the influence of process parameters, environmental performance and fuel properties of hydrochar. *Renewable and Sustainable Energy Reviews*, 123.
57 Kambo, H.S. and Dutta, A., 2015. A comparative review of biochar and hydrochar in terms of production, physico-chemical properties and applications. *Renewable and Sustainable Energy Reviews*, 45, 359–378.
58 Kochermann, J., Gorsch, K., Wirth, B., Muhlenberg, J., Klemm, M., 2018. Hydrothermal carbonization: Temperature influence on hydrochar and aqueous phase composition during process water recirculation. *Journal of Environmental Chemical Engineering*, 6, 5481–5487.
59 Lucian, M., Volpe, M., Gao, L.H., Piro, G., Goldfarb, J.L., Fiori, L., 2018. Impact of hydrothermal carbonization conditions on the formation of hydrochars and secondary chars from the organic fraction of municipal solid waste. *Fuel*, 233, 257–268.
60 Li, Y., Tsend, N., Li, T.K., Liu, H.Y., Yang, R.Q., Gai, X.K., Wang, H.P., Shan, S.D., 2019. Microwave assisted hydrothermal preparation of rice straw hydrochars for adsorption of organics and heavy metals. *Bioresource Technology*, 273, 136–143.
61 Zhu, L., Lei, H.W., Wang, L., Yadavalli, G., Zhang, X.S., Wei, Y., Liu, Y.P., Yan, D., Chen, S.L., Ahring, B., 2015. Biochar of corn stover: Microwave-assisted pyrolysis condition induced changes in surface functional groups and characteristics. *Journal of Analytical and Applied Pyrolysis*, 115, 149–156.
62 Bardestani, R., Kaliaguine, S., 2018. Steam activation and mild air oxidation of vacuum pyrolysis biochar. *Biomass and Bioenergy*, 108, 101–112.
63 Hou, C.H., Liu, N.L., Hsi, H.C., 2015. Highly porous activated carbons from resource-recovered Leucaena leucocephala wood as capacitive deionization electrodes. *Chemosphere*, 141, 71–79.
64 Tuomikoski, S., Kupila, R., Romar, H., Bergna, D., Kangas, T., Runtti, H., Lassi, U., 2019. Zinc adsorption by activated carbon prepared from lignocellulosic waste biomass. *Applied Sciences*, 9.
65 Duan, X.H., Srinivasakannan, C., Peng, J.H., Zhang, L.B., Zhang, Z.Y., 2011. Comparison of activated carbon prepared from Jatropha hull by conventional heating and microwave heating. *Biomass and Bioenergy*, 35, 3920–3926.

66 Cuong, D.V., Matsagar, B.M., Lee, M.S., Hossain, M.S.A., Yamauchi, Y., Vithanage, M., Sarkar, B., Ok, Y.S., Wu, K.C.W., Hou, C.H., 2021. A critical review on biochar-based engineered hierarchical porous carbon for capacitive charge storage. *Renewable and Sustainable Energy Reviews*, 145.

67 Zhang, H.F., Zhang, Y.H., Bai, L.Q., Zhang, Y.G., Sun, L., 2021. Effect of physiochemical properties in biomass-derived materials caused by different synthesis methods and their electrochemical properties in supercapacitors. *Journal of Materials Chemistry A*, 9, 12521–12552.

68 Li, D., Li, Y., Liu, H.J., Ma, J., Liu, Z.G., Gai, C., Jiao, W.T., 2019. Synthesis of biomass tar-derived foams through spontaneous foaming for ultra-efficient herbicide removal from aqueous solution. *Science of the Total Environment*, 673, 110–119.

69 Li, D., Guo, Y.C.A., Li, Y., Liu, Z.G., Chen, Z.L., 2022. Waste-biomass tar functionalized carbon spheres with N/P Co-doping and hierarchical pores as sustainable low-cost energy storage materials. *Renewable Energy*, 188, 61–69.

70 Phiri, J., Dou, J.Z., Vuorinen, T., Gane, P.A.C., Maloney, T.C., 2019. Highly porous willow wood-derived activated carbon for high-performance supercapacitor electrodes. *Acs Omega*, 4, 18108–18117.

71 Cao, J., Jafta, C.J., Gong, J., Rang, Q.D., Lin, X.Z., Felix, R., Wilks, R.G., Bar, M., Yuan, J.Y., Ballauff, M., Lu, Y., 2016. Synthesis of dispersible mesoporous nitrogen-doped hollow carbon nanoplates with uniform hexagonal morphologies for supercapacitors. *ACS Applied Materials & Interfaces*, 8, 29628–29636.

72 Long, C., Zhuang, J.L., Xiao, Y., Zheng, M.T., Hu, H., Dong, H.W., Lei, B.F., Zhang, H.R., Liu, Y.L., 2016. Nitrogen-doped porous carbon with an ultrahigh specific surface area for superior performance supercapacitors. *Journal of Power Sources*, 310, 145–153.

73 Sanchez-Sanchez, A., Izquierdo, M.T., Medjahdi, G., Ghanbaja, J., Celzard, A., Fierro, V., 2018. Ordered mesoporous carbons obtained by conditions soft-templating of tannin in mild conditions. *Microporous and Mesoporous Materials*, 270, 127–139.

74 Wan, C.C., Jiao, Y., Li, J., 2016. Core-shell composite of wood-derived biochar supported MnO_2 nanosheets for supercapacitor applications. *RSC Advances*, 6, 64811–64817.

75 Adorna, J., Borines, M., Dang, V.D., Doong, R.A., 2020. Coconut shell derived activated biochar-manganese dioxide nanocomposites for high performance capacitive deionization. *Desalination*, 492.

76 Yao, F.B., Yang, Q., Yan, M., Li, X.L., Chen, F., Zhong, Y., Yin, H.Y., Chen, S.J., Fu, J., Wang, D.B., Li, X.M., 2020. Synergistic adsorption and electrocatalytic reduction of bromate by Pd/N-doped loofah sponge-derived biochar electrode. *Journal of Hazardous Materials*, 386.

77 Mansoori, S., Davarnejad, R., Ozumchelouei, E.J., Ismail, A.F., 2021. Activated biochar supported iron-copper oxide bimetallic catalyst for degradation of ciprofloxacin via photo-assisted electro-Fenton process: A mild pH condition. *Journal of Water Process Engineering*, 39.

78 Liu, Q., Jiang, S.Q., Su, X.T., Zhang, X.L., Cao, W.M., Xu, Y.F., 2021. Role of the biochar modified with $ZnCl_2$ and $FeCl_3$ on the electrochemical degradation of nitrobenzene. *Chemosphere*, 275.

79 Paul, R., Zhu, L., Chen, H., Qu, J., Dai, L.M., 2019. Recent advances in carbon-based metal-free electrocatalysts. *Advanced Materials*, 31.

80 Zhang, C., Li, H.Q., Yang, X., Tan, X.J., Wan, C.L., Liu, X., 2022. Characterization of electrodes modified with sludge-derived biochar and its performance of electrocatalytic oxidation of azo dyes. *Journal of Environmental Management*, 324.

81 Zhang, C., Li, F., Wen, R.B., Zhang, H.K., Elumalai, P., Zheng, Q., Chen, H.Y., Yang, Y.J., Huang, M.Z., Ying, G.G., 2020. Heterogeneous electro-Fenton using three-dimension NZVI-BC electrodes for degradation of neonicotinoid wastewater. *Water Research*, 182.

82 Zhang, P., Sun, H.W., Ren, C., Min, L.J., Zhang, H.M., 2018. Sorption mechanisms of neonicotinoids on biochars and the impact of deashing treatments on biochar structure and neonicotinoids sorption. *Environmental Pollution*, 234, 812–820.

83 Deng, F.X., Olvera-Vargas, H., Garcia-Rodriguez, O., Zhu, Y.S., Jiang, J.Z., Qiu, S., Yang, J.X., 2019. Waste-wood-derived biochar cathode and its application in electro-Fenton for sulfathiazole treatment at alkaline pH with pyrophosphate electrolyte. *Journal of Hazardous Materials*, 377, 249–258.

84 Deng, F.X., Li, S.X., Zhou, M.H., Zhu, Y.S., Qiu, S., Li, K.H., Ma, F., Jiang, J.Z., 2019. A biochar modified nickel-foam cathode with iron-foam catalyst in electro-Fenton for sulfamerazine degradation. *Applied Catalysis B: Environmental*, 256.

85 Pourzamani, H., Mengelizadeh, N., Hajizadeh, Y., Mohammadi, H., 2018. Electrochemical degradation of diclofenac using three-dimensional electrode reactor with multi-walled carbon nanotubes. *Environmental Science and Pollution Research*, 25, 24746–24763.

86 Soares, C., Correia-Sa, L., Paiga, P., Barbosa, C., Remor, P., Freitas, O.M., Moreira, M.M., Nouws, H.P.A., Correia, M., Ghanbari, A., Rodrigues, A.J., Oliveira, C.M., Figueiredo, S.A., Delerue-Matos, C., 2022. Removal of diclofenac and sulfamethoxazole from aqueous solutions and wastewaters using a three-dimensional electrochemical process. *Journal of Environmental Chemical Engineering*, 10.

87 Zhao, Z.L., Wang, X., Zhu, G.C., Wang, F., Zhou, Y., Dong, W.Y., Wang, H.J., Sun, F.Y., Xie, H.J., 2022. Enhanced removal of Cu-EDTA in a three-dimensional

electrolysis system with highly graphitic activated biochar produced via acidic and K_2FeO_4 treatment. *Chemical Engineering Journal*, 430.

88 Wang, X., Zhao, Z.L., Wang, H.J., Wang, F., Dong, W.Y., 2023. Decomplexation of Cu-1-hydroxyethylidene-1,1-diphosphonic acid by a three-dimensional electrolysis system with activated biochar as particle electrodes. *Journal of Environmental Sciences*, 124, 630–643.

89 Goglio, A., Marzorati, S., Zecchin, S., Quarto, S., Falletta, E., Bombelli, P., Cavalca, L., Beggio, G., Trasatti, S., Schievano, A., 2022. Plant nutrients recovery from agro-food wastewaters using microbial electrochemical technologies based on porous biocompatible materials. *Journal of Environmental Chemical Engineering*, 10.

90 Stephanie, H., Mlsna, T.E., Wipf, D.O., 2021. Functionalized biochar electrodes for asymmetrical capacitive deionization. *Desalination*, 516.

91 Sun, K.G., Wang, C., Tebyetekerwa, M., Zhao, X.S., 2022. Electrocapacitive desalination with nitrogen-doped hierarchically structured carbon prepared using a sustainable salt-template method. *Chemical Engineering Journal*, 446.

92 Wang, B., Zhai, Y.B., Hu, T.J., Niu, Q.Y., Li, S.H., Liu, X.M., Liu, X.P., Wang, Z.X., Li, C.T., Xu, M., 2021. Green quaternary ammonium nitrogen functionalized mesoporous biochar for sustainable electro-adsorption of perchlorate. *Chemical Engineering Journal*, 419.

93 Bergmann, H., 2021. Electrochemical disinfection - State of the art and tendencies. *Current Opinion in Electrochemistry*, 28.

94 Sun, W., Lu, Z.L., Zuo, K.C., Xu, S., Shi, B.Y., Wang, H.B., 2022. High efficiency electrochemical disinfection of pseudomons putida using electrode of orange peel biochar with endogenous metals. *Chemosphere*, 289.

12

Peroxide-Based Biochar-Assisted Advanced Oxidation

Yang Cao[1], Qiaozhi Zhang[2], Yuqing Sun[4], and Daniel C.W. Tsang[3]

[1] Department of Environmental Science and Engineering, Fudan University, Shanghai, China
[2] Department of Civil and Environmental Engineering, The Hong Kong Polytechnic University, Hong Kong, China
[3] State Key Laboratory of Clean Energy Utilization, Zhejiang University, China
[4] School of Agriculture, Sun Yat-Sen University, Guangzhou, Guangdong, China

12.1 Introduction

The increasingly serious water environmental issues urgently require the development of efficient and green wastewater treatment methods [1]. Among existing strategies, advanced oxidation processes (AOPs) are considered to be the most effective method to reduce wastewater pollution [2]. Specifically, AOPs typically employ active species with high redox potentials to achieve mineralization of pollutants in the liquid phase [2, 3]. Peroxides in which two oxygen atoms are linked together by a single covalent bond have been widely explored in AOPs, mainly including hydrogen peroxide (H_2O_2), persulfate like peroxymonosulfate (PMS) and peroxydisulfate (PDS), and percarbonate (Figure 12.1) [4–6]. Since peroxides with a lower redox potential can be easily activated and generate reactive oxidizing species (ROS), various activation techniques have been developed to enhance reaction performance [4–6]. In peroxide-based AOPs, Fenton-like system, persulfate (PS) activation system, and photocatalytic system have been widely studied, and the advantages of the peroxide-based activation mainly include low energy consumption, reduced metal leaching, and high efficiency [5, 7, 8]. Among these reactions, Fenton-like oxidation is one of the most frequently used ones, which can release hydroxyl radical ($^{\bullet}OH$) by the decomposition of H_2O_2 [7, 9]. Recently,

Biochar Applications for Wastewater Treatment, First Edition. Edited by Daniel C.W. Tsang and Yuqing Sun.
© 2023 John Wiley & Sons, Inc. Published 2023 by John Wiley & Sons, Inc.

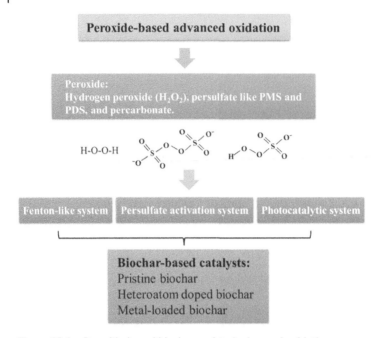

Figure 12.1 Peroxide-based biochar-assisted advanced oxidation.

PS activation system, as a new alternative to AOPs, has attracted more and more attention because it can overcome the strict pH limitation in the Fenton-like reaction and solve the problems of H_2O_2 storage and transportation cost [5, 10, 11]. Photocatalytic oxidation is considered to be an economical and environmentally friendly method for wastewater treatment [12, 13]. In this system, ROS can be generated on the conduction band of the catalyst, and organic pollutants can also be degraded via a direct oxidation by valence-band holes. Different reaction mechanisms and pathways are involved in these oxidation systems, which should be deeply understood for future large-scale applications.

Heterogeneous catalysts play a crucial role in peroxide-based AOPs as the degradation/mineralization efficiency is highly dependent on the activation of peroxides by activators to generate ROS and the high catalytic capacity of various catalysts [14]. Efficient and economical redox metal oxide catalysts have been developed to improve degradation efficiency. However, metal leaching may cause secondary pollution, which limits its environmental application. Recently, many studies have focused on carbonaceous materials for environmental applications due to their high thermal stability, easy synthesis, and high catalytic efficiency [15, 16]. Biochar, a facile synthesized carbon-enriched material from biomass, has been explored as a sustainable and carbon-negative catalyst in environmental catalysis field owing to its cost-effective, multiple functionalities, and tailorable porous structure [17–22]. According to previous studies, reactive sites in biochar-based catalysts mainly include oxygen-containing functional groups, persistent free radicals (PFRs), surface/structural defects induced by heteroatom doping or chemical activation, and rich mineral constituents, which are beneficial to the activation of peroxides and pollutant adsorption and degradation [14, 23]. Among heterogeneous catalysts, metal (e.g., iron, copper, cobalt, manganese, and multi-metallic species) loaded biochar catalysts were frequently used due to their high catalytic performance [9, 24–26]. Various modification strategies were developed to tune the physicochemical properties of biochar-based catalysts and the reactivity of the actual active site and ROS generation. In addition to metal-loaded biochar catalysts, metal-free biochar catalysts synthesized by heteroatom doping like N, S, or B into carbon framework of biochar have been explored as peroxide activators in environmentally friendly applications [27–29]. The introduction of electronegative or electropositive heteroatoms can effectively tailor the charge distribution of the adjacent carbon atoms, thereby promoting the activation of peroxides. Both radical and nonradical reaction pathways were proposed according to the nature of active sites [29–31]. But the real active sites, peroxide activation mechanism, and degradation pathways remain unclear and controversial because of the complexity and variability of biochar-based catalysts. Therefore, it is worthwhile to critically review the applications of biochar-based catalysts as peroxide activators in recent years in order to better understand their catalytic performance and potential applications in wastewater treatment.

This chapter aims to introduce biochar-based catalysts as sustainable and green catalysts from basic components and physicochemical properties to their wide application in peroxide-based AOPs for wastewater treatment. The recent advances in biochar-assisted peroxide activation in terms of catalyst structures and properties, degradation performance, generation of ROS, reaction mechanisms, and future perspectives were summarized and critically discussed.

12.2 Biochar-Based Catalysts

Biochar, a type of carbon-enriched solid material, can be produced from the fast/slow pyrolysis of residual biomass in an oxygen-limited environment, and biochar production shows an attractive property for low carbon footprint of biomass conversion. Biochar catalysts mainly include pristine biochar, redox metal-loaded biochar, and heteroatom-doped biochar catalysts. Heteroatom doping and metal loading are the most commonly used approaches to enhance catalytic efficiency of biochar-based catalysts in peroxide-based advanced oxidation. This section discusses different types of biochar-based catalysts for wastewater treatment.

12.2.1 Pristine Biochar

Pristine biochar catalysts have been explored for peroxide activation due to their facile synthesis and diverse functional structures such as functional groups, inorganic components, and surface/structural defects. Feedstocks properties and pyrolysis parameters (e.g., pyrolysis temperature, time, atmosphere, and activation strategies) can affect its functional structures and physicochemical properties. Dai et al. investigated metal-free biochar (BC) obtained from wheat straw pyrolysis at different temperatures for the degradation of bisphenol A (BPA) [32]. BC_{800} showed a significant activation ability through the enhanced electron-transfer cycles between carbon-activated periodate complexes (electron acceptor) and BPA (electron donor). The pyrolysis of sludge into functional biochar catalysts provides a valuable process for wastewater treatment and the utilization of excess organic solid waste. For instance, high-performance biochar catalysts from sewage sludge were fabricated and applied to the activation of PDS for degrading landfill leachate [33]. Sludge-derived biochar catalysts have high reactivity due to the richness of redox metals. Fu et al. reported PMS can be activated by iron self-doped sludge-derived biochar with a larger specific surface area (SSA) and high dispersion of Fe and N species. These unique characteristics and properties of pristine biochar facilitate the activation of PMS for the degradation of perfluorooctanoic acid [34].

Important physicochemical properties of biochar catalysts for peroxide activation include SSA, functional groups, and surface/structural defects (Figure 12.2)

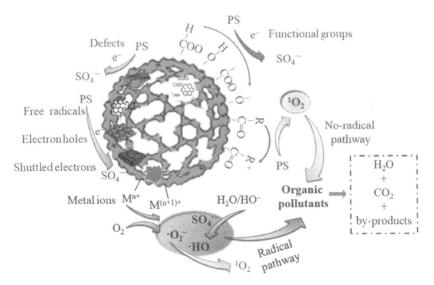

Figure 12.2 The physicochemical properties and active sites of biochar-based catalysts for peroxide activation [35].

[35]. Chemical activation with bases and acids (e.g., KOH, NH_4OH, or HCl treatment) and physical activation with soft/hard templates were employed for biochar modification [36]. The KOH-activated biochar with large SSA and oxygen-containing functional groups can enhance catalytic activity [37]. CO_2 modification was another conversional method to fabricate a hierarchical porous structure with high polarity and acidity [38]. The large SSA of the engineered biochar with tailored surface functionalities promoted the adsorption of 2,4-dichlorophenoxy acetic acid for further degradation under microwave irradiation. So far, many studies have shown that graphitic biochar exhibited excellent performance toward PS activation [39, 40]. As compared to the original BC, graphitized BC derived from wood chips not only promoted the formation of ROS but also facilitated electrons transfer from sulfamethoxazole to PS via nonradical reactions [39]. Zhu et al. further demonstrated the critical role of π-π^* transitions in the BC aromatic ring for the direct activation of peroxides [41]. Future studies are suggested to optimize the rational design of biochar with large SSA, high porosity, and multiple functionalities for peroxide activation.

12.2.2 Redox Metal-Loaded Biochar

Heterogeneous metals and metal oxides are very active in peroxide activation but are limited by their low stability. For metal-loaded biochar catalysts, biochar is typically a support for dispersing and stabilizing active metal species. Until now, iron, copper, cobalt, manganese, and multi-metallic species have been studied in peroxide activation for wastewater treatment [42–44]. Nano-zero-valent iron-loaded biochar (nZVI/BC) catalysts have been widely used in Fenton-like reactions [26, 45]. nZVI played a critical role in the activation and decomposition of H_2O_2 into $^{\bullet}OH$ for pollutant degradation, and BC can cooperate with active metal species to promote the adsorption of pollutants and the activation of peroxides and alleviate nZVI passivation [26]. Li et al. prepared corn cob-derived biochar with abundant functional groups to anchor and stabilize nano-sized nZVI for organic degradation [45]. Fe^0 species and O-containing groups were the major active sites to generate ROS. Despite the vital role of ROS, a nonradical pathway was proposed in Fe/BC systems for degrading and eliminating pollutants. The rapid generation of Fe^{2+}/Fe^{4+} redox cycle in Fe/BC$_{400}$ (low-temperature pyrolysis) could facilitate the continuous degradation of pollutants, and the high-performance Fe/BC$_{700}$ (high-temperature pyrolysis) could promote surface-mediated electron transfer between the formed Fe/BC$_{700}$−PS* reactive complex to degrade pollutants via nonradical pathways [46]. Recently, much effort has been devoted to developing bimetallic catalysts (e.g., CuFe, MnFe, CoFe, and NiFe) to improve catalytic efficiency [9, 24, 47]. For instance, MnOx-Fe$_3$O$_4$/biochar composites showed a high degradation efficacy because the acceleration of Fe^{3+}/Fe^{2+} and Mn^{3+}/Mn^{2+} cycles enhanced H_2O_2 decomposition for pollutant degradation [24].

The synergistic effect of Fe^{3+}/Fe^{2+} and Cu^{2+}/Cu^{+} redox cycles in Fenton-like systems was reported by Xin et al, and high catalytic efficiency was achieved by using biochar-based bimetallic catalysts [9].

There were many studies that used low-cost Cu-loaded biochar catalysts for peroxide activation [48, 49]. Wan et al. designed a Cu-doped graphitic biochar catalyst for the removal of organic contaminants [49]. Cu species can be introduced into the carbon matrix under CO_2 atmosphere, and the stabilized Cu species and formed ketonic groups and unconfined π electrons in the carbon lattice of graphitic biochar could enable electron transfer to activate PDS via nonradical pathways. Zhao et al. prepared CuO@BC for efficient activation of PDS by Cu^{+}/Cu^{2+} redox couples and supposed that the activation primarily occurred over CuO (001) crystal facet [50]. In addition, low-valent Cu species were also reported for the degradation of organic pollutants in wastewater by Yu et al., and both Cu^{0}/Cu^{+} redox cycle and pyridinic N in biochar supports contribute to the enhanced catalytic performance [48]. In addition to copper species, Xue et al. modified biochar by KOH chemical activation to stabilize CuO for the degradation of pollutants, and its porous structure and functional groups can improve the adsorption and activation of PS [43].

12.2.3 Heteroatom-Doped Biochar

Heteroatom doping is an effective strategy to enhance the catalytic activity of biochar-based catalysts, and N-doped biochar catalysts have been widely explored in wastewater treatment via AOPs [51, 52]. Ye et al. prepared N-doped biochar-based catalysts derived from biomass fiber for PS activation by a nonradical pathway, and the catalyst showed a high catalytic efficiency due to the sp^2-hybridized carbon framework and N-induced surface defects as well as ketonic groups [53]. Zuo et al. fabricated N/O co-doped carbon catalysts with two types of active sites, pyridinic N and carbonyl groups, for a Fenton-like reaction [51]. In addition to thermal treatment for heteroatom doping into biochar, ball milling was reported to introduce N into carbon structure with abundant surface defects, mainly including heteroatom doping, vacancy defects, and edge defects (Figure 12.3) [18, 52]. These surface/structure defects could promote the activation of peroxides and degrade pollutants via radical and nonradical reactions.

Besides N-doped biochar-based catalysts, other nonmetallic atoms such as sulfur and boron were explored to modify biochar structure and adjust its physicochemical properties [27, 54]. Sulfurized biochar was facilely synthesized from sewage sludge and applied in the activation of PMS and PDS for PBA degradation [40]. The sulfur can be effectively introduced into sp^2-hybridized carbon networks to form Lewis acid and basic sites for peroxide activation. Cui et al. synthesized an N,S-anchored biochar catalyst derived from coffee grounds, and the prepared catalysts showed high efficiency for peroxide activation and high degradation efficiency for

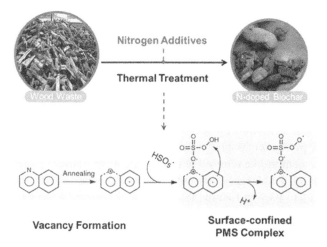

Figure 12.3 N vacancy for peroxide activation [18] (Zhonghao Wan et al., 2021 / from American Chemical Society).

a wide range of organic pollutants [54]. Microwave-assisted pyrolysis was reported to fabricate B, N-codoped biochar with $ZnFe_2O_4$ to enhance the photocatalytic elimination of tetracycline hydrochloride (TCH) [27]. Dou et al. synthesized N/B co-doped biochar catalysts and demonstrated that N and B dopants not only promoted the catalytic efficiency but also switched the radical reaction into a nonradical reaction by enchancing electron transfer on the catalyst surface [55]. Overall, heteroatom doping into the carbon structure of biochar can regulate the electron density, electron distribution, and electron transfer direction, thereby optimizing the catalytic activity and reaction mechanisms, such as radical-based AOPs and nonradical oxidation pathways.

12.3 Peroxide-Based Advanced Oxidation

Peroxide-based advanced oxidation reactions are generally classified into Fenton-like systems, persulfate-activated systems, and photocatalytic systems according to the type of peroxide involved and activation mechanism. This section aims to demonstrate the degradation performance and mechanisms of various biochar-based catalysts based on the review of previously reported studies.

12.3.1 Fenton-Like System

Fenton-like reaction is the most widely studied AOPs. Hydroxyl radicals (·OH) can be produced from H_2O_2 by various heterogenous catalysts, and the produced ROS exhibit high oxidation capacity for pollutant degradation/mineralization.

Fe-based catalyst (e.g., Fe^{2+}, Fe^{3+}, and ZVI) is one of the key parts in traditional Fenton reactions, and its eco-friendly feature shows high potential for large-scale applications (Eq. 2). Recently, ZVI-induced Fenton-like reaction has been widely explored [56, 57]. ZVI has a high reducing capacity ($E^0 = -0.44$) and is considered as an ideal electron donor to promote Fe^{3+} reduction. Specifically, Fe^0 can be easily reduced into Fe^{2+}, which could accelerate the activation of H_2O_2 to produce ˙OH (Eqs. 12.1 and 12.2). While the released Fe^{3+} can react with Fe^0 to sustainably supply Fe^{2+} to participate in Fenton-like processes (Eq. 12.3) [26]. A Fe@BC catalyst was prepared by the pyrolysis of poplar kraft pulp at 900 °C with KOH activation to activate H_2O_2 for Fenton-like removal of rhodamine B (RhB) dyes. Under the reaction conditions of 2 mM H_2O_2 concentration and 50 mg/L RhB, the removal rate can reach up to 100% in 10 min [56]. Bashir et al. reported a biochar-decorated nZVI@Fe_3O_4 catalyst with a core-shell structure, and a high degradation of methylene blue dye was achieved due to the redox transformation of Fe^{3+} to Fe^{2+} [58]. FeMn/BC-H_2O_2 systems were reported to improve degradation efficiency in a Fenton-like system [9, 24]. The decomposition of H_2O_2 into ˙OH and superoxide anion radical (˙O_2^-) in FeMn/BC-H_2O_2-hydroxylamine system was verified. This system has been applied for ciprofloxacin degradation in real wastewater (e.g., industrial wastewater, river water, lake water, and tap water), and the removal rate was 70–82.8% [24]. Fe/N-BC materials with excellent recyclability have also been reported to generate ˙OH and ˙O_2^- from H_2O_2 for wastewater treatment [59]. $CuFeO_2$/BC catalyst also showed high catalytic activity for tetracycline (TC) degradation, and the TC removal rate can reach up to 598 mg/L [9].

$$Fe^0 + H_2O_2 + 2H^+ \rightarrow Fe^{2+} + 2H_2O \tag{12.1}$$

$$Fe^{2+} + H_2O_2 \rightarrow Fe^{3+} + ˙OH + OH^- \tag{12.2}$$

$$2Fe^{3+} + Fe^0 \rightarrow 3Fe^{2+} \tag{12.3}$$

As discussed, the metal component plays an important role in Fenton and Fenton-like reactions through the redox cycle of M^{n+}/M^{n+1+}. In addition to Fe/BC, Cu-loaded biochar catalysts showed a similar H_2O_2 activation mechanism. Cu^0 as a strong reducing agent can accelerate the formation of the Cu^{2+}/Cu^+ cycle. Yu et al. fabricated N-doped biochar encapsulated Cu^0 for the degradation of organic pollutants in a Fenton-like system (Figure 12.4) [48]. In addition to ˙OH radicals, ˙O_2^- and singlet oxygen ($^1O_2$) can be generated and participate in the degradation reactions. Because the Cu^{2+} could react with H_2O_2, forming $HO_2˙$ that can be decomposed subsequently into ˙O_2^-. As shown in Figure 12.5, the dissolved oxygen could also be activated into ˙O_2^- by Cu^+. ˙O_2^- further reacted with $HO_2˙$ to generate $^1O_2$ in this system. Therefore, the degradation efficiency can be enhanced by the redox cycle of Cu^{2+}/Cu^+ and Cu^+/Cu^0. For biochar-based catalysts, other

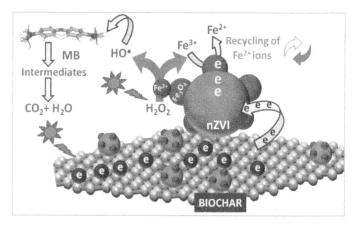

Figure 12.4 The reaction mechanism in Fenton-like reaction by using nZVI@Fe$_3$O$_4$-BC [58].

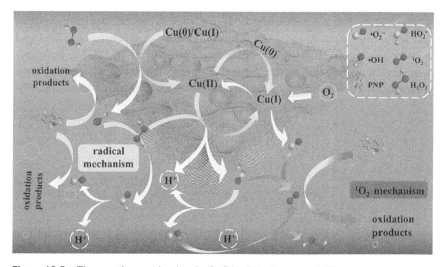

Figure 12.5 The reaction mechanism in Cu@N: C/H$_2$O$_2$ system [48].

active sites derived in carbon matrix such as surface functional groups and surface/structural defects (heteroatom doping, surface vacancies, and edge defects) could also induce H$_2$O$_2$ decomposition for pollutant degradation [51].

12.3.2 Persulfate Activation System

Alternative AOPs using PMS or PDS as an oxidant have emerged based on the same strategy. In persulfate-based AOPs, highly reactive ˙SO$_4^-$ can be produced by cleaving the peroxide bond in PMS and PDS molecular via energy and electron transfer [18, 46]. $\cdot SO_4^-$ is a strong oxidizing agent and shows a better ability to degrade organic pollutants. Compared with $\cdot OH$, $\cdot SO_4^-$ is more stable, and it has a higher redox potential and a longer half-life [49]. Unlike H_2O_2, persulfate-based advanced oxidation includes radical and nonradical reactions [60]. Specifically, the organic pollutants can be directly oxidized by persulfates via electron transfer reactions (Figure 12.6). For nonradical reactions, singlet oxygenation, electron transfer, and high-valent metal-induced oxidation were involved in the persulfate activation system [10]. This section briefly outlines the reaction mechanism and relevant cases, and the next chapter (Chapter 13) will discuss this system in detail.

Similar to a Fenton-like reaction, metal-loaded biochar catalysts can catalyze PMS and PDS by losing one electron for the generation of $\cdot SO_4^-$ [5]. Metal-free biochar catalysts were widely explored in the persulfate activation system. Zhu et al. reported a wood-derived biochar pyrolyzed at 700 °C for the degradation of clofibric acid. The results indicated that BC_{suf}–OH played a crucial role in the activation of PDS to form BC_{suf}–O, $\cdot SO_4^-$, and HSO_4^- species. In addition, the π-π^* transition in the aromatic ring of BC_{700} could contribute to directly decomposing PDS into $\cdot SO_4^-$, thereby promoting the degradation of clofibric acid at these multiple active sites through radical pathways [41]. Wan et al. fabricated a N-doped graphitic biochar and investigated nonradical PMS activation [18]. The positively charged nitrogen vacancies were formed after high-temperature pyrolysis. They can act as active sites for the adsorption of PMS and subsequent activation by

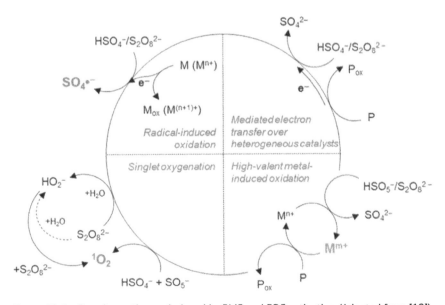

Figure 12.6 Reaction pathways induced by PMS and PDS activation (Adapted from [10]).

forming robust metastable complex (*HSO_5^-). A high removal rate of 4-chlorophenol (90.9%) was achieved through a nonradical reaction. In another case, S-doped biochar showed the potential to effectively remove BPA within 30 min (91.1%), which was mainly attributed to PFRs and vacancy defects [6].

12.3.3 Photocatalytic System

Photocatalysis is considered as a sustainable technique to solve various environmental issues. Typically, the photocatalytic activity is dependent on the sensitivity of photosensitizers to light. Various ROS including $^{\bullet}OH$, $^{\bullet}O_2^-$, H_2O_2, and $^1O_2$ can be generated in the photocatalytic system for the degradation of various organic pollutants (Figure 12.7).

For biochar-based photocatalysts, the effects of heteroatom doping, loaded photosensitizers, functional groups in biochar and its carbon matrix on the catalytic activity have been investigated systematically. Liu et al. fabricated a sewage sludge-derived biochar (SDBC) for the activation of PDS by photothermal conversion under solar light irradiation [33]. SDBC showed an excellent broad-spectrum response that enables photothermal activation of PDS. TiO_2-loaded straw biochar catalysts were synthesized by using a sol-gel method and explored to degrade sulfamethoxazole (SMX) under UV light irradiation [62]. Biochar not only enhanced the adsorption of SMX to promote further degradation but also contributed to

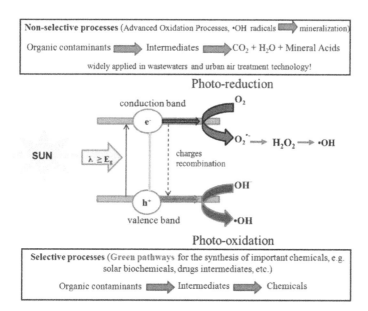

Figure 12.7 Primary reactions and production of ROS in the photocatalytic system [61].

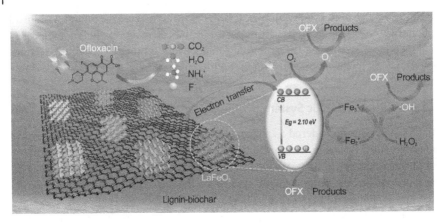

Figure 12.8 Possible mechanism in the photo-Fenton-like system [12].

prolonging the separation lifetime of electrons and valence band hole (h^+) due to its surface-stabilized intermediates. A TiO_2/Fe/Fe_3C-biochar composite with a high mesoporous surface was synthesized for degradation of methylene blue (MB). The prepared catalysts generated ·OH and ·O_2^- radicals through photocatalytic activation of H_2O_2 and exhibited high MB removal capacity (376.9 mg/L) at neutral pH [63]. $ZnFe_2O_4$ binary metal oxides loaded on BN-BC were explored to remove 98.19% of TCH in wastewater [27]. The surface functional groups of BN-BC (e.g., C–OH, C–N, C=O, and C–B–O) played a vital role in the adsorption of amphoteric TCH. In another case, Chen et al. investigated the performance of bimetallic biochar-based materials for the high-efficiency degradation of ofloxacin [12]. As shown in Figure 12.8, $LaFeO_3$/lignin-biochar composites can improve the adsorption capacity of ofloxacin and promote the generation of ·O_2^- radicals through electron transfer on the metal-biochar surface. The improved degradation efficiency of ofloxacin by $LaFeO_3$/lignin-biochar (95.6%) was achieved compared with $LaFeO_3$ (53.4%).

12.4 Conclusion and Future Perspectives

The high carbon content, adjustable porosity, high specific surface area, and stable chemical structure make biochar an attractive catalyst for the degradation of pollutants through peroxide-based AOPs. The physicochemical properties of pristine biochar catalysts can be tuned by heteroatom doping to facilitate the activation of various peroxides. For metal-free biochar, fabricating multiple functionalities and introducing surface defects could help achieve high catalytic activity. Biochar can also serve as a carbon support to disperse and stabilize active

12.4 Conclusion and Future Perspectives

metal species, contributing to the fast electron transfer processes and thus enhancing catalytic efficiency. Metal-loaded biochar catalysts have been widely explored in Fenton-like systems, persulfate activation systems, and photocatalytic systems. Peroxide-based biochar-assisted AOPs are rapidly progressing toward a mature research subject for wastewater treatment. We believe these opportunities can be pursued through a deep understanding of the applicability and limitations of peroxide-based AOPs, as well as a full appreciation of the unique properties of biochar-based materials.

Future research is suggested to focus on developing highly stable and selective biochar-based catalysts and their large-scale application of. Major conclusion and perspectives follow:

1) Various synthetic strategies need to be developed to achieve biochar-based catalysts with low production costs. Biomass properties may need to be critically evaluated as feedstock characteristics will determine/affect the functional structure of biochar-based catalysts. The novel synthesis methods should take advantage of the inherent properties of biomass feedstock to prepare engineered biochar, which do not require further modification.
2) Catalyst stability is a critical issue directly related to industrial applications. Understanding the deactivation mechanisms of biochar-based catalyst in different AOPs may provide new insights into the development of highly stable biochar-based catalysts and improve the selective degradation of targeted pollutants.
3) The feasible production of high-quality biochar with desirable structures/properties and high catalytic activity targeted is the primary goal of the current research for large-scale applications. Both pyrolysis and modification strategies should ensure the homogeneity of biochar catalyst structure, which is of great importance for identifying the real active sites and improving the selective degradation of specific pollutants.
4) The possible reaction mechanism can be inferred from the in situ chemical reactions on the catalyst surface during peroxide activation. In the actual wastewater treatment process, several water parameters such as pH, organic matters, and comm anions will affect the activation mechanism of peroxide and the degradation pathway of pollutants. However, the effects of the aforementioned parameters have not been systematically evaluated in most studies. The properties and constituents of the target wastewater should be fully evaluated.
5) Because biochar materials are derived from organic solid waste, their environmental impacts in various AOPs still need to be explored. For example, dissolution of organics and leaching of metals may occur in Fenton-like systems under acidic conditions, so it is necessary to design and optimize various reaction systems to reduce the impact on the environment.

References

1 Parvulescu, V.I., Epron, F., Garcia, H., Granger, P., 2022. Recent progress and prospects in catalytic water treatment. *Chemical Reviews*, 122, 2981–3121.

2 Zhou, X., Zhu, Y., Niu, Q., Zeng, G., Lai, C., Liu, S., Huang, D., Qin, L., Liu, X., Li, B., Yi, H., Fu, Y., Li, L., Zhang, M., Zhou, C., Liu, J., 2021. New notion of biochar: A review on the mechanism of biochar applications in advanced oxidation processes. *Chemical Engineering Journal*, 416, 129027.

3 Wang, J., Cai, J., Wang, S., Zhou, X., Ding, X., Ali, J., Zheng, L., Wang, S., Yang, L., Xi, S., Wang, M., Chen, Z., 2022. Biochar-based activation of peroxide: Multivariate-controlled performance, modulatory surface reactive sites and tunable oxidative species. *Chemical Engineering Journal*, 428, 131233.

4 Yi, Y., Tu, G., Eric Tsang, P., Fang, Z., 2020. Insight into the influence of pyrolysis temperature on Fenton-like catalytic performance of magnetic biochar. *Chemical Engineering Journal*, 380, 122518.

5 Ren, W., Cheng, C., Shao, P., Luo, X., Zhang, H., Wang, S., Duan, X., 2022. Origins of electron-transfer regime in persulfate-based nonradical oxidation processes. *Environmental Science & Technology*, 56, 78–97.

6 Wan, Z., Xu, Z., Sun, Y., Zhang, Q., Hou, D., Gao, B., Khan, E., Graham, N.J.D., Tsang, D.C.W., 2022. Stoichiometric carbocatalysis via epoxide-like C-S-O configuration on sulfur-doped biochar for environmental remediation. *Journal of Hazardous Materials*, 428, 128223.

7 Fang, G., Gao, J., Liu, C., Dionysiou, D.D., Wang, Y., Zhou, D., 2014. Key role of persistent free radicals in hydrogen peroxide activation by biochar: Implications to organic contaminant degradation. *Environmental Science & Technology*, 48, 1902–1910.

8 Lee, J., von Gunten, U., Kim, J.-H., 2020. Persulfate-based advanced oxidation: Critical assessment of opportunities and roadblocks. *Environmental Science & Technology*, 54, 3064–3081.

9 Xin, S., Liu, G., Ma, X., Gong, J., Ma, B., Yan, Q., Chen, Q., Ma, D., Zhang, G., Gao, M., Xin, Y., 2021. High efficiency heterogeneous Fenton-like catalyst biochar modified CuFeO2 for the degradation of tetracycline: Economical synthesis, catalytic performance and mechanism. *Applied Catalysis B: Environmental*, 280, 119386.

10 Liang, J., Duan, X., Xu, X., Chen, K., Zhang, Y., Zhao, L., Qiu, H., Wang, S., Cao, X., 2021. Persulfate oxidation of sulfamethoxazole by magnetic iron-char composites via nonradical pathways: Fe(IV) versus surface-mediated electron transfer. *Environmental Science & Technology*, 55, 10077–10086.

11 Li, F., Lu, Z., Li, T., Zhang, P., Hu, C., 2022. Origin of the excellent activity and selectivity of a single-atom copper catalyst with unsaturated Cu-N2 sites via peroxydisulfate activation: Cu(III) as a dominant oxidizing species. *Environmental Science & Technology*, 56, 8765–8775.

12. Chen, X., Zhang, M., Qin, H., Zhou, J., Shen, Q., Wang, K., Chen, W., Liu, M., Li, N., 2022. Synergy effect between adsorption and heterogeneous photo-Fenton-like catalysis on LaFeO3/lignin-biochar composites for high efficiency degradation of ofloxacin under visible light. *Separation and Purification Technology*, 280, 119751.
13. Shi, J., Wang, J., Liang, L., Xu, Z., Chen, Y., Chen, S., Xu, M., Wang, X., Wang, S., 2021. Carbothermal synthesis of biochar-supported metallic silver for enhanced photocatalytic removal of methylene blue and antimicrobial efficacy. *Journal of Hazardous Materials*, 401, 123382.
14. Gupta, A.D., Singh, H., Varjani, S., Awasthi, M.K., Giri, B.S., Pandey, A., 2022. A critical review on biochar-based catalysts for the abatement of toxic pollutants from water via advanced oxidation processes (AOPs). *Science of the Total Environment*, 849, 157831.
15. Lee, J., Kim, K.-H., Kwon, E.E., 2017. Biochar as a Catalyst. *Renewable and Sustainable Energy Reviews*, 77, 70–79.
16. Luo, H., Fu, H., Yin, H., Lin, Q., 2022. Carbon materials in persulfate-based advanced oxidation processes: The roles and construction of active sites. *Journal of Hazardous Materials*, 426, 128044.
17. He, X., Zheng, N., Hu, R., Hu, Z., Yu, J.C., 2020. Hydrothermal and pyrolytic conversion of biomasses into catalysts for advanced oxidation treatments. *Advanced Functional Materials*, 31, 2006505.
18. Wan, Z., Xu, Z., Sun, Y., He, M., Hou, D., Cao, X., Tsang, D.C.W., 2021. Critical impact of nitrogen vacancies in nonradical carbocatalysis on Nitrogen-doped graphitic biochar. *Environmental Science & Technology*, 55, 7004–7014.
19. Wan, Z., Sun, Y., Tsang, D.C.W., Xu, Z., Khan, E., Liu, S.-H., Cao, X., 2020. Sustainable impact of tartaric acid as electron shuttle on hierarchical iron-incorporated biochar. *Chemical Engineering Journal*, 395, 125138.
20. Zhang, Q., Xu, S., Cao, Y., Ruan, R., Clark, J.H., Hu, C., Tsang, D.C.W., 2022. Sustainable production of gluconic acid and glucuronic acid via microwave-assisted glucose oxidation over low-cost Cu-biochar catalysts. *Green Chemistry*, 24, 6657–6670.
21. Cao, Y., He, M., Dutta, S., Luo, G., Zhang, S., Tsang, D.C.W., 2021. Hydrothermal carbonization and liquefaction for sustainable production of hydrochar and aromatics. *Renewable and Sustainable Energy Reviews*, 152, 111722.
22. Chen, S.S., Cao, Y., Tsang, D.C.W., Tessonnier, J.-P., Shang, J., Hou, D., Shen, Z., Zhang, S., Ok, Y.S., Wu, K.C.W., 2020. Effective dispersion of MgO nanostructure on biochar support as a basic catalyst for glucose isomerization. *ACS Sustainable Chemistry & Engineering*, 8, 6990–7001.
23. Wu, W., Zhu, S., Huang, X., Wei, W., Jin, C., Ni, B.J., 2021. Determination of instinct components of biomass on the generation of persistent free radicals (PFRs) as critical redox sites in pyrogenic chars for persulfate activation. *Environmental Science & Technology*, 55, 7690–7701.

24 Li, L., Liu, S., Cheng, M., Lai, C., Zeng, G., Qin, L., Liu, X., Li, B., Zhang, W., Yi, Y., Zhang, M., Fu, Y., Li, M., Long, M., 2021. Improving the Fenton-like catalytic performance of MnO(x)-Fe(3)O(4)/biochar using reducing agents: A comparative study. *Journal of Hazardous Materials*, 406, 124333.

25 Li, L., Lai, C., Huang, F., Cheng, M., Zeng, G., Huang, D., Li, B., Liu, S., Zhang, M., Qin, L., Li, M., He, J., Zhang, Y., Chen, L., 2019. Degradation of naphthalene with magnetic bio-char activate hydrogen peroxide: Synergism of bio-char and Fe-Mn binary oxides. *Water Research*, 160, 238–248.

26 Deng, J., Dong, H., Zhang, C., Jiang, Z., Cheng, Y., Hou, K., Zhang, L., Fan, C., 2018. Nanoscale zero-valent iron/biochar composite as an activator for Fenton-like removal of sulfamethazine. *Separation and Purification Technology*, 202, 130–137.

27 Peng, H., Li, Y., Wen, J., Zheng, X., 2021. Synthesis of $ZnFe_2O_4$/B,N-codoped biochar via microwave-assisted pyrolysis for enhancing adsorption-photocatalytic elimination of tetracycline hydrochloride. *Industrial Crops and Products*, 172, 114066.

28 Xu, L., Wu, C., Liu, P., Bai, X., Du, X., Jin, P., Yang, L., Jin, X., Shi, X., Wang, Y., 2020. Peroxymonosulfate activation by nitrogen-doped biochar from sawdust for the efficient degradation of organic pollutants. *Chemical Engineering Journal*, 387, 124065.

29 Hung, C.M., Chen, C.W., Huang, C.P., Dong, C.D., 2022. Metal-free single heteroatom (N, O, and B)-doped coconut-shell biochar for enhancing the degradation of sulfathiazole antibiotics by peroxymonosulfate and its effects on bacterial community dynamics. *Environmental Pollution*, 311, 119984.

30 Yu, J., Tang, L., Pang, Y., Zeng, G., Wang, J., Deng, Y., Liu, Y., Feng, H., Chen, S., Ren, X., 2019. Magnetic nitrogen-doped sludge-derived biochar catalysts for persulfate activation: Internal electron transfer mechanism. *Chemical Engineering Journal*, 364, 146–159.

31 Ortiz-Medina, J., Wang, Z., Cruz-Silva, R., Morelos-Gomez, A., Wang, F., Yao, X., Terrones, M., Endo, M., 2019. Defect engineering and surface functionalization of nanocarbons for Metal-free catalysis. *Advanced Materials*, 31, e1805717.

32 Dai, J., Wang, Z., Chen, K., Ding, D., Yang, S., Cai, T., 2023. Applying a novel advanced oxidation process of biochar activated periodate for the efficient degradation of bisphenol A: Two nonradical pathways. *Chemical Engineering Journal*, 453, 139889.

33 Liu, Y., Huang, R., Hu, W., Lin, L., Liu, J., Wang, Q., Wang, D., Wu, Z., Zhang, J., 2022. High-performance photothermal conversion of sludge derived biochar and its potential for peroxydisulfate-based advanced oxidation processes. *Separation and Purification Technology*, 303, 122214.

34 Fu, S., Zhang, Y., Xu, X., Dai, X., Zhu, L., 2022. Peroxymonosulfate activation by iron self-doped sludge-derived biochar for degradation of perfluorooctanoic acid: A singlet oxygen-dominated nonradical pathway. *Chemical Engineering Journal*, 450, 137953.

35 Zhao, Y., Yuan, X., Li, X., Jiang, L., Wang, H., 2021. Burgeoning prospects of biochar and its composite in persulfate-advanced oxidation process. *Journal of Hazardous Materials*, 409, 124893.

36 Mian, M.M., Liu, G., 2020. Activation of peroxymonosulfate by chemically modified sludge biochar for the removal of organic pollutants: Understanding the role of active sites and mechanism. *Chemical Engineering Journal*, 392, 123681.

37 Li, J., Liu, Y., Ren, X., Dong, W., Chen, H., Cai, T., Zeng, W., Li, W., Tang, L., 2021. Soybean residue based biochar prepared by ball milling assisted alkali activation to activate peroxydisulfate for the degradation of tetracycline. *Journal of Colloid and Interface Science*, 599, 631–641.

38 Sun, Y., Yu, I.K.M., Tsang, D.C.W., Fan, J., Clark, J.H., Luo, G., Zhang, S., Khan, E., Graham, N.J.D., 2020. Tailored design of graphitic biochar for high-efficiency and chemical-free microwave-assisted removal of refractory organic contaminants. *Chemical Engineering Journal*, 398, 125505.

39 Du, L., Xu, W., Liu, S., Li, X., Huang, D., Tan, X., Liu, Y., 2020. Activation of persulfate by graphitized biochar for sulfamethoxazole removal: The roles of graphitic carbon structure and carbonyl group. *Journal of Colloid and Interface Science*, 577, 419–430.

40 Wang, H., Guo, W., Liu, B., Si, Q., Luo, H., Zhao, Q., Ren, N., 2020. Sludge-derived biochar as efficient persulfate activators: Sulfurization-induced electronic structure modulation and disparate nonradical mechanisms. *Applied Catalysis B: Environmental*, 279, 119361.

41 Zhu, K., Wang, X., Geng, M., Chen, D., Lin, H., Zhang, H., 2019. Catalytic oxidation of clofibric acid by peroxydisulfate activated with wood-based biochar: Effect of biochar pyrolysis temperature, performance and mechanism. *Chemical Engineering Journal*, 374, 1253–1263.

42 Ye, G., Zhou, J., Huang, R., Ke, Y., Peng, Y., Zhou, Y., Weng, Y., Ling, C., Pan, W., 2022. Magnetic sludge-based biochar derived from Fenton sludge as an efficient heterogeneous Fenton catalyst for degrading Methylene blue. *Journal of Environmental Chemical Engineering*, 10, 107242.

43 Xue, Y., Guo, Y., Zhang, X., Kamali, M., Aminabhavi, T.M., Appels, L., Dewil, R., 2022. Efficient adsorptive removal of ciprofloxacin and carbamazepine using modified pinewood biochar – A kinetic, mechanistic study. *Chemical Engineering Journal*, 450, 137896.

44 Fang, G., Li, J., Zhang, C., Qin, F., Luo, H., Huang, C., Qin, D., Ouyang, Z., 2022. Periodate activated by manganese oxide/biochar composites for antibiotic degradation in aqueous system: Combined effects of active manganese species and biochar. *Environmental Pollution*, 300, 118939.

45 Li, Z., Sun, Y., Yang, Y., Han, Y., Wang, T., Chen, J., Tsang, D.C.W., 2020. Biochar-supported nanoscale zero-valent iron as an efficient catalyst for organic degradation in groundwater. *Journal of Hazardous Materials*, 383, 121240.

46 Jiang, N., Xu, H., Wang, L., Jiang, J., Zhang, T., 2020. Nonradical oxidation of pollutants with single-atom-Fe(III)-activated persulfate: Fe(V) Being the possible intermediate oxidant. *Environmental Science & Technology*, 54, 14057–14065.

47 Li, Z., Sun, Y., Yang, Y., Han, Y., Wang, T., Chen, J., Tsang, D.C.W., 2020. Comparing biochar- and bentonite-supported Fe-based catalysts for selective degradation of antibiotics: Mechanisms and pathway. *Environmental Research*, 183, 109156.

48 Yu, C., Huang, R., Xie, Y., Wang, Y., Cong, Y., Chen, L., Feng, L., Du, Q., Sun, W., Sun, H., 2022. In-situ synthesis of N-doped biochar encapsulated Cu(0) nanoparticles with excellent Fenton-like catalytic performance and good environmental stability. *Separation and Purification Technology*, 295, 121334.

49 Wan, Z., Sun, Y., Tsang, D.C.W., Yu, I.K.M., Fan, J., Clark, J.H., Zhou, Y., Cao, X., Gao, B., Ok, Y.S., 2019. A sustainable biochar catalyst synergized with copper heteroatoms and CO2 for singlet oxygenation and electron transfer routes. *Green Chemistry*, 21, 4800–4814.

50 Zhao, Y., Yu, L., Song, C., Chen, Z., Meng, F., Song, M., 2022. Selective degradation of electron-rich organic pollutants induced by CuO@Biochar: The key role of outer-sphere interaction and singlet oxygen. *Environmental Science & Technology*, 56, 10710–10720.

51 Zuo, S., Zhu, S., Wang, J., Liu, W., Wang, J., 2022. Boosting Fenton-like reaction efficiency by co-construction of the adsorption and reactive sites on N/O co-doped carbon. *Applied Catalysis B: Environmental*, 301, 120783.

52 Annamalai, S. and Shin, W.S., 2022. Efficient degradation of trimethoprim with ball-milled nitrogen-doped biochar catalyst via persulfate activation. *Chemical Engineering Journal*, 440, 135815.

53 Ye, S., Zeng, G., Tan, X., Wu, H., Liang, J., Song, B., Tang, N., Zhang, P., Yang, Y., Chen, Q., Li, X., 2020. Nitrogen-doped biochar fiber with graphitization from Boehmeria nivea for promoted peroxymonosulfate activation and non-radical degradation pathways with enhancing electron transfer. *Applied Catalysis B: Environmental*, 269, 118850.

54 Cui, P., Yang, Q., Liu, C., Wang, Y., Fang, G., Dionysiou, D.D., Wu, T., Zhou, Y., Ren, J., Hou, H., Wang, Y., 2021. An N,S-anchored single-atom catalyst derived from domestic waste for environmental remediation. *ACS ES&T Engineering*, 1, 1460–1469.

55 Dou, J., Cheng, J., Lu, Z., Tian, Z., Xu, J., He, Y., 2022. Biochar co-doped with nitrogen and boron switching the free radical based peroxydisulfate activation into the electron-transfer dominated nonradical process. *Applied Catalysis B: Environmental*, 301, 120832.

56 Xia, J., Shen, Y., Zhang, H., Hu, X., Mian, M.M., Zhang, W.-H., 2022. Synthesis of magnetic nZVI@biochar catalyst from acid precipitated black liquor and Fenton

sludge and its application for Fenton-like removal of rhodamine B dye. *Industrial Crops and Products*, 187, 115449.
57 Yan, J., Qian, L., Gao, W., Chen, Y., Ouyang, D., Chen, M., 2017. Enhanced Fenton-like degradation of trichloroethylene by hydrogen peroxide activated with nanoscale zero valent iron loaded on biochar. *Scientific Reports*, 7, 43051.
58 Bashir, A., Pandith, A.H., Qureashi, A., Malik, L.A., Gani, M., Perez, J.M., 2022. Catalytic propensity of biochar decorated with core-shell nZVI@Fe_3O_4: A sustainable photo-Fenton catalysis of methylene blue dye and reduction of 4-nitrophenol. *Journal of Environmental Chemical Engineering*, 10, 107401.
59 Li, H.-C., Ji, X.-Y., Pan, X.-Q., Liu, C., Liu, W.-J., 2020. Ionothermal carbonization of biomass to construct Fe, N-Doped biochar with prominent activity and recyclability as cathodic catalysts in heterogeneous electro-Fenton. *ACS ES&T Engineering*, 1, 21–31.
60 Zhu, S., Huang, X., Ma, F., Wang, L., Duan, X., Wang, S., 2018. Catalytic removal of aqueous contaminants on N-doped graphitic biochars: Inherent roles of adsorption and nonradical mechanisms. *Environmental Science & Technology*, 52, 8649–8658.
61 Colmenares, J.C. and Luque, R., 2014. Heterogeneous photocatalytic nanomaterials: Prospects and challenges in selective transformations of biomass-derived compounds. *Chemical Society Reviews*, 43, 765–778.
62 Zhang, H., Wang, Z., Li, R., Guo, J., Li, Y., Zhu, J., Xie, X., 2017. TiO(2) supported on reed straw biochar as an adsorptive and photocatalytic composite for the efficient degradation of sulfamethoxazole in aqueous matrices. *Chemosphere*, 185, 351–360.
63 Mian, M.M. and Liu, G., 2019. Sewage sludge-derived TiO(2)/Fe/Fe(3) C-biocharcomposite as an efficient heterogeneous catalyst for degradation of methylene blue. *Chemosphere*, 215, 101–114.

13

Persulfate-Based Biochar-Assisted Advanced Oxidation

Mengdi Zhao[1], Zibo Xu[2], and Daniel C.W. Tsang[3]

[1] EIT Institute for Advanced Study, Ningbo, China
[2] Department of Civil and Environmental Engineering, The Hong Kong Polytechnic University, Hong Kong, China
[3] State Key Laboratory of Clean Energy Utilization, Zhejiang University, China

13.1 Introduction

Conventional advanced oxidation technologies have shown exemplary performance in wastewater treatment, especially for refractory organics in wastewater. The advanced oxidation technology based on hydroxyl radicals (HR-AOP) with hydrogen peroxide has been widely studied, but it contains certain limitations, such as a narrow working pH range (2.0–4.0), low stability, and difficulties in long-term storage and transportation of hydrogen peroxide. In recent years, sulfate radical-based advanced oxidation technology (SR-AOP) has also been widely used in environmental remediation with advantages including similar or higher redox potential (E^0 = 2.5 V–3.1 V), wider pH range (pH = 2.0–8.0), longer half-life time (3–4 × 10^{-5} s) and higher stability, compared with hydroxyl radicals (•OH). Therefore, advanced oxidation technology based on sulfate radicals has received more and more attention.

As the primary source of sulfate radical generation, persulfate (PS) can be divided into two types: peroxomonosulfate (PMS) and peroxydisulfate (PDS). Among them, PDS has a symmetrical structure, its peroxygen bond length is 1.50 Å, and its bond energy is 140 kJ/mol. Besides, PMS has an asymmetric structure with a shorter bond length (1.46 Å), resulting in higher activation energy. Due to structural differences, PDS and PMS exhibit differences in specific activation processes. Polar PMS is more active than PDS when subjected to nucleophilic attack by

Biochar Applications for Wastewater Treatment, First Edition. Edited by Daniel C.W. Tsang and Yuqing Sun.
© 2023 John Wiley & Sons, Inc. Published 2023 by John Wiley & Sons, Inc.

organic pollutants with electron-rich moieties [1]. PMS and PDS are relatively stable under normal conditions and need to be activated under certain conditions to generate sulfate radicals. Physical (thermal, photoactivation, ultrasonic, microwave), chemical (alkali, quinone, metal ion, metal or metal oxide, carbon material), and combined activation methods are widely used.

PS activation can be divided into the free radical and non-radical pathways for pollutant degradation. In the free radical path, the O-O bonds of PS are activated to produce $SO_4\cdot^-$ and $\cdot OH$, which can effectively oxidize and mineralize organic pollutants. In addition, the process may also involve nonradical pathways such as surface activation, electron transfer, and singlet oxygen ($^1O_2$). In real wastewater, the nonradical pathway is more resistant to the influence of inorganic ions and natural organic matter (NOMs) than the radical pathway [2], which makes the non-free radical pathway more advantageous in complex wastewater treatment.

Biomass-derived, environmentally friendly carbon material has attracted extensive attention for its application in persulfate (PS) activation. Biochar has a wide range of sources and is easy to prepare. Due to the advantages of large specific surface area and stable structure, biochar is an excellent platform for the preparation of carbon-based catalysts. Persistent free radicals (PFR), oxygen-containing functional groups (especially C-OH), defective structure, endogenous transition metals and minerals in biochar may promote the activation of PS. Biochar can also be modified by acid, alkali, metal and heteroatom doping to improve its activation effect on PS.

In this chapter, we systematically review and analyze the application of biochar-based catalysts for PS activation: First, the activation pathway and reaction mechanism of biochar-based catalysts are discussed. Second, the activation mechanisms of biochar doped with different metals and their oxides on PS are reviewed respectively. Third, the related research on the activation of PS by nitrogen, sulfur, and other heteroatom doping biochar is introduced. Finally, the concluding section discusses the challenges and possible future directions for the biochar/PS system to degrade pollutants. The contents of this chapter provide reference and help for the application of biochar-based catalysts in advanced oxidation.

13.2 Activation Pathway and Reaction Mechanism of Persulfate by Biochar

13.2.1 Distinction between Different Pathways

The SR-AOP with biochar can be divided into free-radical controlled, non-free-radical controlled, and mixed reactions based on the contribution of free-radical pathways.

The electron paramagnetic resonance (EPR) technique, probe experiment and free radical quenching experiment are commonly used to identify the free radicals in SR-AOP. The radical scavengers used in the quenching experiment are ethanol (EtOH) and tert-butyl alcohol (TBA) because of the difference of the second-order reaction rate constants with $SO_4 \bullet^-$ or $\bullet OH$. The results of quenching experiment may combine with probe experiments and EPR measurements to analysis. For example, in the study of the activation pathway of carbon structure defects to PS, the EPR results show that the signal intensity differences of $^1O_2$ were consistent with the degradation results [3]. Then, furfuryl alcohol (FA), a typical $^1O_2$ quencher, was used to further verify the dominant role of $^1O_2$ in this degradation system.

13.2.2   Properties Necessitating the Generation of Radicals with PS

The active sites on biochar-based catalysts include persistent free radicals, oxygen-containing groups, structural defects, oxygen vacancies, and metals and their oxides, which can activate PS to generate free radicals generated ($SO_4 \bullet^-$, $\bullet OH$ and $O_2 \bullet^-$). Therefore, it is necessary to promote the generation of free radicals by increasing the active sites on the biochar catalyst. An important factor affecting the physicochemical properties of biochar is the pyrolysis temperature. With the increase of pyrolysis temperature, the specific surface area of biochar will gradually increase and the graphitized structure will increase, which improves the degradation effect. However, too-high temperature will also have some negative effects, such as reduced yield and reduced cation exchange capacity of biochar. At the same time, the choice of pyrolysis temperature will change according to the difference of raw materials.

13.2.3   Nonradical Degradation with Biochar

The nonradical pathways of PS activation by biochar are mainly $^1O_2$ and surface electron transfer. Related studies have shown that the graphitized structure enhances the charge transfer ability of organisms, which facilitates PS activation through nonradical pathways.

In addition to the graphitic structure, the PFRs on biochar also participate in the electron transfer process. During the pyrolysis process, the phenolic lignin present produces a large amount of phenolic or quinone species, which then transfer electrons to PFRs in biochar. Such PFRs can promote the generation of reactive oxygen species (ROS, such as $\bullet OH$, $SO_4 \bullet^-$, $O_2 \bullet^-$, $^1O_2$), which is beneficial to the catalytic degradation of organic pollutants [4, 5]. However, the formation of PFRs mainly occurs at low or moderate pyrolysis temperatures, since high pyrolysis temperatures cause carbon clusters to condense and form graphitic-like structures [6].

13.2.4 Modifying Biochar for Enhanced Properties Related to the Degradation Process

13.2.4.1 Thermal Modification

There are two main thermal modification methods of biochar; one is to change the temperature of biochar preparation and the other is to perform additional heat treatment, that is, secondary pyrolysis.

The pyrolysis temperature range for biochar production is 300–900 °C. The three conditions of temperature, residence time and heating rate in this process will affect the physical and chemical properties of biochar. Among them, the pyrolysis temperature is the most influential condition on biochar. As the pyrolysis temperature increases, the specific surface area (SSA) and pore volume of biochar increase, which is due to the gasification of volatile substances in the feedstock and the removal of surface functional groups [7–9]. If the pyrolysis temperature is too high, the carbon skeleton will collapse, and the SSA and porosity of biochar will decrease instead [10, 11]. Increasing SSA and porosity can also be achieved by increasing residence time, allowing more VOCs to be released. When selecting the heating rate, it is better to be lower than 10 °C/min to prevent the rapid temperature rise from destroying the pore wall and causing the porous structure to collapse [12]. As the pyrolysis temperature increased, the graphitization degree, defects, surface groups, porosity and PFR, surface area, and porous structure of biochar change, as shown in Figure 13.1.

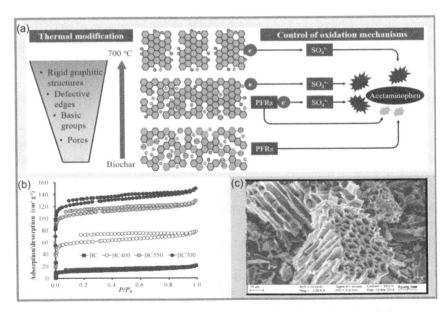

Figure 13.1 (a) Diagram of the effect of thermal modification on the degradation mechanism, (b) N_2 adsorption-desorption isotherms, and (c) SEM image of BC700 [13] (Chongqing Wang et al., 2021 / with permission from Elsevier).

The secondary pyrolysis method will affect the oxygen functional group, carbon structure, and specific surface area of biochar, thereby affecting the activation effect on PS. Zhang et al. (2022) [14] thermally modified biochar from coffee grounds by secondary pyrolysis at 550 °C. The results showed that the secondary pyrolysis reduced the specific surface area of biochar, increased the average pore size, and decreased the micropores. At the same time, the –OH and aromatic C=O contained in the biochar after secondary pyrolysis increased. The modified biochar can activate PMS better because of the introduction of inorganic anions and the improvement of electrical conductivity (EC).

13.2.4.2 Acid/Alkaline Modification Alkaline Modification

Acid–base modification is a common method to improve the performance of biochar. Acid–base treatment can effectively increase the surface area and introduce functional groups [15]. Acid modification uses various acids, including sulfuric acid (H_2SO_4), hydrochloric acid (HCl), nitric acid (HNO_3), and phosphoric acid (H_3PO_4), to remove impurities and introduce functional groups on the biochar surface. The alkali modification generally uses sodium hydroxide (NaOH) and potassium hydroxide (KOH). Alkali modification can not only effectively expand SSA and introduce hydroxyl groups but also change the aromaticity of biochar.

Acid/alkali modification can be divided into pretreatment or post-treatment, according to the treatment time. Among them, postprocessing is milder than preprocessing. Pretreatment may present strong corrosion problems.

13.3 Metal-Biochar Composites in Persulfate Activation System

Biochar can be used as catalyst support based on its porous structure, high surface area, and rich surface functionality. The mesoporous structure of biochar benefits the immobilization of nanoparticles and reduces/avoids particle aggregation due to intraparticle interactions and the complexation with surface functional groups [16, 17]. Therefore, the leaching of metal ions in wastewater was significantly reduced or eliminated during the wastewater treatment [18–20].

Moreover, the prepared biochar composites usually have a larger surface area than the mineral particles, which offer more area for attracting both PS and pollutants. This causes the accumulation of PS and pollutants at the near-surface area, leading to enhanced catalytic activity and ultimately the high degradation rate of the target pollutants [21–23].

Metal-modified biochar catalysts, especially transition metal-modified biochar catalysts, have been widely used in PS activation systems due to their high catalytic activity.

13.3.1 Iron-Biochar

Iron-based catalysts have attracted much attention due to their high activity, low toxicity, good magnetism, and favorable recycling. Therefore, a variety of catalytic iron species have been utilized, including zero-valent iron (Fe^0), metal oxides (e.g., Fe_2O_3, Fe_3O_4), and metal salts (in the form of Fe^{2+} or Fe^{3+}) for doping with biochar.

13.3.1.1 Fe^0

Zero-valent iron (ZVI) and nanoscale zero-valent iron (nZVI) have been successfully used to activate PS to degrade various pollutants. However, nZVI suffers from the problems of easy aggregation and oxidation, which will reduce its activation performance. nZVI is rapidly oxidized when exposed to air, and also aggregates to form micron-sized particles due to van der Waals force and magnetic attraction. In addition, the difficulty of recycling also hinders its practical application. Due to the advantages of low cost, high specific surface area, porosity, and stability, biochar is widely used as a carrier material to improve the dispersion and reactivity of nZVI. [24, 25, 26]. For example, when nZVI is loaded on biochar, ferric iron (Fe^{3+}) or ferrous iron (Fe^{2+}) ions are combined with –OH through bidentate complexation and iron oxide layer around nZVI reacts with biochar surface functionalities to form stable bonds.

Some researchers studied the removal of trichloroethylene (TCE), a common organic pollutant in groundwater, by thhe nZVI/BC/PS system [26]. The as-synthesized Fe/C composites are characterized by good dispersion of Fe nanoparticles on the biochar surface, and the material has hierarchical porosity. This material exhibits excellent catalytic performance in PMS activation for TCE degradation and complete mineralization, outperforming freshly prepared nZVI and other fabricated Fe/C catalysts. The results indicated that the porous biochar support enhanced the PMS activity and facilitated the removal of TCE by reducing the aggregation of iron nanoparticles.

13.3.1.2 Fe_3O_4

Magnetite (Fe_3O_4) can effectively activate PS, but its magnetic properties will lead to easy aggregation and reduce its specific surface area, thereby weakening the catalytic activity. Biochar can be used as a suitable carrier to disperse Fe_3O_4 particles. There are two main synthesis methods – iron ion impregnation method and chemical co-precipitation method. In the PDS/PMS system of the synthesized Fe_3O_4-BC, Fe(II) on the surface of the catalyst will activate PS to form free radicals [27].

13.3.1.3 Fe_3C

Iron carbide (Fe_2C, Fe_5C_2, Fe_3C, etc.) is a composition consisting of carbon atoms occupying the interstices of the close-packed iron lattice. This configuration leads to less oxidation and aggregation during the preparation of FeC than ZVI, and higher catalytic activity toward PS [28].

For example, 3D magnetic Fe$_3$C/Fe doped biochar was successfully synthesized by a one-step pyrolysis method using loofah sponges as raw materials, and the results of its activation PMS experiment illustrated the synergistic effect of Fe$_3$C and biochar [29]. Biochar promotes Fe^{2+} regeneration by accelerating electron transfer, and Fe$_3$C effectively activates PMS and acts as an intermediate layer efficiently catalyzed electron transfer from the Fe to the biochar. This reaction mechanism involves SO$_4$•$^-$, •OH and ^{1}O$_2$.

13.3.2 Copper-biochar

The redox properties of copper (Cu) are similar to those of iron, and the activation effect of PS using copper-based biochars has also been extensively studied. The methods of copper loaded onto biochar are also impregnation method and co-precipitation method.

Copper is often supported on biochar in the form of copper oxide (CuO), which can be an effective PS activator. Some studies have synthesized CuO-BC through co-precipitation, calcination and other steps using corn stalks as raw materials, and applied it to activate PMS [3]. The results showed that ^{1}O$_2$ played a dominant role in the catalytic degradation in the BC-CuO/PMS system, while the oxygen-containing functional groups on the biochar and the hydroxylation on the CuO surface contributed to the generation of ^{1}O$_2$.

In copper-based biochar materials, the existing copper species are not only CuO but also Cu or Cu$_2$O. For example, N-Cu/biochar with good activation effect can be synthesized by Cu and N co-doping [30]. The copper species in the material mainly exist in the form of copper. In the reaction of PS activated by N-Cu/biochar, the degradation pathway is mainly the free radical pathway, and the dominant free radical is •OH.

13.3.3 Cobalt Biochar

Besides Fe and Cu, cobalt (Co) has also been widely used to modify biochar to degrade pollutants in SR-AOP systems. The Co^{2+}-based catalyst Co$_3$O$_4$ exhibits the advantages of high efficiency and high stability in a wide pH range and is widely used to activate PS. For example, in some studies, biochar-supported Co$_3$O$_4$ composites were synthesized to activate PS to degrade antibiotics [31]. In these experiments, the degradation pathway of the reaction is usually the free radical pathway, acting through SO$_4$•$^-$ or •OH. Among them, there is a synergistic effect between biochar and Co species. In the degradation reaction, BC can promote electron transfer between the Co species and HSO$_5^-$, provide a large surface area for the dispersal of Co$_3$O$_4$ and accelerate the Co^{3+}/Co^{2+} redox cycle.

In addition, CoO, Co^0, and Co$_9$S$_8$ can also be combined with biochar to activate PS. Luo et al. (2020) loaded Co onto the surface of biochar by carbonizing Co-impregnated goat manure wastes [32]. There are heteroatoms such as phosphorus and nitrogen in the biomass raw material, and they will also have an activation effect. In the synthesized material, Co mainly exists in the form of Co^0, and also contains a small amount of CoO. A catalyst is used to activate PMS to degrade ciprofloxacin (CIP). The experimental results show that Co^0 is not only easier to generate free radicals in PMS than Co^{2+} and Co^{3+}, but also can accelerate the conversion of Co^{2+} to Co^{3+} and enhance the catalytic effect. Other researchers obtained biochar with Co_9S_8 as the main cobalt species by pyrolyzing cobalt-impregnated biochar [33]. The carbon material can effectively activate PMS to degrade atrazine (ATZ), and the main active substance in this process is $SO_4 \bullet^-$. PFRs in biochar can promote the conversion of Co(III) to Co(II), accelerate the Co(III)/Co(II) cycle, and thus improve the degradation effect.

13.3.4 Biochar of Other Metal and Mixed Metal

Compared with single-metal-based biochars, polymetallic biochars were also used in the activation of PSs. Current research on bimetallic catalysts focuses on iron. For example, the researchers successfully combined biochar-supported $CuO-Fe_3O_4$ catalysts via co-precipitation and high-temperature calcination, and performed well in PDS activation [34]. The results of EPR and quenching experiments show that there are free radical pathways and non-free radical pathways, and the non-free radical pathway of $^1O_2$ is dominant. In another study, the difference in the activation effect of iron-cobalt co-doped biochar (FeCo-BC) and iron-copper co-doped biochar (FeCu-BC) on PMS was compared [35]. The results show that the catalytic activity of FeCo-BC is higher than that of FeCu-BC. In addition, the catalytic activity of FeCu-BC is also higher than that of Fe-BC and Cu-BC, showing a bimetallic synergistic effect.

Other transition metals, such as titanium (Ti), nickel (Ni), and manganese (Mn), have also been used to modify biochar [36]. For example, Lin et al. (2020) prepared biochar by microwave irradiation, and then synthesized FeMn-co-doped biochar FeMn-BC by a combination of impregnation and microwave depyrolysis [37]. The characterization results showed that the bimetallic Fe-Mn oxides were uniformly distributed on the surface of the biochar in the form of Fe(II), Fe(III), Mn(II) and Mn(III). In the FeMn-BC/PS system, $SO_4 \bullet^-$ played a major role in this system for the pollutant degradation.

13.4 Heteroatom-Doped Biochar for PS Activation

Heteroatom doping is an effective strategy to enhance the biochar reactivity for PS activation. By changing the heteroatom doping method and synthesis conditions, the type and content of active sites can be adjusted, thereby affecting the catalytic activity of PS.

13.4.1 Nitrogen-doped Biochar

N doping is an effective way to improve the catalytic performance of inert carbonaceous materials towards PS [38, 39]. N-doped biochar is usually prepared by the pyrolysis method, which was achieved by the co-pyrolysis of biochar/biomass and nitrogen-rich chemicals mixed through liquid-phase and solid-phase doping. Various nitrogen-containing chemicals can be used as nitrogen sources, including urea, polyaniline, indole, melamine, dicyandiamide, ethylenediamine, thiourea, etc.

The prepared nitrogen-doped biochar has the characteristics of a larger specific surface area, richer defects, and a much more porous structure. More importantly, the doped nitrogen as the form of nitrogen functionalities can increase the surface alkalinity of biochar to facilitate the adsorption of PMS and accelerate electron transfer by activating adjacent sp^2 carbon atoms [23]. Therefore, it is considered an effective method to improve the catalytic ability of biochar. For example, Zhu et al.(2018) evaporated the solvent by uniformly mixing ammonium nitrate and treated reed biomass at 85 °C precipitation temperature for 4 h [40]. Then the mixture was heated under nitrogen (N_2) for 90 min at 400–1000 °C to synthesize N-doped biochar. The nitrogen-doped biochar synthesized under the condition of pyrolysis temperature of 900 °C has the best performance on the activation of PDS. In the free radical pathway, the electronegative quaternary N leads to asymmetric spin density and low electron density of adjacent carbons, which helps to activate PDS to generate $SO_4{}^{\bullet-}$. At the same time, high temperature also leads to the increase of graphite structure and structural defects in biochar, which promotes the transport of unpaired electrons to generate free radicals. In the nonradical pathway, the porous structure, large specific surface area, and N doping of biochar all facilitate the adsorption of target pollutants, which in turn facilitates the direct transfer of electrons from organics to PDS-carbon complexes.

The three most frequently constructed types of N-bond configurations in carbon structures are pyridinic N, pyrrolic N, and graphitic N, and the speciation can be controlled by the thermal treatment. Qi et al. (2020) used Enteromorpha with high N content as a raw material to generate carbonized Enteromorpha through the first pyrolysis [41]. It is then mixed with potassium carbonate and pyrolyzed a second time, which helps to form a different form of porous structure. For the final synthesized nitrogen-doped biochar materials, the fraction of graphitic N increased from 0.497% to 0.692% with the increase of the second pyrolysis temperature. Due to the good thermal stability of graphitic N, the conversion of pyridinic N or pyrrolic N to graphitic N increases with increasing activation temperature. This may lead to the distortion of the carbon network and the formation of defect edges that contribute to the activation of PS. Besides, graphitic N also increases the adsorption of pollutants sulfamethoxazole (SMX) by biochar and the binding of biochar to PS. Both free radical pathways ($O_2{}^{\bullet-}$) and nonradical pathways ($^1O_2$ and surface electron transfer) exist during this degradation process.

13.4.2 Sulfur-Doped Biochar

Besides nitrogen, other heteroatoms such as sulfur can also be doped into biochar. Since the electronegativity of sulfur atoms (2.58) is close to that of carbon atoms (2.55), S-doping will destroy the carbon network of carbon materials and form thiophene sulfur -CSC-. The incorporation of sulfur species leads to the adjustment of the carbon planar electronic structure, which enhances the catalytic activity [42]. Similar to N-doped biochar, S-doped biochar is mainly prepared through the high-temperature pyrolysis of sulfur-containing chemicals with biomass/biochar. For example, Wang et al. (2020) used sludge as a raw material to synthesize biochar by pyrolyzing a mixture of a pyrolytically treated sludge-derived biochar sample and sulfuric acid at 650 °C for 5 h. It has shown good performance in activating both PMS and PDS to remove Bisphenol A (BPA). This is due to the combined effect of zigzag-edge S atoms and graphitic/pyridinic N, greatly altering the electronic properties of SB, resulting in more Lewis acidic and basic sites, which facilitate persulfate activation. However, the activation process for PMS and PDS is different. The activation of PMS occurs at Lewis acid sites and generates $^1O_2$, while the activation of PDS occurs at Lewis basic sites and generates a surface-bound complex (PDS-SSB), followed by electron transfer from BPA to PDS-SSB. [34].

There are also studies using rice straw to prepare biochar and doping S atoms to activate PMS to degrade metolachlor [26]. The results show that S-doped biochar harms the degradation process and the synergistic effect produced by doping does not exist. The inconsistency might come from the fact that the incorporated N preferentially replaces C atoms as graphitic N, while S would preferentially replace oxygen-containing groups. The calculation results based on density functional theory (DFT) also indicated that N doping increases the positively charged C atoms on the biochar, thereby promoting the interaction between PMS and the catalyst surface. •OH and $^1O_2$ play a crucial role in the degradation process. Overall, Further research is needed on S-doped biochar.

13.5 Conclusion and Perspectives

This chapter outlines the principles and application forms of biochar-based catalysts in the advanced persulfate oxidation process, summarizes the activation pathways and differentiation methods of persulfate, and explains the role of biochar in the persulfate activation system. However, some aspects must be further considered for applying biochar-based catalysts in SR-AOP.

1) The overall economic concern of biochar/modified biochar materials should be considered. When preparing biochar activators, some high-cost approaches and chemical feedstock were used, which will improve the overall cost of the prepared biochar activators. Before the field application of biochar activators,

more efforts should be put into the green and low-cost preparation of efficient biochar activators.
2) More mechanisms-based study is still needed. The nonradical pathway of biochar-based catalysts has attracted more and more attention due to stability in real wastewater. The nonradical path has a more vital anti-interference ability to inorganic ions and NOMs and has better selectivity for the degradation of specific pollutants. However, the determination of the reaction pathway is still complicated, and the detection of free radicals is not ideal, and further research on the catalytic mechanism is needed. How to modify biochar with a desired degradation pathway needs more studies.
3) The application of BCs/SR-AOP in the catalytic degradation of pollutants in an aqueous solution has been widely studied, but its practical application in wastewater is very lacking. Its feasibility and practicability in real environments need to be further evaluated. In particular, the influence of inorganic ions and organic compounds in the actual wastewater on its catalytic performance is still insufficiently studied.
4) The potential environmental risk of biochar-based catalysts needs more evaluation. The possibility of releasing heavy metal ions and polycyclic aromatic hydrocarbons from biochar should be studied for the environmental risk when applied to the actual environment for PS activation. In this context, finding more sustainable and effective ways to modify catalysts for applications under complex conditions is necessary.

References

1 Lee, J., Von Gunten, U., Kim, J.-H., 2020. Persulfate-based advanced oxidation: Critical assessment of opportunities and roadblocks. *Journal of Environmental Science & Technology*, 54(6), 3064–3081.
2 Yu, J., Tang, L., Pang, Y., Zeng, G., Feng, H., Zou, J., et al., 2020. Hierarchical porous biochar from shrimp shell for persulfate activation: A two-electron transfer path and key impact factors. *Applied Catalysis B: Environmental*, 260, 118160.
3 Li, H., Liu, Y., Jiang, F., Bai, X., Li, H., Lang, D.,Wang, L., Pan, G., 2022. Persulfate adsorption and activation by carbon structure defects provided new insights into ofloxacin degradation by biochar. *Science of the Total Environment*, 806, 150968. https://doi.org/10.1016/j.scitotenv.2021.150968.
4 Qin, Y., Li, G., Gao, Y., Zhang, L., Ok, Y.S., An, T., 2018. Persistent free radicals in carbon-based materials on transformation of refractory organic contaminants (ROCs) in water: A critical review. *Journal of Water Research*, 137, 130–143.

5 Huang, D., Zhang, Q., Zhang, C., Wang, R., Deng, R., Luo, H., et al., 2020. Mn doped magnetic biochar as persulfate activator for the degradation of tetracycline. *Chemical Engineering Journal*, 391, 123532.

6 Duan, X., Zhang, C., Wang, S., Ren, N.-q., Ho, S.-H., 2020. Graphitic biochar catalysts from anaerobic digestion sludge for nonradical degradation of micropollutants and disinfection. *Journal of Chemical Engineering Journal*, 384, 123244.

7 Nidheesh, P., Gopinath, A., Ranjith, N., Akre, A.P., Sreedharan, V., Kumar, M.S., 2021. Potential role of biochar in advanced oxidation processes: A sustainable approach. *Journal of Chemical Engineering Journal*, 405, 126582.

8 Meng, Y., Li, Z., Tan, J., Li, J., Wu, J., Zhang, T., et al., 2022. Oxygen-doped porous graphitic carbon nitride in photocatalytic peroxymonosulfate activation for enhanced carbamazepine removal: Performance, influence factors and mechanisms. *Chemical Engineering Journal*, 429, 130860.

9 Zhu, K., Wang, X., Geng, M., Chen, D., Lin, H., Zhang, H., 2019. Catalytic oxidation of clofibric acid by peroxydisulfate activated with wood-based biochar: Effect of biochar pyrolysis temperature, performance and mechanism. *Journal of Chemical Engineering Journal*, 374, 1253–1263.

10 Ding, D., Yang, S., Qian, X., Chen, L., Cai, T., 2020. Nitrogen-doping positively whilst sulfur-doping negatively affect the catalytic activity of biochar for the degradation of organic contaminant. *Applied Catalysis B: Environmental*, 263, 118348.

11 Wang, S., Wang, J., 2021. Nitrogen doping sludge-derived biochar to activate peroxymonosulfate for degradation of sulfamethoxazole: Modulation of degradation mechanism by calcination temperature. *Journal of Hazardous Materials*, 418, 126309.

12 Li, R., Lu, X., Yan, B., Li, N., Chen, G., Cheng, Z., et al., 2022. Sludge-derived biochar toward sustainable peroxymonosulfate activation: Regulation of active sites and synergistic production of reaction oxygen species. *Chemical Engineering Journal*, 440, 135897.

13 Wang, C., Huang, R., Sun, R., Yang, J., Sillanpää, M., 2021. A review on persulfates activation by functional biochar for organic contaminants removal: Synthesis, characterizations, radical determination, and mechanism. *Journal of Environmental Chemical Engineering*, 9(5), 106267.

14 Zhang, X., Yang, Y., Ngo, H.H. et al. 2022. Enhancement of urea removal from reclaimed water using thermally modified spent coffee ground biochar activated by adding peroxymonosulfate for ultrapure water production. *Bioresource Technology*, 349, 126850.

15 Wang, J., Wang, S., 2019. Preparation, modification and environmental application of biochar: A review. *Journal of Cleaner Production*, 227, 1002–1022.

16 Fu, H., Ma, S., Zhao, P., Xu, S., Zhan, S., 2019. Activation of peroxymonosulfate by graphitized hierarchical porous biochar and $MnFe_2O_4$ magnetic nanoarchitecture for organic pollutants degradation: Structure dependence and mechanism. *Journal of Chemical Engineering Journal*, 360, 157–170.

17 Gan, L., Zhong, Q., Geng, A., Wang, L., Song, C., Han, S., et al., 2019. Cellulose derived carbon nanofiber: A promising biochar support to enhance the catalytic performance of $CoFe_2O_4$ in activating peroxymonosulfate for recycled dimethyl phthalate degradation. *Science of the Total Environment*, 694, 133705.

18 Yang, M.-T., Tong, W.-C., Lee, J., Kwon, E., Lin, K.-Y.A., 2019. CO_2 as a reaction medium for pyrolysis of lignin leading to magnetic cobalt-embedded biochar as an enhanced catalyst for oxone activation. *Journal of Colloid and Interface Science*, 545, 16–24.

19 Du, J., Kim, S.H., Hassan, M.A., Irshad, S., Bao, J., 2020. Application of biochar in advanced oxidation processes: Supportive, adsorptive, and catalytic role. *Journal of Environmental Science and Pollution Research*, 27(30), 37286–37312.

20 Nguyen, V.-T., Nguyen, T.-B., Chen, C.-W., Hung, C.-M., Huang, C., Dong, C.-D., 2019. Cobalt-impregnated biochar (Co-SCG) for heterogeneous activation of peroxymonosulfate for removal of tetracycline in water. *Journal of Bioresource Technology*, 292, 121954.

21 Palansooriya, K.N., Yang, Y., Tsang, Y.F., Sarkar, B., Hou, D., Cao, X., et al., 2020. Occurrence of contaminants in drinking water sources and the potential of biochar for water quality improvement: A review. *Critical Reviews in Environmental Science and Technology*, 50(6), 549–611.

22 He, J., Xiao, Y., Tang, J., Chen, H., Sun, H., 2019. Persulfate activation with sawdust biochar in aqueous solution by enhanced electron donor-transfer effect. *Journal of Science of the Total Environment*, 690, 768–777.

23 Liu, C., Chen, L., Ding, D., Cai, T., 2019. From rice straw to magnetically recoverable nitrogen doped biochar: Efficient activation of peroxymonosulfate for the degradation of metolachlor. *Journal of Applied Catalysis B: Environmental*, 254, 312–320.

24 Wang, B., Li, Y.-N., Wang, L., 2019. Metal-free activation of persulfates by corn stalk biochar for the degradation of antibiotic norfloxacin: Activation factors and degradation mechanism. *Chemosphere*, 237, 124454.

25 Gao, J., Han, D., Xu, Y., Liu, Y., Shang, J., 2020. Persulfate activation by sulfide-modified nanoscale iron supported by biochar (S-nZVI/BC) for degradation of ciprofloxacin. *Separation and Purification Technology*, 235, 116202.

26 Li, Z., Sun, Y., Yang, Y., Han, Y., Wang, T., Chen, J., et al., 2020. Biochar-supported nanoscale zero-valent iron as an efficient catalyst for organic degradation in groundwater. *Journal of Hazardous Materials*, 383, 121240.

27 Cui, X., Zhang, S.-S., Geng, Y., Zhen, J., Zhan, J., Cao, C., Ni, S.-Q., 2021. Synergistic catalysis by Fe_3O_4-biochar/peroxymonosulfate system for the removal

of bisphenol a. *Separation and Purification Technology*, 276, 119351. ISSN 1383-5866, https://doi.org/10.1016/j.seppur.2021.119351.

28 Lyu, S., Wang, L., Li, Z., et al., 2020. Stabilization of ε-iron carbide as high-temperature catalyst under realistic Fischer–Tropsch synthesis conditions. *Nature Communications*, 11(1), 6219.

29 Gou, G., Huang, Y., Wang, Y., et al., 2023. Peroxymonosulfate activation through magnetic Fe_3C/Fe doped biochar from natural loofah sponges for carbamazepine degradation. *Separation and Purification Technology*, 306, 122585.

30 Zhong, Q., Lin, Q., Huang, R., et al., 2020. Oxidative degradation of tetracycline using persulfate activated by N and Cu codoped biochar. *Chemical Engineering Journal*, 380, 122608.

31 Xu, H., Zhang, Y., Li, J., et al., 2020. Heterogeneous activation of peroxymonosulfate by a biochar-supported Co_3O_4 composite for efficient degradation of chloramphenicols. *Environmental Pollution*, 257, 113610.

32 Luo, J., Bo, S., Qin, Y., et al., 2020. Transforming goat manure into surface-loaded cobalt/biochar as PMS activator for highly efficient ciprofloxacin degradation. *Chemical Engineering Journal*, 395, 125063.

33 Liu, B., Guo, W., Wang, H., et al., 2020. Activation of peroxymonosulfate by cobalt-impregnated biochar for atrazine degradation: The pivotal roles of persistent free radicals and ecotoxicity assessment. *Journal of Hazardous Materials*, 398, 122768.

34 Cai, S., Wang, T., Wu, C., et al., 2023. Efficient degradation of norfloxacin using a novel biochar-supported CuO/Fe_3O_4 combined with peroxydisulfate: Insights into enhanced contribution of nonradical pathway. *Chemosphere*, 138589.

35 Wang, S., Wang, J., 2023. Bimetallic and nitrogen co-doped biochar for peroxymonosulfate (PMS) activation to degrade emerging contaminants. *Separation and Purification Technology*, 307, 122807.

36 Yao, B., Luo, Z., Du, S., Yang, J., Zhi, D., Zhou, Y., 2022. Magnetic $MgFe_2O_4$/biochar derived from pomelo peel as a persulfate activator for levofloxacin degradation: Effects and mechanistic consideration. *Bioresource Technology*, 346, 126547.

37 Chen, L., Jiang, X., Xie, R., et al., 2020. A novel porous biochar-supported Fe-Mn composite as a persulfate activator for the removal of acid red 88. *Separation and Purification Technology*, 250, 117232.

38 Wang, H., Guo, W., Liu, B., Wu, Q., Luo, H., Zhao, Q., et al., 2019. Edge-nitrogenated biochar for efficient peroxydisulfate activation: An electron transfer mechanism. *Water Research*, 160, 405–414.

39 Xu, L., Fu, B., Sun, Y., Jin, P., Bai, X., Jin, X., et al., 2020. Degradation of organic pollutants by Fe/N co-doped biochar via peroxymonosulfate activation: Synthesis, performance, mechanism and its potential for practical application. *Chemical Engineering Journal*, 400, 125870.

40 Zhu, S., Huang, X., Ma, F., et al. 2018. Catalytic removal of aqueous contaminants on N-doped graphitic biochars: inherent roles of adsorption and nonradical mechanisms. *Environmental Science & Technology*, 52(15), 8649–8658.

41 Qi, Y., Ge, B., Zhang, Y., et al., 2020. Three-dimensional porous graphene-like biochar derived from Enteromorpha as a persulfate activator for sulfamethoxazole degradation: Role of graphitic N and radicals transformation. *Journal of Hazardous Materials*, 399, 123039.

42 Liu, B., Guo, W., Wang, H., Si, Q., Zhao, Q., Luo, H., et al., 2020. Activation of peroxymonosulfate by cobalt-impregnated biochar for atrazine degradation: The pivotal roles of persistent free radicals and ecotoxicity assessment. *Journal of Hazardous Materials*, 398, 122768.

43 Wang, H., Guo, W., Liu, B., Si, Q., Luo, H., Zhao, Q., et al., 2020. Sludge-derived biochar as efficient persulfate activators: Sulfurization-induced electronic structure modulation and disparate nonradical mechanisms. *Applied Catalysis B: Environmental*, 279, 119361.

14

Biochar-Enhanced Ozonation for Sewage Treatment

Dongdong Ge[1], Nanwen Zhu[1], Mingjing He[2], and Daniel C.W. Tsang[3]

[1] School of Environmental Science & Engineering, Shanghai Jiao Tong University, Shanghai, China
[2] Department of Civil and Environmental Engineering, The Hong Kong Polytechnic University, Hung Hom, Kowloon, Hong Kong, China
[3] State Key Laboratory of Clean Energy Utilization, Zhejiang University, China

14.1 Introduction

The chemically oxidative treatment of recalcitrant pollutants has attracted wide attention, and developing an innovative process for efficient wastewater handling has had success in the last decade (Oh and Nguyen, 2022). Ozone has been successfully employed in sewage treatment due to the high oxidation potential (2.08 V) without secondary pollutants (Babar et al., 2022). The single ozonation process fails in the breakdown and mineralization of refractory organic pollutants, which is ascribed to the limited oxidative capacity, selective degradation characteristic, and low ozone utilization efficiency (Ge et al., 2022a). It is noteworthy that the heterogeneous catalytic ozonation reaction arouses the extensive attentions of investigators engaged in emerging pollutant treatment due to the small footprint, convenient operation, and high mineralization efficiency, and simple separation of catalysts (Li et al., 2022, Li, S. et al., 2021). Carbon materials and metal oxides have been widely adopted as the efficient catalysts for the heterogeneous catalytic ozonation (Liu et al., 2021). However, the practical application of these potential catalytic materials are always prevented for the typical disadvantages such as poor stability, high cost, and metal ion leaching (Moussavi and Khosravi, 2012). Hence, developing the stable, highly active, and cost-efficient catalysts for ozonation has been essential.

Biochar has aroused widespread environmental concerns, involving climate change, soil remediation, gas storage, adsorption, and catalyzation for its porous carbon structure, huge surface area, and numerous functional groups (Singh et al.,

Biochar Applications for Wastewater Treatment, First Edition. Edited by Daniel C.W. Tsang and Yuqing Sun.
© 2023 John Wiley & Sons, Inc. Published 2023 by John Wiley & Sons, Inc.

2017; Xiao et al., 2021; Zhu et al., 2022). These enhance its versatile physicochemical characteristics. The feedstocks for biochar preparation are widely derived from agricultural residuals, yard wastes, animal manures, and municipal wastes (Li et al., 2022; Zhu et al., 2022), meanwhile reducing their resultant secondary pollutants. For the high catalytic activity, the metal oxides are commonly considered to be loaded into the biochar to reinforce ozonation for more yield of reactive oxygen species (ROS), such as iron oxides, manganese oxides, and titanium oxides (Dabuth et al., 2022; Xiao et al., 2021; Yang et al., 2022). The promising biochar-based catalysts not only show high catalytic capacity but also protect the active components from losing and aggregating. Biochar-enhanced ozonation processes have been employed in the handling of various kinds of sewage containing dyestuffs (Babar et al., 2022), herbicides (Tian et al., 2021), endocrine disruptors (Li S. et al., 2021), etc., which currently pose the serious threats to human life and environmental safety.

14.2 Preparation of Biochar-Based Catalyst for Ozonation

The feedstock is commonly crushed into power first, and then the pretreatments were performed for the better physicochemical characteristics of prepared biochar. Afterward, the mixed materials undergo further pyrolysis procedure at 300–900 °C for 0.5–3 h under inert gas protection (Zhang et al., 2018; Zhu et al., 2022). For example, with the alkali (NaOH) solution pretreatment, the pyrolysis rice straw–based biochar showed the larger surface area of 703 m^2/g, 2.73-fold higher than the raw straw-based biochar, and also the pore volume of the bicohar with pretreatment rose by 5.8-fold to 0.99 cm^3/g (Zhu et al., 2022). The KOH pretreatment could improve the porous structure and micro-pore contents by the gasification and oxidation reactions from the K_2CO_3 decomposition under high temperature (Eqs. 14.1–14.5) (Cha et al., 2022). The pretreatments normally enhanced the specific surface area and pore structure of biochar for the satisfactory contaminant adsorption effect and catalytic ozonation performance (Chen et al., 2021). For instance, after the *Arundo donax* activated by KOH, the specific surface area and micropore volume increased to 1122 m^2/g and 0.5 cm^3/g, respectively (Singh et al., 2017).

$$2KOH \leftrightarrow K_2O + H_2O \tag{14.1}$$

$$6KOH + 2C \rightarrow 2K + 3H_2 + 2K_2CO_3 \tag{14.2}$$

$$K_2CO_3 \leftrightarrow K_2O + CO_2 \tag{14.3}$$

$$K_2CO_3 + 2C \rightarrow 2K + CO \tag{14.4}$$

$$K_2O + C \rightarrow 2K + CO \tag{14.5}$$

It is worth noting that a series of decomposition of various organic components occur during the thermally chemical reactions. First, the moisture is evaporated from the feedstock surface; then, some functional groups and the major carbonaceous structure are damaged and the hemicellulose and cellulose star to disorganize simultaneously; the decomposition of lignin with the various branched aromatic rings commonly occurred at the final stage (Yang et al., 2007). The pyrolysis temperature apparently affected the biochar microstructure, and the high temperature would stimulate the surface roughness of the formed biochar, which is caused by the destruction of cellulose and lignin, the increment of the ash, and the generation of crystalline mineral components (Kloss et al., 2012; Wu et al., 2016). With the increased high pyrolysis temperature, the elements of Ca and Mg would be accumulated highly due to their nonvolatile characteristics. The O/C ratio was decreased, reflecting the less hydrophilicity of the resultant biochar (Li et al., 2022).

Large surface area is favorable for the reactants adsorption and the improved mass transfer for the targeted degradation of pollutants (Liu et al., 2021). As the pyrolysis temperature rose, the polar functional groups were lowered, and the aromaticity increased (Gaskin et al., 2008). The representative C–C/C=C, C–O, C=O, and O–C=O groups are presented commonly, and especially, the oxygen-containing functional groups such as O=C=O, C–O–C, and –OH groups are focused on for the enhanced adsorption and catalytic performance (Oh and Nguyen, 2022; Zhang et al., 2018; Zhu et al., 2022). Notably, O 1s peak might be deconvoluted into four components including hydroxyl oxygen (O_w) at 533.3 eV, vacancy oxygen (O_v) at 532.2 eV, adsorbed oxygen (O_{ads}) at 531.1–532.0 eV, and lattice oxygen (O_L) at 529.0–530.0 eV (He et al., 2020). The high O_{ads}/O_L ratio calculated from the XPS spectra suggests the plentiful absorbed oxygen (O_2^-, O^-, and O_2^{2-}) on catalyst surface, and also the ratio of ($O_{ads} + O_v$)/O_L is an indicator of the amount of the surface active oxygen (Cha et al., 2022).

Microwave is another accessible pretreatment method to prepare the sludge-based biochar. Li et al. (2021a) mixed the sludge with iron and cobalt salts and treated the mixture under 200 W and 2450 MHz at 70 °C for 30 min, and then the collected black powder was subjected to thermal treatment by a tube-furnace at 800 °C for 120 min, and finally, the $CoFe_2O_4$@biochar was obtained (Li, S. et al., 2021). The raw sludge possessed smooth appearance, and there are no signs of porosity and agglomeration on its surface. The prepared $CoFe_2O_4$@biochar has smaller particle size with the specific surface area of 71.1 m^2/g and superior crystal structure, and the nanoparticles are distributed evenly. However, compared with the raw biochar, the metal oxides modified biochar has bigger or smaller surface area, and its microporous and mesoporous volume changed differently (Cha et al., 2022; Chen et al., 2019; Singh et al., 2017). The pH of the point of zero charge (pHpzc) of biochar is also worthy of concerning. Dabuth et al. observed that the pH_{pzc} of the TiO_2 coated coal-based, coconut shell-based, and wood-based biochar materials ranged from 5.76 to 6.01, which indicated that these loaded catalysts would work as the cation exchangers with negative charges via protonation (Dabuth et al., 2022).

14.3 Efficacy of Biochar-Catalytic Ozonation on Sewage Treatment

The single ozonation only oxidized 48% atrazine (ATZ) with 2.5 mg/L O_3 for 30 min, but the additives MnOx/biochar and FeOx/biochar combined with ozonation treatments removed 83% and 100% ATZ, respectively (Tian et al., 2021). With the rice straw–based biochar added, the 2,4-dichlorophenoxyacetic acid (2,4-D) removal effect was boosted and an abrupt decrement of 2,4-D occurred within the 1–2 min (Zhu et al., 2022). When the solution pH is above the pHpzc of the bichar, the negatively charged biochar surface would facilitate O_3 decomposition to produce more reactive oxygen radicals, which is attributed to the electrophilic behavior of ozone (Moussavi and Khosravi, 2012). It does not mean the high pH is the optimal condition for organic decomposition. The pH would adjust the adsorption behavior of biochar, which facilitate TOC removal at the pH below pHpzc (Zhu et al., 2022). The biochar catalyst surface showed porosity and disorder morphology before catalyzing O_3 reaction, and after catalytic reaction, the particulate pollutants might deposit on the catalyst surface (Figure 14.1).

Figure 14.1 SEM images of the Fe-loaded catalyst: (a) before and (b) after the catalytic ozonation reaction (Muhammad Babar et al., 2022 / with permission from Elsevier).

14.4 Effects of Process Conditions on Biochar-Enhanced Ozonation Sewage Treatment

The various process parameters have distinctive effects on biochar-enhanced ozonation performance, such as biochar dosage, ozone dosage, initial concentration of pollutants, initial pH, natural organic matter (NOM), and coexisting anions.

The biochar dosage is of great significance for catalyzing ozonation and facilitating organic pollutant disintegration. When the biochar dosage of 1–5 g/L, the decolorization efficiency was elevated from 92.88% to 95.82% (Babar et al., 2022). The reason lies in the broader catalyst surface available for the adsorption of pollutants and generation of hydroxyl radicals. With more biochar addition, there are more active sites for interfacial reactions, enhancing the interactions of reactants and the removal of targeted pollutants (Zhu et al., 2022).

The increased ozone dosage would facilitate the improved ROS, which enhances decontamination effect. Babar et al. (2022) stated that with the increased O_3 dosage from 0.1 to 0.3 mg/mL, the decolorization rate of methylene blue was elevated from 82.5% to 96.33% in the catalytic ozonation process, which was ascribed to the more ozone molecules and hydroxyl radicals, accelerating the decomposition and transformation of contaminants (Babar et al., 2022).

Generally speaking, the higher initial pollutant concentration would cause remarkable decrement of the decontamination effect in both single and catalytic ozonation processes due to the increased organic load (Issaka et al., 2022). Comparatively, the biochar-enhanced ozonation process has better sewage treatment performance over the single ozonation process. For instance, with the increased initial concentration of BPA from 20 to 100 mg/L, the TOC removal efficiency declined from 80.6% to 32.6% for the individual ozonation system, and 73.5% TOC removal was still retained in $CoFe_2O_4$@biochar combined ozonation process (Li, S. et al., 2021), which indicated the efficacious degradation and transformation of BPA intermediates in the biochar-enhanced ozonation process probably due to the boosted reaction of O_3 molecule and organic molecule absorbed on the catalyst surface. Similarly, when the initial 2,4-D concentration was decreased from 2000 to 1000, 500, 200, and 100 μg/L, the removal efficiency of 2,4-D was elevated from 22.7% to 26.1%, 32.1%, 48.1%, and 53.1%, correspondingly, in the 20 mg/L catalytic ozonation process (Zhu et al., 2022).

The solution pH is a critical parameter influencing the catalytic ozonation effect via adjusting the O_3 decomposition, the physicochemical characteristics of catalyst, and the degradation of pollutants in aqueous liquid. Babar et al. (2022) found that the color-removal efficiency rose from 84.42% to 96.5% corresponding acidic (pH 4.0) and basic (pH 10.0) condition, respectively, in the Fe-loaded biochar catalytic ozonation process (Babar et al., 2022). Finally, if the pH values increased after ozonation and catalytic ozonation treatments, the results were ascribed to the generation of OH⁻ from the •OH scavenged by carbonate and bicarbonate (Dabuth et al., 2022). Interestingly, as the solution pH increased from 3.5 to 11.0, the preadsorption removal effect of 2,4-D by biochar dropped from 57.4% to 18.2%, and the terminal 2,4-D removal rate declined from 94.0% to 76.0% (Zhu et al., 2022). Totally, the pollutant elimination was decided by the organic matter structure, catalysts, and oxidants. In detail, the pH_{pzc} of bubbles is approximate 3.0 (Temesgen et al., 2017), indicating the negatively charged surface in the aqueous liquid of pH of 3.5; also, the H of the –COOH group of 2,4-D structure (pK_a = 2.64) dissociate and functional groups carry negative charges. However, the surface of the biochar with the pHpzc of 4.0 would be protonated and charged positively. Thus, the 2,4-D pollutants and O_3 bubbles are rapidly absorbed on the biochar surface via electrostatic effects, further improving the interactions of reactants and the removal efficiency of contaminants (Zhu et al., 2022). On the contrary, the increased pH results in the electrostatic exclusion of reactants, thereby restricting their interfacial reactions.

NOM is one of the crucial constituents in organic matters, influencing the contaminant disintegration to some extent (Feng et al., 2019; Li, W. et al., 2021). Often, the investigation took humic acid as the model substance representing NOM. Humic acid in the aqueous liquid has no competitive impact on oxalic acid, but the competitive adsorption between NOM and O_3 would greatly restrict the decomposition of O_3 molecules into •OH (Dabuth et al., 2022). When humic acid concentration ranged from 0.5 to 3.0 mg/L, the removal rate of 2,4-D decreased from 95.5% to 77.0% noticeably. The role of humic acid can be concluded by two aspects. First, the contained –OH groups of humic acid are favorable for catalyzing ozone reaction. Second, the excessive humic acid shows adverse effects on organics degradation by competing O_3 and reactive radicals (Zhu et al., 2022).

The coexisting anions, such as Cl^-, SO_4^{2-}, NO_3^-, HCO_3^-, and PO_4^{3-}, are pervasive in natural water (Ahmad et al., 2022; Chen et al., 2021). The interactions between them and hydroxyl radical occur with various reactive rate constants (Eqs. 14.6–14.11). Zhu et al. (2022) found that no significant influence on the process performance occurred in the catalytic ozonation system with the anions' concentration less than 1.0 mM (Zhu et al., 2022). Also, for Cl^-, the reverse reaction happened after it combined with •OH. Similar reports show that Cl^- and SO_4^{2-} could affect the adsorption of organic pollutants on the $MnFe_2O_4$/carbon microsphere nanocatalyst (CMS-$MnFe_2O_4$) but show little influence on the final TOC reduction (Jin et al., 2021), which reflects that CMS-$MnFe_2O_4$ can resist the disturbance of anions on ozonation sewage treatment. NO_3^- and HCO_3^- react with •OH at the rate constants of 5×10^9 $M^{-1} \cdot s^{-1}$ and 8.5×10^6 $M^{-1} \cdot s^{-1}$, respectively. However, HCO_3^- negatively affects the wastewater handling in the catalytic ozonation. Reportedly, with the increased HCO_3^- concentration from 1 to 5 mM, the removal efficiencies of ATZ were decreased by 16.6% and 19.9%, respectively (Tian et al., 2021), due to the competition for free radicals between HCO_3^- and ATZ. Phosphate as a strong Lewis base is able to impede ozone to react with catalysts, inhibiting O_3 decomposition and competing with organic contaminants via trapping $^\bullet OH$ with the reactive rate constants of $K^\bullet OH$, $H_2PO_4^-$ of 2×10^4 $M^{-1} \cdot s^{-1}$ and $K^\bullet OH$, HPO_4^{2-} of 1.5×10^5 $M^{-1} \cdot s^{-1}$ (Yang et al., 2022).

$$^\bullet OH + SO_4^{2-} \rightarrow SO_4^{\bullet -} + OH^- \quad k = 3.5 \times 10^5 M^{-1} \cdot s^{-1} \tag{14.6}$$

$$^\bullet OH + Cl^- \rightarrow ClOH^{\bullet -} \quad k = 4.3 \times 10^9 M^{-1} \cdot s^{-1} \tag{14.7}$$

$$ClOH^{\bullet -} \rightarrow {^\bullet OH} + Cl^- \quad k = 6.1 \times 10^9 M^{-1} \cdot s^{-1} \tag{14.8}$$

$$^\bullet OH + NO_3^- \rightarrow NO_3^\bullet + OH^- \quad k = 5 \times 10^9 M^{-1} \cdot s^{-1} \tag{14.9}$$

$$^\bullet OH + HCO_3^- \rightarrow HCO_3^\bullet + OH^- \quad k = 8.5 \times 10^6 M^{-1} \cdot s^{-1} \tag{14.10}$$

$$^\bullet OH + CO_3^{2-} \rightarrow CO_3^{\bullet -} + OH^- \quad k = 3.9 \times 10^8 M^{-1} \cdot s^{-1} \tag{14.11}$$

14.5 Technical Mechanism and Implementation Prospects

The biochar could efficiently improve the O_3 utilization efficiency, and the ozone decomposition rate constant in the biochar-enhance ozonation process was 0.31 min^{-1}, 1.82-fold higher than that in the single ozonation process (Zhu et al., 2022). Notably, carbon microsphere shell can work as insulation, which hinders the collision and agglomeration of particles, enhancing the chemical stability and catalytic ability of the biochar-based catalyst (Jin et al., 2021). In single ozonation process, the produced intermediates would result in the resistance on O_3 decomposition, thereby inducing the inadequate decomposition of contaminants (Ge et al., 2022b). However, the tendency of the oxalic acid degradation is consistent with that of TOC removal, suggesting no stabilized intermediates in the CMS-$MnFe_2O_4$ catalyzed ozonation process (Jin et al., 2021).

Tert-butyl alcohol (TBA), chloroform (CF), and NaN_3 were commonly employed as the radical scavengers for hydroxyl radical ($^\bullet OH$) (5.9×10^8 $M^{-1} \cdot s^{-1}$), superoxide radical ($O_2^{\bullet -}$) (3×10^{10} $M^{-1} \cdot s^{-1}$), and singlet oxygen ($^1O_2$) (Sun et al., 2021), respectively. However, some researchers reported that the TBA mainly scavenged the $^\bullet OH$ in bulk solution, whereas the generated $^\bullet OH$ on the catalyst surface could continue to react with the targeted pollutants (Ledjeri et al., 2017; Li, S. et al., 2021). The dominant ROS responsible for the degradation and mineralization of pollutants can be judged by the conjugate data analysis of the quenching experiments and EPR spectra. The oxygen-containing groups (e.g., −OH and C=O groups) and C=C bonds probably serve as the reactive sites for catalyzing ozone decomposition (Moussavi and Khosravi, 2012; Zhang et al., 2018). Zhu et al. (2022) used the Boehm titration method to determine the quantity of various chemical groups before and after the catalytic ozonation, and confirmed that the −OH groups were the dominant active sites on the biochar surface for O_3 decomposition (Zhu et al., 2022).

The interfacial active mode of −OH group and O_3 molecule can be further explained by DFT calculation. In detail, the bond length of O−O of the stable ozone structure is 1.277 Å and the bond angle of O−O−O bond angle is 117.660°; various adsorption energies are computationally acquired for O_3 on −OH (0.31 eV), −COOH (0.16 eV), and −COOR (0.21 eV) (Figure 14.2a) (Zhu et al., 2022). Hence, the interaction effect of −OH group and O_3 molecule is stronger than that of other group. Especially, the distance from O_3 molecule to −OH group is 2.687 Å optimally, which is shorter in comparison with 2.880 Å of −COOR and 3.309 Å of −COOH (Figure 14.2b). The potential mechanism was put forward in Figure 14.2c.

The delocalized π electron can be characterized by the density of C=C structure, which is able to react with H_2O to generate hydronium (H_3O^+) and hydroxide (OH^-) (Eq. 12), then the chain reactions are initiated (Eqs. 14.13–14.17) (Li et al., 2022). The speculated main mechanism of the catalytic ozonation oxidation of ketoprofen by peanut shell-based biochar is presented in Figure 14.3.

$$\text{carbon} - \pi + 2H_2O \rightarrow \text{carbon} - H_3O^+ + OH^- \tag{14.12}$$

$$O_3 + OH^- \rightarrow HO_2^- + O_2 \tag{14.13}$$

$$O_3 + HO_2^- \rightarrow HO_2^\bullet + O_3^{\bullet -} \tag{14.14}$$

$$HO_2^\bullet \rightarrow O_2^{\bullet -} + H^+ \tag{14.15}$$

$$O_3^{\bullet -} \rightarrow O_2 + O^{\bullet -} \tag{14.16}$$

$$O^{\bullet -} + H_2O \rightarrow {}^\bullet OH + OH^- \tag{14.17}$$

Noticeably, the chemical state of N show influential role on O_3 decomposition. Reportedly, the various N components consisting of 34.25% graphitic N, 25.00% pyrrolic N, and 34.25% pyridinic N were shown in CMS-$MnFe_2O_4$ (Jin et al., 2021). Especially, the pyrrolic N and pyridinic N have been verified to be active site for O_3 decomposition (Chen et al., 2022). The negatively charged organic pollutants were easily absorbed on Lewis acid site via electrostatic interaction, which is

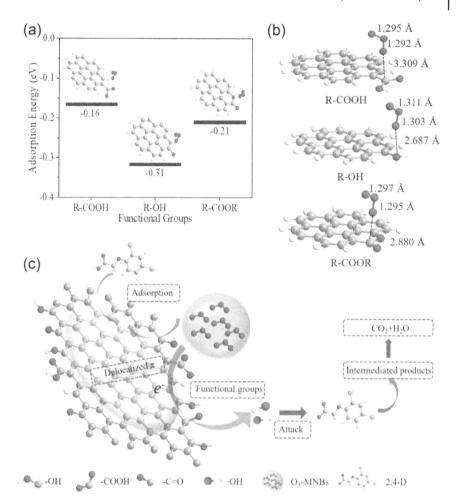

Figure 14.2 (a) The calculated adsorption energy, (b) the optimized configuration of ozone adsorbed by R–OH, R–COOH, and R–COOR on graphene surface, and (c) interfacial mechanism of adsorption and catalytic ozonation by biochar for 2,4-D removal (Zhu et al., 2022 / with permission from Elsevier).

helpful for their degradation. Lewis acid sites enhance the adsorption of O_3 molecule on biochar surface and also participate O_3 decomposition.

The electron transfer of the metal ion redox couples could boost the generation of ROS in the catalyzed ozonation reaction, such as Fe^{2+}/Fe^{3+}, Co^{2+}/Co^{3+}, and Mn^{3+}/Mn^{4+} redox couples (Cha et al., 2022; Tian et al., 2021; Yang et al., 2022).

Oxygen vacancy has the high absorption capacity on ozone molecule (Cha et al., 2022; He et al., 2020). The defects from carbonaceous materials can result in

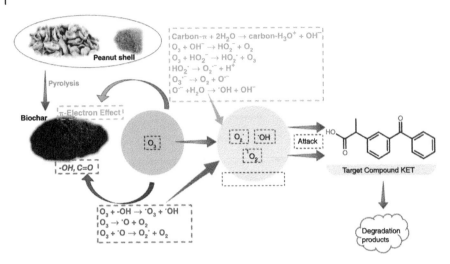

Figure 14.3 The main mechanism of the catalytic ozonation oxidation of ketoprofen by peanut shell-based biochar (Haiquan Li et al., 2020 / from Taylor and Francis group).

delocalized electrons, which are useful active sites for O_3 decomposition (Zhu et al., 2022). The I_D/I_G ratio of biochar material declined, indicating that the defects might join the O_3 decomposition reaction (Wang et al., 2015).

Moreover, the persistent free radicals on biochar surface are reported to boost the ROS generation via a single-electron transfer process (Guo et al., 2020). It is also worth noting that the addition of biochar could increase the electric current dramatically, thereby boosting the electron transfer behavior (Li, S. et al., 2021). Totally speaking, the efficient pollutant disintegration was caused by the synergistic effects of free radical oxidation and electron transfer (Li, S. et al., 2021). Finally, the deactivation of the catalyst was caused by the accumulation of the organic intermediates in the catalytic ozonation process (He et al., 2020). Additionally, the hydroxyl groups on the biochar surface transformed into carbonyl and further into carboxyl groups after catalytic ozonation reaction (Zhang et al., 2018; Zhu et al., 2022), causing the deactivation of biochar in reutilization.

The stability and reusability of a biochar catalyst should be concerned emphatically. The stability of the catalyst decides the catalyst's fate in practical implementation. Biochar keeps its outstanding catalytic performance for the initial cyclized catalyzation, whereas the consecutive reutilization causes a distinct attenuation of catalytic performance. The pore structure of the used biochar was different from that of fresh biochar, and collapse of pore walls occurred distinctly (Figure 14.4) (Zhu et al., 2022). Only a 3% reduction in efficiency was shown after the third cycle, reflecting the stability of the Fe-loaded biochar (Babar et al., 2022). Moreover, the formed Mn(IV) and Fe(III) after would accumulate gradually (Tian et al., 2021), reducing their catalytic activity for ozone decomposition.

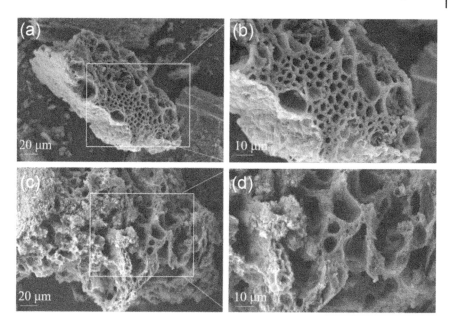

Figure 14.4 SEM images of the fresh biochar (a and b) and the used biochar (c and d) (Xinwei Zhu et al., 2022 / with permission from Elsevier).

The leached amounts of various metals from biochar materials after their use are worth exploring for effluent quality security (Jin et al., 2021). For Mn- and Fe-loaded biochar, the leached concentrations were 1.8 μg/L and 1.2 μg/L, respectively, after the first cycling use, and were 1.7 μg/L and 1.4 μg/L, respectively, after the fourth reuse (Tian et al., 2021), which satisfied the quality standard for drinking water of various countries and areas.

Further investigations on shortening reaction time, reducing operation cost, enhancing mass transfer efficiency, and reinforcing reusability should be comprehensively performed to stimulate the development of the biochar-enhanced ozonation sewage treatment process in practical engineering.

References

Ahmad, A., Priyadarshini, M., Yadav, S., Ghangrekar, M.M., Surampalli, R.Y., 2022. The potential of biochar-based catalysts in advanced treatment technologies for efficacious removal of persistent organic pollutants from wastewater: A review. *Chemical Engineering Research and Design*, 187, 470–496.

Babar, M., Munir, H.M.S., Nawaz, A., Ramzan, N., Azhar, U., Sagir, M., Tahir, M.S., Ikhlaq, A., Mubashir, M., Khoo, K.S., 2022. Comparative study of ozonation and ozonation catalyzed by Fe-loaded biochar as catalyst to remove methylene blue from aqueous solution. *Chemosphere*, 307, 135738.

Cha, J.S., Kim, Y.-M., Choi, Y.J., Rhee, G.H., Song, H., Jeon, B.-H., Lam, S.S., Khan, M.A., Lin, K.-Y.A., Chen, W.-H., 2022. Mitigation of hazardous toluene via ozone-catalyzed oxidation using MnOx/Sawdust biochar catalyst. *Environmental Pollution*, 312, 119920.

Chen, C., Yan, X., Xu, Y., Yoza, B.A., Wang, X., Kou, Y., Ye, H., Wang, Q., Li, Q.X., 2019. Activated petroleum waste sludge biochar for efficient catalytic ozonation of refinery wastewater. *Science of the Total Environment*, 651, 2631–2640.

Chen, J., Tu, Y., Shao, G., Zhang, F., Zhou, Z., Tian, S., Ren, Z., 2022. Catalytic ozonation performance of calcium-loaded catalyst (Ca-C/Al$_2$O$_3$) for effective treatment of high salt organic wastewater. *Separation and Purification Technology*, 301, 121937.

Chen, X., Fu, L., Yu, Y., Wu, C., Li, M., Jin, X., Yang, J., Wang, P., Chen, Y., 2021. Recent development in sludge Biochar-based catalysts for advanced oxidation processes of wastewater. *Catalysts*, 11(11), 1275.

Dabuth, N., Thuangchon, S., Prasert, T., Yuthawong, V., Phungsai, P., 2022. Effects of catalytic ozonation catalyzed by TiO$_2$ activated carbon and biochar on dissolved organic matter removal and disinfection by-product formations investigated by Orbitrap mass spectrometry. *Journal of Environmental Chemical Engineering*, 10(2), 107215.

Feng, J., Xing, B., Chen, H., 2019. Catalytic ozonation of humic acid in water with modified activated carbon: Enhancement and restoration of the activity of an activated carbon catalyst. *Journal of Environmental Management*, 237, 114–118.

Gaskin, J., Steiner, C., Harris, K., Das, K., Bibens, B., 2008. Effect of low-temperature pyrolysis conditions on biochar for agricultural use. *Transactions of the ASABE*, 51(6), 2061–2069.

Ge, D., Huang, S., Cheng, J., Han, Y., Wang, Y., Dong, Y., Hu, J., Li, G., Yuan, H., Zhu, N., 2022a. A new environment-friendly polyferric sulfate-catalyzed ozonation process for sludge conditioning to achieve deep dewatering and simultaneous detoxification. *Journal of Cleaner Production*, 359, 132049.

Ge, D., Wu, W., Li, G., Wang, Y., Li, G., Dong, Y., Yuan, H., Zhu, N., 2022b. Application of CaO$_2$-enhanced peroxone process to adjust waste activated sludge characteristics for dewaterability amelioration: Molecular transformation of dissolved organic matters and realized mechanism of deep-dewatering. *Chemical Engineering Journal*, 437, 135306.

Guo, J., Jia, X., Gao, Q., 2020. Insight into the improvement of dewatering performance of waste activated sludge and the corresponding mechanism by biochar-activated persulfate oxidation. *Science of the Total Environment*, 744, 140912.

He, C., Wang, Y., Li, Z., Huang, Y., Liao, Y., Xia, D., Lee, S., 2020. Facet engineered α-MnO$_2$ for efficient catalytic ozonation of odor CH$_3$SH: Oxygen vacancy-induced active centers and catalytic mechanism. *Environmental Science & Technology*, 54(19), 12771–12783.

Issaka, E., Amu-Darko, J.N.-O., Yakubu, S., Fapohunda, F.O., Ali, N., Bilal, M., 2022. Advanced catalytic ozonation for degradation of pharmaceutical pollutants—A review. *Chemosphere*, 289, 133208.

Jin, X., Wu, C., Tian, X., Wang, P., Zhou, Y., Zuo, J., 2021. A magnetic-void-porous $MnFe_2O_4$/carbon microspheres nano-catalyst for catalytic ozonation: Preparation, performance and mechanism. *Environmental Science and Ecotechnology*, 7, 100110.

Kloss, S., Zehetner, F., Dellantonio, A., Hamid, R., Ottner, F., Liedtke, V., Schwanninger, M., Gerzabek, M.H., Soja, G., 2012. Characterization of slow pyrolysis biochars: Effects of feedstocks and pyrolysis temperature on biochar properties. *Journal of Environmental Quality*, 41(4), 990–1000.

Ledjeri, A., Yahiaoui, I., Kadji, H., Aissani-Benissad, F., Amrane, A., Fourcade, F., 2017. Combination of the Electro/Fe^{3+}/peroxydisulfate (PDS) process with activated sludge culture for the degradation of sulfamethazine. *Environmental Toxicology and Pharmacology*, 53, 34–39.

Li, H., Liu, S., Qiu, S., Sun, L., Yuan, X., Xia, D., 2022. Catalytic ozonation oxidation of ketoprofen by peanut shell-based biochar: Effects of the pyrolysis temperatures. *Environmental Technology*, 43(6), 848–860.

Li, S., Wu, Y., Zheng, Y., Jing, T., Tian, J., Zheng, H., Wang, N., Nan, J., Ma, J., 2021a. Free-radical and surface electron transfer dominated bisphenol A degradation in system of ozone and peroxydisulfate co-activated by $CoFe_2O_4$-biochar. *Applied Surface Science*, 541, 147887.

Li, W., Zhu, N., Yuan, H., Shen, Y., 2021b. Influence of sludge organic matter on elimination of polycyclic aromatic hydrocarbons (PAHs) from waste activated sludge by ozonation: Controversy over aromatic compounds. *Science of the Total Environment*, 797, 149232.

Liu, Y., Chen, C., Duan, X., Wang, S., Wang, Y., 2021. Carbocatalytic ozonation toward advanced water purification. *Journal of Materials Chemistry A*, 9(35), 18994–19024.

Moussavi, G., Khosravi, R., 2012. Preparation and characterization of a biochar from pistachio hull biomass and its catalytic potential for ozonation of water recalcitrant contaminants. *Bioresource Technology*, 119, 66–71.

Oh, S.-Y., Nguyen, T.-H.A., 2022. Ozonation of phenol in the presence of biochar and carbonaceous materials: The effect of surface functional groups and graphitic structure on the formation of reactive oxygen species. *Journal of Environmental Chemical Engineering*, 10(2), 107386.

Singh, G., Kim, I.Y., Lakhi, K.S., Srivastava, P., Naidu, R., Vinu, A., 2017. Single step synthesis of activated bio-carbons with a high surface area and their excellent CO_2 adsorption capacity. *Carbon*, 116, 448–455.

Sun, Z., Ma, J., Liu, Y., Wang, H., Cao, W., Zhu, N., Lou, Z., 2021. Mineralization of refractory organics in oil refinery wastewater by the catalytic ozonation with magnetic praseodymium-catalysts: Catalytic performances and mechanisms. *Separation and Purification Technology*, 277, 119506.

Temesgen, T., Bui, T.T., Han, M., Kim, T.-i., Park, H., 2017. Micro and nanobubble technologies as a new horizon for water-treatment techniques: A review. *Advances in Colloid and Interface Science*, 246, 40–51.

Tian, S.-Q., Qi, J.-Y., Wang, Y.-P., Liu, Y.-L., Wang, L., Ma, J., 2021. Heterogeneous catalytic ozonation of atrazine with Mn-loaded and Fe-loaded biochar. *Water Research*, 193, 116860.

Wang, X., Qin, Y., Zhu, L., Tang, H., 2015. Nitrogen-doped reduced graphene oxide as a bifunctional material for removing bisphenols: Synergistic effect between adsorption and catalysis. *Environmental Science & Technology*, 49(11), 6855–6864.

Wu, Y., Zhang, P., Zeng, G., Ye, J., Zhang, H., Fang, W., Liu, J., 2016. Enhancing sewage sludge dewaterability by a skeleton builder: Biochar produced from sludge cake conditioned with rice husk flour and $FeCl_3$. *ACS Sustainable Chemistry & Engineering*, 4(10), 5711–5717.

Xiao, T., Dai, X., Wang, X., Chen, S., Dong, B., 2021. Enhanced sludge dewaterability via ozonation catalyzed by sludge derived biochar loaded with $MnFe_2O_4$: Performance and mechanism investigation. *Journal of Cleaner Production*, 323, 129182.

Yang, H., Yan, R., Chen, H., Lee, D.H., Zheng, C., 2007. Characteristics of hemicellulose, cellulose and lignin pyrolysis. *Fuel*, 86(12–13), 1781–1788.

Yang, X., Zeng, L., Huang, J., Mo, Z., Guan, Z., Sun, S., Liang, J., Huang, S., 2022. Enhanced sludge dewaterability by a novel $MnFe_2O_4$-biochar activated peroxymonosulfate process combined with tannic acid. *Chemical Engineering Journal*, 429, 132280.

Zhang, F., Wu, K., Zhou, H., Hu, Y., Sergei, P., Wu, H., Wei, C., 2018. Ozonation of aqueous phenol catalyzed by biochar produced from sludge obtained in the treatment of coking wastewater. *Journal of Environmental Management*, 224, 376–386.

Zhu, X., Wang, B., Kang, J., Shen, J., Yan, P., Li, X., Yuan, L., Zhao, S., Cheng, Y., Li, Y., 2022. Interfacial mechanism of the synergy of biochar adsorption and catalytic ozone micro-nano-bubbles for the removal of 2, 4-dichlorophenoxyacetic acid in water. *Separation and Purification Technology*, 299, 121777.

15

Biochar-Supported Odor Control

Jingyi Gao[1], Zibo Xu[2], and Daniel C.W. Tsang[3]

[1] EIT Institute for Advanced Study, Ningbo, Zhejiang, China
[2] Department of Civil and Environmental Engineering, The Hong Kong Polytechnic University, Hong Kong, China
[3] State Key Laboratory of Clean Energy Utilization, Zhejiang University, China

An odor (UK spelling: odour) is also called a "smell" or a "scent," which can be pleasant or unpleasant. Odors are actually caused by one or more volatile compounds, which are perceived by humans and animals through the sense of smell (Odor, 2022). When odors act as an annoying smell, odor pollution becomes an environmental nuisance that affects public facilities and the quality of life in the community (Hu et al., 2020; Piccardo et al., 2022). A complex mixture of odorant substances, usually sulfur, nitrogen, and volatile organic compounds (VOCs), enables the perception of an odor whose intensity is mainly related to the concentration of the odorant substance and whose properties are mainly related to its chemical structure (Hawko et al., 2021). The different chemicals contained in the smell affect its odor profile; for example, hydrogen sulfide is described as rotten eggs, and methyl mercaptan is described as rotten cabbage and garlic (Piccardo et al., 2022). Sulfur, nitrogen, and volatile organic compounds (VOCs) that make up the odor will be perceived by the sense of smell at quite low concentrations, thereby causing a certain degree of physical and psychological harm to the public (Toledo et al., 2019).

Biochar Applications for Wastewater Treatment, First Edition. Edited by Daniel C.W. Tsang and Yuqing Sun.
© 2023 John Wiley & Sons, Inc. Published 2023 by John Wiley & Sons, Inc.

15.1 Causes and Treatment of Odor

Sources of odor pollution are divided into natural sources and man-made sources. Natural sources include odors produced by algae and plants in stagnant lakes, the stench produced by the decomposition of animal waste, dead animals, and plants by microorganisms, and natural phenomena such as lightning and volcanic eruptions. The manmade source is the main source of odor pollution, which is further divided into domestic sources of waste disposal and industrial production sources (Gallego et al., 2012).

With the rapid development of livestock industry, the pressure on animal manure management is increasing. Aerobic composting and anaerobic digestion are considered common biological treatments for animal manure to produce biofertilizers and recover biomethane as a resource (Akdeniz, 2019; Sardar et al., 2021). Aerobic composting is more efficient at converting livestock manure into humus-rich biofertilizers for agricultural applications than anaerobic digestion. However, odor pollutants such as ammonia and volatile sulfur compounds (VSCs) are produced during composting, compromising the environmental benefits of livestock manure treatment (Gao et al., 2022). Besides, digestate from FW (DFW) produced after anaerobic digestion (AD) process used to treat food waste may also produce a large amount of ammonia and volatile sulfur compounds (VSCs) to generate odor during the composting process (N. Wang et al., 2022). Landfills are also high incidences of odor pollution, with typical odor sources identified as landfill working faces (WF), leachate treatment areas, landfill gas wells, and waste haulage vehicles (Ying et al., 2012). Municipal solid waste (MSW) will release volatile trace compounds and cause odor pollution, especially in developing countries where the amount of waste is increasing rapidly (Liu et al., 2019; Zhao et al., 2015). Most odor compounds emitted from MSW facilities are present in very low concentrations and are therefore considered volatile trace compounds (Du et al., 2023). These odorous substances generally include both volatile organic compounds and inorganic compounds, where VOCs include sulfur compounds, oxygenates (alcohols, aldehydes, ketones, acids, and esters), aromatics, terpenes, halogenated compounds, and saturated and unsaturated hydrocarbons (Lim et al., 2018; Zhao et al., 2015), and inorganic compounds mainly include ammonia (NH_3) and hydrogen sulfide (H_2S) (Shen et al., 2019). In addition to the two waste disposal technologies of composting and landfilling, waste incineration can also cause odorous pollution.

VOCs emitted from municipal solid waste incineration power plants (MSWIPP) play an important role in odor pollution (Sun et al., 2023). Sludge waste from wastewater treatment plants (WWTPs) produces odorous compounds such as organic sulfide (methyl mercaptan) and inorganic sulfide (hydrogen sulfide and sulfur dioxide), ammonia, amines, volatile fatty acids, aromatic compounds, terpenes, acetone, phenols (Lebrero et al., 2011). In terms of industrial production, oil refineries, chemical plants, textile mills, and leather factories emit odorous emissions during the production process because they require a large amount of organic substances and odorous inorganic substances as materials and the chemical processes they perform (Piccardo et al., 2022).

Odor emission is evaluated by two methods of chemical analysis and sensory analysis (Conti et al., 2020). Chemical analysis is widely used to determine the molecules present in the air and their chemical concentrations, such as using gas chromatography-mass spectrometry (GC–MS) to separate components from the odor mixture based on their affinity to the stationary phase in the column (Sadowska-Rociek et al., 2009). The sensory measurement method is to use the human olfactory organs to reflect the degree of malodorous substances to evaluate their properties (Blanco-Rodríguez et al., 2018).

In general, odor abatement technologies can be divided into three categories: physical, chemical, and biological, depending on the base of the abatement process (Talaiekhozani et al., 2016). Physical and chemical technologies, such as dry adsorption using activated carbon and other adsorbents and chemical scrubbers, have been widely used because of their low empty bed residence time (EBRT), extensive experience in design and operation, and rapid startup (Le-Minh et al., 2018). On the other hand, biotechnologies have been marketed as low-cost, environmentally friendly odor abatement methods, which include biofilters, bioscrubbers, and suspended-growth bioreactors (Capelli et al., 2019).

In recent years, due to its good physical and chemical properties, biochar has been used as a compost additive to reduce the emission of volatile organic compounds and is an effective means of absorbing odor (Nguyen et al., 2022; Toledo et al., 2020a). Moreover, adding biochar to biofilters has been reported to be a suitable alternative for the efficient adsorption of hydrogen sulfide (Das et al., 2019). Biochar application can be used not only on an industrial scale but also on a domestic scale, such as removing volatile organic sulfur compounds (VOSCs) from cat urine by adding biochars to the cat litter formula (S. Wang et al., 2022). Biochar can be produced from low-cost biomass waste pyrolysis, hardwood chips, poultry litter, husks, corncobs and bagasse, for example, and the production and application of biochar have been proven for multiple advantages, including carbon sequestration and waste upcycling (Xu et al., 2021). Therefore, using biochar for odor pollution control has great potential in terms of economic and environmental benefits (Hwang et al., 2018; Marrakchi et al., 2017).

15.2 Odor Pollutants

Odor pollution belongs to sensory pollution. The public can perceive the existence of peculiar smell through smell, which will seriously affect the quality of the public's living environment (Y. Zhang et al., 2021). There are many kinds of odor pollutants; according to incomplete statistics, there are more than 4,000 kinds, including sulfur-containing compounds, nitrogen-containing compounds, aldehydes, ketones, esters, acids, phenols, aromatic hydrocarbons, terpenes, and other substances. They have low olfactory threshold and strong odor, which can lead to the occurrence of malodorous pollution. Moreover, the types, concentrations, and emissions of odorous pollutants emitted by different industries vary greatly (Ministry of Ecology and Environment of the People's Republic of China, 2018). Table 15.1 shows the odor profiles and odor thresholds of common odor pollutants.

Odor pollution comes from a wide range of sources, including chemical, petroleum refining, pharmaceutical, coating, papermaking, food processing, flavors

Table 15.1 Odor profiles and odor thresholds of common odor compounds (Piccardo et al., 2022; Talaiekhozani et al., 2016).

Category	Compounds	Odor threshold value (ppm)	Odor profiles
Sulfide compounds	H_2S hydrogen sulfide	$0.4E^{-1}$–4	Rotten egg
	CS_2 Carbon disulfide	0.016–32	Vegetable, sulfide, medicinal
	$(CH_3)_2S_2$ Dimethyl disulfide	$0.29E^{-3}$–1.45	Cabbage, sulfurous
	$(CH_3)_2S$ Dimethyl sulfide	$0.12E^{-3}$–8.11	Disagreeable, putridus asparagus
	C_2H_6S Ethyl mercaptan	$0.87E^{-5}$–18	Rotten cabbage, flatulence, skunk-like
	SO_2 Sulfur dioxide	0.33–8	Pungent; similar to a just-struck match
Nitrogen compounds	NH_3 Ammonia	0.043–60.3	Strong pungent odor
	CH_5N Methylamine	$0.75E^{-3}$–4.8	Fishy
	C_2H_7N Dimethylamine	$0.76E^{-3}$–4.2	Ammoniacal, rotten fish
	$C_6H_{15}N$ Trimethylamine	$0.2E^{-4}$–1.82	Fishy, pungent
Organic acids	CH_2O_2 Formic acid	0.52–3.40	Sharp
	$C_2H_4O_2$ Acetic acid	$0.4E^{-3}$–204	Pungent, vinegar
	$C_3H_6O_2$ Propionic acid	$0.99E^{-3}$–4.65	Sour

Table 15.1 (Continued)

Category	Compounds	Odor threshold value (ppm)	Odor profiles
Aldehydes	C_2H_4O Acetaldehyde	$0.15E^{-2}$–1000	Pungent, fruity, suffocating, fresh, green
Ketones	$C_5H_{10}O$ Methyl isopropyl ketone	0.51–4.8	Sweet, sharp
	$C_5H_{10}O$ Methyl propyl ketone	0.028–65	Fingernail polish

and fragrances and other factories and enterprises, as well as nonpoint sources such as sewage treatment plants, landfills, and other municipal facilities. The effects of odor often lead to complaints from the to authorities, so elected officials have substantial motivation to regulate odor pollution (Brancher et al., 2017). Therefore, countries around the world have formulated odor policies and regulations to control odor emissions. Asian countries such as China, Japan, and South Korea have a more detailed classification of odor-causing gases than Europe and the United States, limiting the concentration of each pollutant. Japan was one of the first countries to conduct research on odor pollution and the current Odor Prevention Act was revised in 2000 (Ministry of the Environment, Government of Japan, 2006). The perimeter concentration limits in Korea are all taken from the lower limit of the Japanese limits and are therefore more stringent than the Japanese standards (Seoul Solution, 2010). China promulgated the Emission Standards for Odor Pollutants (GB 14554–93) in 1993, but with the continuous progress of science and technology, environmental management and the improvement of people's living standards, the previous standards were no longer strong enough, and in 2018 the Emission Standards for Odor Pollutants (Draft for Public Comments) (GB 14554–201X) was released, which tightened the limits for ammonia, trimethylamine hydrogen sulphide, methyl mercaptan, methyl sulphide, dimethyl sulphide, carbon disulphide, and styrene perimeter concentration limits (Ministry of Ecology and Environment of the People's Republic of China, 2018). Table 15.2 summarizes the standards for odor in China, Japan, and Korea.

15.3 Properties of Biochar for the Removal of Odor Pollutants

The properties of biochar are the key to its application in the removal of odor pollutants, such as its surface area, porosity, element composition, functional group, and pH. The adsorption of both organic and inorganic odor pollutants on biochar

Table 15.2 Odor emission standards in China, Japan and Korea.

Pollutant	China			Japan			Korea	
	Unit	GB 14554-93 Boundary standard values	GB 14554-201X Boundary standard values	Offensive odor control law			Malodor Prevention Act	
				Unit	Industry Area	Non-Industry Area	Unit	Boundary standard values
Azane (ammonia)	mg/m³	1.0–5.0	0.2	mg/m³	1.5–3.8	0.76–1.5	mg/m³	0.76
N,N-dimethylmethanamine (trimethylamine)		0.05–0.80	0.05		0.053–0.18	0.013–0.053		0.013
Sulfane (hydrogen sulfide)		0.03–0.06	0.02		0.09–0.3	0.03–0.09		0.03
Methanethiol (methyl mercaptan)		0.0004–0.035	0.002		0.009–0.02	0.004–0.009		0.004
Methylsulfanylmethane (Dimethyl sulfide)		0.03–1.10	0.02		0.14–0.55	0.028–0.14		0.028
(Methyldisulfanyl)methane (Dimethyl disulfide)		0.03–0.71	0.05		0.13–0.42	0.038–0.13		0.038
Carbon disulfide		2.0–10	0.5					
Styrene (vinyl benzene)		33.0–19	1.0		3.7–9.3	1.9–3.7		1.9
Odor concentration		20			10–126			15–20

is a combination of physical adsorption and chemical adsorption. The physical adsorption part is mainly determined by the texture characteristics of biochar, such as specific surface area (SSA), pore volume (PV), and pore size distribution, while the chemical adsorption is mainly affected by the type and quantity of surface functional groups. The following describes several vital properties of biochar associated with the removal of odorous odor pollutants, mainly volatile organic pollutants, including organic and inorganic odor.

15.3.1 Surface Area and Total Pore Volume

Gas filtration by biochar often requires large surface areas and pore volumes to be effective (Maziarka et al., 2021). For organic gaseous pollutants, in the study by Xueyang Zhang et al. (2021), it was found that the adsorption capacity of VOCs was significantly correlated with SSA and PV. SSA can provide necessary sites for VOCs adsorption, leading to a higher adsorption capacity with large SSA. In addition, the physical adsorption of VOCs often relies on the gas–liquid phase transition and liquid-like capillary condensation in porous materials, so the PV strongly affects the adsorption amount (Zhang et al., 2017a). However, the biochar material with the highest SSA and PV does not always mean the best adsorption capacity for organic compounds. Vikrant et al. (2020) found no clear correlation between SSA and maximum adsorption capacity for benzene and methyl ethyl ketone, since the overall adsorption mechanism mainly involves a combination of physisorption and partitioning processes, where the physical adsorption process is usually controlled by SSA, while the partition process is mainly controlled by the volatile matter content. For inorganic odor pollutants, it is undeniable that the larger the SSA and PV, the more the number of active sites capable of removing air pollutants, but the positive correlation with the removal capacity is also not always established (Zhao et al., 2022). Braghiroli et al. (2019) indicated that the amount of SO_2 adsorbed by activated biochar in the dynamic test has no linear relationship with the surface area and pore volume. In addition, in the study by Cui et al. (2021), the adsorption capacity of H_2S is not positively related to the specific surface area of some of adsorbents. It was also found that the surface area of biochar is not easy to control the adsorption of gaseous NH_3 during the adsorption process of NH_3, and the adsorption capacity of steam-activated biochar with relatively high specific surface area was not improved (Ro et al., 2015).

Large specific surface area and total pore volume have a positive effect on physical adsorption, but physical adsorption is determined by specific surface area, pore structure, surface properties, and adsorbate properties at the same time, so blindly increasing specific surface area and pore volume does not necessarily achieve better removal effect.

15.3.2 Pore Size Distribution

Pores of different sizes have certain removal biases for target pollutants (He et al., 2021). The molecular diameter of VOCs determines the effective pores for VOCs to enter the adsorbent. Theoretically, pores with a pore size larger than the molecular diameter of VOCs can become effective adsorption sites. However, when the pore size is much larger than the molecular diameter of VOCs, the adsorption force between the adsorbent and the VOCs molecules is too weak, and the pores can only function as channels (Li et al., 2020). The pores in the nanometer range are divided into three types: micropores (<2 nm), mesopores (2–50 nm) and macropores (>50 nm). It is generally believed that micropores can provide adsorption sites for VOCs, mesopores can accelerate the diffusion of VOCs, and macropores can promote the transport and diffusion of VOCs (Gan et al., 2021). For example, toluene is mainly adsorbed on the micropores of the biochar, while the developed mesoporous structure is conducive to the mass transfer and diffusion of toluene, thereby improving adsorption capacity and the regeneration performance (Cheng et al., 2020; Hossein Tehrani et al., 2020). Mass transfer plays an important role in the initial adsorption kinetics, but once the mesopores are plugged, the mass transfer is continuously reduced, at which point macropores are needed to enhance mass transfer by reducing the diffusion length of the reactants (Im et al., 2021). In the study of Yang et al. (2022), hierarchically structured porous carbons with more mesopores and macropores had a faster toluene adsorption rate, and the reason was attributed to the fact that mesopores and macropores sped up the adsorption rate by providing mass transfer channels.

For inorganic odor pollutants, the effect of pore size distribution on its removal efficiency is generally about the removal mechanism. In terms of the removal of SO_2, the effect of microporous structure is more significant than mesoporous structure, the reason is that the pore size of 0.7 nm is the best pore size to facilitate the oxidation of SO_2 to SO_3. The increase of pore width decreases the conversion rate of SO_2 to SO_3, thus decreasing the total amount of SO_2 retained by carbon samples (Raymundo-Piñero et al., 2000). Therefore, some scholars used carbon dioxide activation to create more micropores on biochar. In a study by Iberahim et al. (2022), biochars activated by CO_2 had more micropores to obtain better SO_2 adsorption capacity. Xiong Zhang et al. (2021) also used CO_2 activation to improve the microporous structure of biochar, which significantly improved the ability to adsorb SO_2 at low temperatures. It was also found that the microporous volume plays an important role in the introduction of nitrogen-containing functional groups during the MDEA impregnation process.

However, the highest SO_2 adsorption capacity does not necessarily exist at the maximum micropore volume (Zhu et al., 2012). Braghiroli et al. (2019) indicated that the reason for this is that for SO_2 adsorption, the role of pore structure is not as clear as that of surface oxidation groups, and large pores reduce the conversion of SO_2 to SO_3, although porosity is important for storing sulfuric acid as an oxidation product (Bagreev et al., 2002).

In terms of H_2S removal, H_2S oxidation is less likely to occur in very small micropores and more likely to occur in micropores larger than 0.7 nm in size, which act as nanoreactors for H_2S dissociation and oxidation, while mesoporosity is important for higher diffusion rates of H_2S to the active site and oxidation product deposition (Surra et al., 2019). Su et al. (2021) found that biochars with microporous structure had better H_2S removal performance characteristics than those with large and medium pore structure, where the micropores were responsible for most of the H_2S adsorption while the macropores were responsible for H_2S transport to the micropores. Dou et al. (2022) used microwave-assisted KOH to obtain a higher microporosity of biochar, making it favorable for H_2S adsorption. Chen et al. (2022) obtained that the desulfurization mechanism concerns the retention of S and SO_4^{2-} in the micropores, but also the transfer to the mesopores through the cross-linked porous structure, demonstrating the importance of the mesopores.

For NH_3 adsorption, sufficient investigations have not been conducted on the effect of pore size. Xiao et al. (2023) hypothesized that NH_3 adsorbed on the surface and pores (including micro-, meso- and macro-pores) of their constructed Prussian blue-impregnated waste teak peel-derived biochar through van der Waals forces. However, studies have shown that pore size is related to the distribution of functional groups (Zhang and Calo, 2001). Nitrogen-containing basic functional groups are mainly distributed in micropores, oxygen-containing acidic (polar) groups mostly exist in mesopores/macropores, and oxygen-containing nonpolar functional groups are widely distributed. Feng et al. (2021a) found that the adsorption sites of NH_3 are mainly nitrogen-containing functional groups in micropores (d < 0.7 nm), while the mesoporous structure promotes the adsorption of NH_3. Therefore, when it comes to biochar modification and modification of functional groups, it is important to consider pore size.

Pore size distribution is very important for the removal process of organic and inorganic odor pollutants, and as mentioned above, each size of pore has its special role. Generally, the adsorption reaction sites of pollutants are in micropores and rely on mesopores and macropores for transport, and a suitable pore size distribution can be constructed according to the characteristics of the target pollutants. At present, the difference between the role of macro- and mesopores for inorganic odor pollutants needs further exploration. The role of pore size distribution does not completely determine the adsorption capacity, just like surface area and pore volume, because physical adsorption and chemical adsorption constitute the adsorption process at the same time.

15.3.3 Chemical Functional Group

The surface functional groups of biochar materials are responsible for chemical adsorption. Therefore, the type and quantity of chemical functional groups on the surface of the adsorption material have a great influence on the adsorption capacity of organic and inorganic odor pollutants.

Among the surface functional groups of biochar, oxygen-containing groups are the most abundant species (Zhu et al., 2020). Most oxygen-containing groups (e.g., carboxyl, hydroxyl, carbonyl, anhydride, and lactone) are the source of acidity on the surface of carbon materials, which will enhance the adsorption of hydrophilic VOCs, while the presence of oxygen-containing groups may inhibit specific interactions between hydrophobic VOCs and π-electron-rich regions on carbonaceous adsorbents (Zhao et al., 2018). Li et al. (2011) found that the adsorption capacity of activated carbon materials for the hydrophobic compound, 1,2-dimethylbenzene decreased with the increase of the total surface oxygen-containing groups. Similarly, Cheng et al. (2022) found that the oxygen-containing functional groups on biochar promoted the adsorption of strongly polar ethyl acetate, but weakly promoted the adsorption of weakly polar toluene. Therefore, in order to improve the adsorption performance of hydrophobic VOCs, carbonaceous adsorbents can be modified with weak alkali such as ammonia to reduce the oxygen-containing groups on the surface. In addition to oxygen groups, nitrogen groups on porous carbon are also considered to be important adsorption species, which were introduced into the biochar surface by some modification means (Zhang et al., 2017a). Xueyang Zhang et al. (2021) introduced nitrogen functional groups into biochar using ammonium hydroxide ball milling and enhanced the adsorption capacity of benzene, m-xylene, o-xylene, and p-xylene, which was attributed to the substitution of oxygen-containing groups on the carbon surface by amine groups. Feng et al. (2021b) used the ammonia hydrothermal method to construct biochar containing O and N groups to achieve good adsorption of phenol. The nitrogen-containing functional group can change the surface electrostatic potential of biochar and contribute to the adsorption stability, thus promoting the adsorption of phenol. The promotion effect of oxygen-containing functional groups on the adsorption of phenol is mainly manifested in the offset effect of the electron pairs of adjacent carbon atoms, which increases the electrostatic effect.

For inorganic odor pollutants, the functional groups that play a role are different for gases with different properties. Zhang et al. (2022) used phytic acid as an activator material to synthesize $CuSiO_3$/P-doped porous biochar with excellent NH_3 bulk adsorption capacity, attributed to the presence of acidic active sites (HPO_4^{2-} and $-COOH$ functional groups) on the surface of the material that can combine with NH_3 to generate NH_4^+. Xiao et al. (2023) utilized acidification treatment and Prussian blue impregnation to increase the acidic functional groups on the surface of biochar, endowing more acidic adsorption sites, resulting in enhanced NH_3

capture capacity. For H_2S removal, the basic surface functional groups on the carbon surface are important, mainly N-containing functional groups, such as pyridine N (PdN) and pyrrole N (PyN). Ma et al. (2021) chose urea phosphate as an activator and N source to prepare N-doped biochar adsorbents, which introduced alkaline sites into the carbon and thus improved the chemical reactivity of acid-base chemistry and exhibited good H_2S adsorption capacity. The effective removal of H_2S was closely related to the adsorption and catalytic reactions of the prepared biochar, and more N species were converted to PdN and PyN functional groups at pyrolysis temperatures above 600 °C, where PdN was responsible for the H_2S oxidation in catalytic activity. Chen et al. (2022) synthesized N-doped interconnected mesoporous biochar by H_2O_2 assisted hydrothermal carbonization, containing PdN, PyN, $-C=O$, and π-π* as basic sites, and the aqueous film showed strong local basicity, which facilitated the adsorption and dissociation of H_2S. Meanwhile, PdN and PyN provided electrons for O_2 activation, and the active O species, i.e., O*, provided strong oxidation for the deep oxidation of H_2S, thus effectively improving the sulfur capacity. The alkaline nitrogen-containing functional groups also exhibit a favorable role in SO_2 removal. Braghiroli et al. (2019) used methyldiethanolamine to impregnate biochar to introduce N-containing groups, and to significantly improve the SO_2 adsorption capacity of the adsorbent by reacting N functional groups, such as CN groups and NH- groups, with SO_2. Xiong Zhang et al. (2021) introduced nitrogen functional groups (e.g., PdN and PyN) on the surface of biochar by high temperature ammonification with ammonia, and found that N-containing functional groups play a major role in high-temperature adsorption.

15.3.4 Noncarbonized Organic Matter

Noncarbonized organic matter (NOM) is an important factor affecting the adsorption of VOCs on biochar, which is about the adsorption mechanism of VOCs on biochar. The adsorption of VOCs on biochar is dominated by the dual mechanism of physical adsorption and distribution (Wang et al., 2016). Carbonized organic matter (COM) in biochar behaves as an adsorbent, and NOM behaves as a distribution phase (Chen et al., 2008). Adsorption would be the main mechanism for biochar with higher surface area and lower NOM content produced at higher temperatures, while partitioning would be the main mechanism for lower temperature pyrolytic biochar with higher NOM content and lower surface area (Zhang et al., 2017b).

15.3.5 Mineral constituents

In addition to a rich carbon component, biochar usually has a high content of mineral (ash) components, such as alkali metals (K, Na, etc.) or alkaline earth metals (Ca, Mg, etc.), which usually exist as carbonates, phosphates, or oxides.

Biochar extracted from mineral-rich biomass such as grass, industrial waste, and municipal biosolids has a high ash content, while biochar extracted from woody biomass usually has a low ash content content (Premarathna et al., 2019). The low ash content of wood biochar is attributed to the fact that wood biomass contains a large amount of cellulose, hemicellulose, and lignin, but little inorganic components, while the opposite is true for straw biomass (Cao et al., 2022).

Alkali metals or alkaline earth metals in the form of carbonates are the main alkaline substances responsible for the high pH of biochar from crop residues and chicken manure (Xu et al., 2017), which can promote inorganic acid odor removal of pollutants. In the study of Xu et al. (2016), the SO_2 adsorption capacity of cow manure and sewage sludge with more mineral components was higher than that of rice husk biochar with less mineral components. Minerals in biochar may play an important role in the SO_2 adsorption of biochar because the abundant mineral components in biochar mainly achieve high retention of SO_2 through the formation of stable sulfite and sulfate minerals (Iberahim et al., 2018; Papurello et al., 2020). For H_2S, abundant minerals also play an important role in its high H_2S adsorption capacity and sulfur species conversion (Xu et al., 2014).

For organic odorous pollutants, the mineral composition also influences the adsorption process. Pongkua et al., 2018 found that bamboo and rice husk biochar with similar surface area and pore volume had higher Qmax due to the higher ash content of rice husk biochar (23.50%) compared to bamboo biochar (0.67%). The speculated reason is that ash reduces the porosity of the material and the contact between VOCs molecules, resulting in less binding to VOCs molecules. Different minerals have been reported to have different effects on the adsorption of various types of organic pollutants (e.g., polar and nonpolar, ionic and nonionic organic pollutants) by biochar (Xu et al., 2017). For example, de-ashed biochar resulted in enhanced adsorption of hydrophobic organics phenanthrene (Sun et al., 2013) and propiconazole (Sun et al., 2016) as the removal of minerals caused by favorable and hydrophobic adsorption sites, but some minerals in crop straw-derived biochar can enhance the adsorption of hydrophobic organic tetracycline (Ji et al., 2011) due to ion-π interactions, surface complexation, and ion-exchange reactions. There is insufficient research on the influence of mineral composition on the adsorption of gaseous organic pollutants, and future research is needed to investigate the interaction between biochar minerals and VOCs.

15.4 Application of Biochar in Odor Control

15.4.1 Biochar as Adsorbent

Similar to activated carbon, biochar has a high adsorption capacity for VOCs and other gaseous pollutants because of its physicochemical properties (such as large surface area and porosity) and surface chemistry (Le-Minh et al., 2018). Therefore, biochar can reduce the odor produced by many sources, such as animal manure or urine, vehicle emissions, and anaerobic digestion, with its high adsorption capacity.

Maurer et al. (2017) applied pine biochar directly to the surface of manure stored in deep pits or lagoon storage environments and found that biochar could float on pig manure for at least one month. It was found that a thin layer (thicknesses of 0.375, 0.75, and 1.5 cm) of biochar effectively reduced NH_3 with a 13–23% decrease, while no significant reduction was found for H_2S. The significant removal of NH_3 by biochar may be due to the formation of a semiporous shell on the manure surface, affecting mass transfer to the headspace, while the lack of significant effect on H_2S may be due to the inconsistent nature of swine manure H_2S emissions and the complex mixture of swine manure emissions that interferes with the adsorption of H_2S on biochar.

Thereafter, Papurello et al. (2018) used biochar from wood pyrolysis to achieve the effective removal of several odor components from biogas produced during anaerobic digestion, with Q_{max} value of 3.39, 272.01, 13.15 and 101.59 mg g^{-1} for H_2S, 2-butanone, hexamethylcyclotrisiloxane, and limonene, respectively, indicating biochar is more effective at removing 2-butanone, toluene, and limonene than sulfur and siloxane compounds. For the removal of VOSCs, Hwang et al. (2018) used oak, solid separated pig manure, coconut shells, and poultry litter as biochar feedstock and selected two pyrolysis temperatures (350 °C and 500 °C) to investigate their ability to adsorb the two most odor-causing VOCs (dimethyl disulfide (DMDS) and dimethyl trisulfide (DMTS)) from pig manure emissions. The best adsorption capacity was found in oak biochar at under 500 °C pyrolysis due to the relatively higher SSA compared with 350 °C. It could be concluded that SSA determined the adsorption of odors. However, the high SSA of livestock-manure biochar did not significantly affect the adsorption of DMDS and DMTS, and the effect was not as good as that of the low SSA of plant-biomass biochar.

In addition to pig manure odor control, biochar can be used as an adsorbent to remove cat urine odors for cat litter production and has significant commercial potential. The components of cat urine identified as causing unpleasant odors are mainly VOSCs, including 3-mercapto-3-methyl-1-butanol (MMB), 3-mercapto-3-methylbutyl-formate, 3-methyl-3-methylthio-1- butanol and 3-methyl-3-methylthio-1-butanol. butanol and 3-methyl-3 -(2-methyldisul-fanyl)-1-butanol (Miyazaki et al., 2006). In a study by Vaughn et al. (2020), a litter formulation containing a combination of 10% Eastern red cedar (Juniperus virginiana L.; ERC) wood fibers and biochar was found to have the best adsorption effect on MMB, reducing it by 59.3% owing to the high SSA and porosity of biochar derived from hardwoods. To obtain better performance for removing VOSCs, different modification methods were conducted to produce biochar with desired textural properties and desired functional group. For instance, sugarcane bagasse-based biochar was prepared using ultrasonic-assisted HNO_3 and H_3PO_4 impregnation modification, and a higher (99.87%) removal of MMB was achieved due to its high specific surface areas and high concentrations of functional groups, particularly C=O, –OH, –COOH, PdN, and PyN (S. Wang et al., 2022).

In addition to the domestic application for odor from biogas, livestock manure, and cat urine, biochar has been applied to remove the odors from vehicle emissions (e.g., methyl tert-butyl ether (MTBE), an odor from incomplete combustion of gasoline). Sugarcane bagasse biochar has also been applied to the removal of MTBE, while bone biochar had the lowest (3.51 mg g^{-1}). The reason is that sugarcane bagasse-based biochar has CO and C=O functional groups that can be effectively combined with MTBE and has less ash content that reduce the inhibition of ash content on the binding of materials to MTBE molecules, indicating that the functional groups on the material and the ash content affect the removal of VOCs (Pongkua et al., 2018).

The adsorption of odorous pollutants by biochar is divided into physical and chemical adsorption processes, so the biochar's physical and chemical properties affect the adsorption process. From the above cases, it can seen the physical structure characteristics of biochar do not fully determine the adsorption capacity. The chemical properties of the adsorbent, such as the surface functional groups and ash content of the material, play an important role in gas adsorption simultaneously (Pongkua et al., 2018). The excellent structural properties of biochar provide a good basis for odor adsorption, and the selection and modification of chemical properties of biochar according to the properties of odor can improve its upper limit of adsorption, especially for VOSCs, which are more difficult to be removed by unmodified biochar. Table 15.3 demonstrates the adsorption of different types of biochar for various odor pollutants and their specific surface area (SSA).

15.4.2 Biochar as Additives

Composting is a widely used method for converting organic waste into a nontoxic, pathogen-free stable product that can be used as a suitable fertilizer and soil conditioner to support plant growth (Toledo et al., 2020). Manure from poultry and livestock farming contains a large amount of trace elements such as nitrogen, phosphorus, and potassium, which improves its suitability as an excellent organic fertilizer and has numerous advantages as a compostable substrate (Keck et al., 2018). However, the high temperature conditions (> 45 °C), high pH (pH >7.5), low C/N ratio, and high humidity in the compost pile favor the volatilization of NH_3 and, due to the low C/N ratio and also the high humidity conditions (Gil et al., 2018; Gonzalez et al., 2019), some VOCs and sulfur-containing compounds with unpleasant odors are produced, emitting strong odors of about 25,000 ou E /m^3 (Keck et al., 2018). Therefore, composting of this waste requires the addition of fillers aimed at providing appropriate physical, chemical and biological properties, aerated porosity and a proper C/N ratio (range: 20–40) for the good operation of the process (Duan et al., 2019). The adsorption of odor by biochar by

Table 15.3 Adsorption of different types of biochar for various odor pollutants and specific surface area of biochar.

Biochar feedstocks	Target pollutants	C_{in}^a (mg mL^{-1})	Q_{max}^b (mg g^{-1})	Removal efficiency (%)	Specific surface area (SSA) (m^{2}g^{-1})	Reference
Oak	DMDS	18.9	0.448		9.8	(Hwang et al., 2018)
		202	5.748			
	DMTS	4.8	0.107			
		61.9	1.427			
Swine manure	DMDS	6.8	0.150		164.3	
		275	1.312			
	DMTS	1.1	0.0109			
		1.2	0.034			
ERC flakes	MMB	1		59.3	484	(Vaughn et al., 2020)
Sugarcane bagasse	MMB	1.09	204.48	94.02	1603.15	(S. Wang et al., 2022)
			191.48	87.99	1414.58	
			132.88	61.06	898.63	
Sugarcane bagasse	MTBE	0.25	8.26	100	169.13	(Pongkua et al., 2018)
Sawdust			6.99	89	174.98	
Bamboo			6.13	75	61.18	
Corncob			4.52	72	160.74	
Rice husk			4.16	63	43.36	
Bone			3.51	61	127.99	
Wood	H$_2$S		3.39		75	(Papurello et al., 2018)
	2-butanone		272.01			
	Hexamethyl cyclotrisiloxane		13.15			
	Limonene		101.59			
Pistachio shells	Toluene		169.9	92	983.35	(Cheng et al., 2022)
	Ethyl acetate		96.77	87		
	Toluene		123	64	758.25	
	Ethyl acetate		66.6	52		
Fruit waste	hexane	0.0001		100	8.3	(Kaikiti et al., 2021)
	DMTS	0.00001	0.021	100		
	p-cresol	0.00001	0.019	100		

[a] C_{in}: The initial VOC concentration.
[b] Q_{max}: Maximum adsorption capacity.

virtue of its high adsorption capacity has been widely reported, but beyond that, biochar can be used as an additive or bulking agent in the composting process of organic waste such as poultry manure (Sánchez-Monedero et al., 2019). However, the main reason for achieving odor control in this process is not the adsorption capacity of biochar, but the improvement of aeration in the compost pile by virtue of its good nitrogen fixation capacity, which promotes the organic matter humification process (Toledo et al., 2020).

The addition of 3% holm oak-derived biochar to a mixture of poultry manure and barley straw for composting has been shown to lead to a reduction in the total amount of VOCs produced during the thermophilic and thermophilic phases, such as phenols, ketones and volatile fatty acids (VFA) that affect odor perception in livestock operations (Sánchez-Monedero et al., 2019). This is because the thermophilic phase is when the release of VOCs is the largest. For VFA, Duan et al. (2019) found that bamboo biochar added at 10% mixed with wheat straw and chicken manure compost was the most practical to suppress their odor. And the application of biochar significantly enhanced the degree of humification, thus enhancing the odor elimination effect. In the study by Li et al. (2022), biochar made from charcoal was added to the compost mixture of peanut straw and chicken manure, and the amount of biochar added was chosen to be 10%. Compared with the control group without biochar, the addition of biochar reduced NH_3, H_2S, and total VOCs (TVOC) by 20.04%, 16.18%, and 17.55%, respectively. The effective reduction of NH_3 volatilization by biochar is based on the decomposition and adsorption characteristics of biochar. At the same time, the alkaline biochar has a high dispersion of catalytic centers in the pore system, which is conducive to the effective oxidation of H_2S, and the removal of TVOC is through the surface adsorption and distribution mechanism. Table 15.4 shows the biochar used as additives for composting.

The amount of biochar additive is important to the composting process. According to studies, the addition of 10% biochar seems to be most suitable for optimal composting performance (Duan et al., 2019; Liu et al., 2021). And the use of biochar above 10% is discouraged because excessive additions can lead to severe water loss and heat dissipation, which adversely affects the composting process (Liu et al., 2017). The use of excessive amounts of biochar is also costly (Akdeniz, 2019). Biochar can adsorb odors from compost piles due to its original large specific surface area, but biochar can rapidly saturate in composting systems that produce high concentrations of odors, which means that the adsorption effect is limited (Hwang et al., 2018). Moreover, the high moisture, temperature and organic content of the compost pile are not conducive to the adsorption of gaseous pollutants on the surface of the biochar, which may also limit its adsorption capacity (Zhang et al., 2017b). Therefore, the mechanism of biochar application as a compost additive should be based on a combination of factors.

Table 15.4 Biochar for composting additives.

Substrates	Biochar feedstock	Addition rate	Composting time	Impact on odor control	References
Chicken manure and straw	charcoal	10%	40 days	NH_3, H_2S, and TVOC emissions decreased by 20.04%, 16.18%, and 17.55%, respectively.	(Li et al., 2022)
Poultry manure and barley straw	Holm oak	3%	20 weeks	The addition of biochar resulted in a reduction of TVOC production in the thermophilic and thermophilic phases, while no effect was observed in the mature compost.	(Sánchez-Monedero et al., 2019)
Chicken manure and wheat straw	Bamboo	0%, 2%, 4%, 6%, 8% and 10%	42 days	The amount of volatile fatty acids and odor production decreases with the increase of biochar addition.	(Duan et al., 2019)
Pig manure and apple sawdust		5%	64 days	The NH_3 emissions from the compost with biochar were lower than other treatments, probably because NH_3 was absorbed by the acidic functional groups on the surface of biochar.	(X. Wang et al., 2022)
Poultry manure and wheat straw	Bamboo	0%, 2%, 4%, 6%, 8% and 10%	42 days	NH_3 volatilization rate and cumulative value decreased significantly with the increase of biochar concentration.	(Awasthi et al., 2020a)
Pig manure and sawdust	Bamboo	5%	36 days	Additives of biochar led to lower emission of NH_3.	(Mao et al., 2019)
Cow manure and wheat straw	Wood and wheat straw	12%	42 days	The addition of biochar suppresses NH3 emissions and reduces odor pollution.	(Awasthi et al., 2020b)
Chopped straw, wheat bran, rotten vegetables, and contaminated soil	Corn cob	7%	45 days	The capacity of activated biochar reduces the effective nitrogen and retains more ammonia and nitrate ions.	(Ye et al., 2019)

With its porous structure and large specific surface area, biochar possesses a good water-retention capacity, which can provide a suitable habitat for microorganisms, especially as the larger pore size is more suitable for microbial colony colonization (Sanchez-Monedero et al., 2018). For example, the inhibition of NH_3 emissions from biochar is usually explained by two mechanisms: (1) adsorption of NH_3 and NH_4^+ on the surface of biochar, which leads to lower NH_3 losses, and (2) creation of favorable conditions for nitrifying bacteria, which convert ammonia to nitrate and thus retain nitrogen in the compost (Godlewska et al., 2017). In addition, biochar acts as an effective bulking agent, reduces the bulk density of the compost pile, helps to aerate the compost pile, inhibits the growth of anaerobic sites in the pile (Steiner et al., 2011), and also promotes the movement of microorganisms and improves the decomposition process (Behera and Samal, 2022; Chen et al., 2018). Moreover, the use of biochar improves the particle size distribution of the compost mixture, accelerates organic matter degradation and improves the physical properties of the compost pile, preventing the formation of agglomerates larger than 70 mm (Sánchez-García et al., 2015). Therefore, these favorable conditions in compost piles containing biochar may lead to a reduction in the production of VOCs, especially oxygenated volatile compounds produced in the absence of aerobic degradation processes (Sánchez-Monedero et al., 2019).

H_2S in compost is mainly produced through two pathways: anaerobic decomposition of proteins or other sulfur-containing organic matter and reduction of sulfate by organic compounds under anoxic conditions, meaning that anaerobic conditions, for example inside poorly aerated material masses, promote the formation of H_2S during composting. Therefore, biochar also has a role in reducing H_2S emissions from composting (Sanchez-Monedero et al., 2018).

15.5 Conclusion and Perspective

Due to its good properties, such as high surface area, high porosity, and rich surface functional groups, biochar has great potential in the control of gas pollutants. In terms of odor pollutant control, biochar can be used not only as an adsorbent to remove the gas after the odor is generated but also as an additive to suppress the odor during the composting process. Although biochar already has good physical and chemical properties, to achieve an ideal effect, biochar needs to be modified according to the type of odor pollutants to be treated. In the adsorption-based removal mechanism, the structural properties and surface chemistry of biochar are key factors. When biochar is used as a compost additive, in addition to the removal of generated odor by adsorption, it can improve aeration by reducing the bulk density of the compost, thus creating a favorable microbial environment, enhancing the nitrification process, reducing NH_3 emissions, as well as avoiding the formation of H_2S due to poor aeration within the material masses.

Future studies of biochar for odor control should focus on the following aspects:

1) *Biochar modification for better removal performance.* Adjusting the pore distribution of biochar, enhancing its chemical adsorption capacity through doping and loading, and overcoming the early saturation can be the potential practical approach for biochar to enhance the adsorption impact. Meanwhile, in terms of biochar for composting additives, reports on the use of modified biochar are relatively scarce, probably because they are costly.
2) *Removal performance and mechanisms of biochar when targeted as multi-component odor pollutants.* In-depth research is needed on the adsorption mechanism of multi-odor pollutants, how the odor components interact with each other, and how the use of these mechanisms can guide the preparation of biochar adsorbents.
3) *The economic and environmental impact.* Studies of biochar application should consider how to achieve balance among cost, environmental friendliness, and function.
4) *Size of study applications.* Studies based on large-scale applications are still lacking.
5) *Renewability.* The renewability and lifespan of biochar determine its utilization potential, but relevant data is relatively lacking under the industrial scale.
6) *Disposal of spent biochar.* The disposal of biochar after reaching the end of its life needs further consideration. An ideal approach is reusing this spent biochar as a soil conditioner to improve soil quality and crop yield, while the potential environmental impact should be double-checked before application.

References

Akdeniz, N., 2019. A systematic review of biochar use in animal waste composting. *Waste Management*, 88, 291–300. https://doi.org/10.1016/j.wasman.2019.03.054.

Awasthi, M.K., Duan, Y., Awasthi, S.K., Liu, T., Zhang, Z., 2020a. Influence of bamboo biochar on mitigating greenhouse gas emissions and nitrogen loss during poultry manure composting. *Bioresource Technology*, 303, 122952. https://doi.org/10.1016/j.biortech.2020.122952.

Awasthi, M.K., Duan, Y., Awasthi, S.K., Liu, T., Zhang, Z., 2020b. Effect of biochar and bacterial inoculum additions on cow dung composting. *Bioresource Technology*, 297, 122407. https://doi.org/10.1016/j.biortech.2019.122407.

Bagreev, A., Bashkova, S., Bandosz, T.J., 2002. Adsorption of SO_2 on activated carbons: The effect of nitrogen functionality and pore sizes. *Langmuir*, 18, 1257–1264. https://doi.org/10.1021/la011320e.

Behera, S., Samal, K., 2022. Sustainable approach to manage solid waste through biochar assisted composting. *Energy Nexus*, 7, 100121. https://doi.org/10.1016/j.nexus.2022.100121.

Blanco-Rodríguez, A., Camara, V.F., Campo, F., Becherán, L., Durán, A., Vieira, V.D., de Melo, H., Garcia-Ramirez, A.R., 2018. Development of an electronic nose to characterize odors emitted from different stages in a wastewater treatment plant. *Water Research*, 134, 92–100. https://doi.org/10.1016/j.watres.2018.01.067.

Braghiroli, F.L., Bouafif, H., Koubaa, A., 2019. Enhanced SO_2 adsorption and desorption on chemically and physically activated biochar made from wood residues. *Industrial Crops and Products*, 138, 111456. https://doi.org/10.1016/j.indcrop.2019.06.019.

Brancher, M., Griffiths, K.D., Franco, D., de Melo Lisboa, H., 2017. A review of odor impact criteria in selected countries around the world. *Chemosphere*, 168, 1531–1570. https://doi.org/10.1016/j.chemosphere.2016.11.160.

Cao, L., Zhang, X., Xu, Y., Xiang, W., Wang, R., Ding, F., Hong, P., Gao, B., 2022. Straw and wood based biochar for CO_2 capture: Adsorption performance and governing mechanisms. *Separation and Purification Technology*, 287, 120592. https://doi.org/10.1016/j.seppur.2022.120592.

Capelli, L., Bax, C., Diaz, C., Izquierdo, C., Arias, R., Salas Seoane, N., 2019. Review on odor pollution, odor measurement, abatement techniques. *D-NOSES*, H2020-SwafS-23-2017-789315.

Chen, B., Zhou, D., Zhu, L., 2008. Transitional adsorption and partition of nonpolar and polar aromatic contaminants by biochars of pine needles with different pyrolytic temperatures. *Environmental Science Technology*, 42, 5137–5143. https://doi.org/10.1021/es8002684.

Chen, H., Awasthi, M.K., Liu, T., Zhao, J., Ren, X., Wang, M., Duan, Y., Awasthi, S.K., Zhang, Z., 2018. Influence of clay as additive on greenhouse gases emission and maturity evaluation during chicken manure composting. *Bioresource Technology*, 266, 82–88. https://doi.org/10.1016/j.biortech.2018.06.073.

Chen, L., Jiang, X., Chen, W., Dai, Z., Wu, J., Ma, S., Jiang, W., 2022. H2O2-assisted self-template synthesis of N-doped biochar with interconnected mesopore for efficient H2S removal. *Separation and Purification Technology*, 297, 121410. https://doi.org/10.1016/j.seppur.2022.121410.

Cheng, H., Sun, Y., Wang, X., Zou, S., Ye, G., Huang, H., Ye, D., 2020. Hierarchical porous carbon fabricated from cellulose-degrading fungus modified rice husks: Ultrahigh surface area and impressive improvement in toluene adsorption. *Journal of Hazardous Materials*, 392, 122298. https://doi.org/10.1016/j.jhazmat.2020.122298.

Cheng, T., Li, J., Ma, X., Zhou, L., Wu, H., Yang, L., 2022. Alkylation modified pistachio shell-based biochar to promote the adsorption of VOCs in high humidity environment. *Environmental Pollution*, 295, 118714. https://doi.org/10.1016/j.envpol.2021.118714.

Conti, C., Guarino, M., Bacenetti, J., 2020. Measurements techniques and models to assess odor annoyance: A review. *Environment International*, 134, 105261. https://doi.org/10.1016/j.envint.2019.105261.

Cui, S., Zhao, Y., Liu, Y., Huang, R., Pan, J., 2021. Preparation of straw porous biochars by microwave-assisted KOH activation for removal of gaseous H_2S. *Energy Fuels*, 35, 18592–18603. https://doi.org/10.1021/acs.energyfuels.1c02241.

Das, J., Rene, E.R., Dupont, C., Dufourny, A., Blin, J., van Hullebusch, E.D., 2019. Performance of a compost and biochar packed biofilter for gas-phase hydrogen sulfide removal. *Bioresource Technology*, 273, 581–591. https://doi.org/10.1016/j.biortech.2018.11.052.

Dou, Z., Chen, H., Liu, Y., Huang, R., Pan, J., 2022. Removal of gaseous H_2S using microalgae porous carbons synthesized by thermal/microwave KOH activation. *Journal of the Energy Institute*, 101, 45–55. https://doi.org/10.1016/j.joei.2021.12.007.

Du, W., Lü, F., Zhang, H., Shao, L., He, P., 2023. Odor emission rate of a municipal solid waste sanitary landfill during different operation stages before final closure. *Science of The Total Environment*, 856, 159111. https://doi.org/10.1016/j.scitotenv.2022.159111.

Duan, Y., Awasthi, S.K., Liu, T., Zhang, Z., Awasthi, M.K., 2019. Response of bamboo biochar amendment on volatile fatty acids accumulation reduction and humification during chicken manure composting. *Bioresource Technology*, 291, 121845. https://doi.org/10.1016/j.biortech.2019.121845.

Feng, D., Guo, D., Zhang, Y., Sun, S., Zhao, Y., Chang, G., Guo, Q., Qin, Y., 2021a. Adsorption-enrichment characterization of CO_2 and dynamic retention of free NH_3 in functionalized biochar with $H_2O/NH_3 \cdot H_2O$ activation for promotion of new ammonia-based carbon capture. *Chemical Engineering Journal*, 409, 128193. https://doi.org/10.1016/j.cej.2020.128193.

Feng, D., Guo, D., Zhang, Y., Sun, S., Zhao, Y., Shang, Q., Sun, H., Wu, J., Tan, H., 2021b. Functionalized construction of biochar with hierarchical pore structures and surface O-/N-containing groups for phenol adsorption. *Chemical Engineering Journal*, 410, 127707. https://doi.org/10.1016/j.cej.2020.127707.

Gallego, E., Roca, F.J., Perales, J.F., Sánchez, G., Esplugas, P., 2012. Characterization and determination of the odorous charge in the indoor air of a waste treatment facility through the evaluation of volatile organic compounds (VOCs) using TD–GC/MS. *Waste Management*, 32, 2469–2481. https://doi.org/10.1016/j.wasman.2012.07.010.

Gan, F., Cheng, B., Jin, Z., Dai, Z., Wang, B., Yang, L., Jiang, X., 2021. Hierarchical porous biochar from plant-based biomass through selectively removing lignin carbon from biochar for enhanced removal of toluene. *Chemosphere*, 279, 130514. https://doi.org/10.1016/j.chemosphere.2021.130514.

Gao, X., Yang, F., Cheng, J., Xu, Z., Zang, B., Li, G., Xie, X., Luo, W., 2022. Emission of volatile sulphur compounds during swine manure composting: Source identification, odor mitigation and assessment. *Waste Management*, 153, 129–137. https://doi.org/10.1016/j.wasman.2022.08.029.

Gil, A., Toledo, M., Siles, J.A., Martín, M.A., 2018. Multivariate analysis and biodegradability test to evaluate different organic wastes for biological treatments: Anaerobic co-digestion and co-composting. *Waste Management*, 78, 819–828. https://doi.org/10.1016/j.wasman.2018.06.052.

Godlewska, P., Schmidt, H.P., Ok, Y.S., Oleszczuk, P., 2017. Biochar for composting improvement and contaminants reduction. A review. *Bioresource Technology*, Special Issue on Biochar: Production, Characterization and Applications – Beyond Soil Applications. 246, 193–202. https://doi.org/10.1016/j.biortech.2017.07.095.

González, D., Colón, J., Sánchez, A., Gabriel, D., 2019. A systematic study on the VOCs characterization and odourodor emissions in a full-scale sewage sludge composting plant. *Journal of Hazardous Materials*, 373, 733–740. https://doi.org/10.1016/j.jhazmat.2019.03.131.

Hawko, C., Verriele, M., Hucher, N., Crunaire, S., Leger, C., Locoge, N., Savary, G., 2021. A review of environmental odor quantification and qualification methods: The question of objectivity in sensory analysis. *Science of The Total Environment*, 795, 148862. https://doi.org/10.1016/j.scitotenv.2021.148862.

He, M., Xu, Z., Sun, Y., Chan, P.S., Lui, I., Tsang, D.C.W., 2021. Critical impacts of pyrolysis conditions and activation methods on application-oriented production of wood waste-derived biochar. *Bioresource Technology*, 341, 125811. https://doi.org/10.1016/j.biortech.2021.125811.

Hossein Tehrani, N.H.M., Alivand, M.S., Rashidi, A., Rahbar Shamskar, K., Samipoorgiri, M., Esrafili, M.D., Mohammady Maklavany, D., Shafiei-Alavijeh, M., 2020. Preparation and characterization of a new waste-derived mesoporous carbon structure for ultrahigh adsorption of benzene and toluene at ambient conditions. *Journal of Hazardous Materials*, 384, 121317. https://doi.org/10.1016/j.jhazmat.2019.121317.

Hu, R., Liu, G., Zhang, H., Xue, H., Wang, X., Lam, P.K.S., 2020. Odor pollution due to industrial emission of volatile organic compounds: A case study in Hefei, China. *Journal of Cleaner Production*, 246, 119075. https://doi.org/10.1016/j.jclepro.2019.119075.

Hwang, O., Lee, S.-R., Cho, S., Ro, K.S., Spiehs, M., Woodbury, B., Silva, P.J., Han, D.-W., Choi, H., Kim, K.-Y., Jung, M.-W., 2018. Efficacy of Different Biochars in Removing Odorous Volatile Organic Compounds (VOCs) Emitted from Swine

Manure. *ACS Sustainable Chemistry & Engineering*, 6, 14239–14247. https://doi.org/10.1021/acssuschemeng.8b02881.

Iberahim, N., Sethupathi, S., Bashir, M.J.K., 2018. Optimization of palm oil mill sludge biochar preparation for sulfur dioxide removal. *Environmental Science and Pollution Research*, 25, 25702–25714. https://doi.org/10.1007/s11356-017-9180-5.

Iberahim, N., Sethupathi, S., Bashir, M.J.K., Kanthasamy, R., Ahmad, T., 2022. Evaluation of oil palm fiber biochar and activated biochar for sulphur dioxide adsorption. *Science of The Total Environment*, 805, 150421. https://doi.org/10.1016/j.scitotenv.2021.150421.

Im, E., Seo, H.J., Kim, D.I., Hyun, D.C., Moon, G.D., 2021. Bimodally-porous alumina with tunable mesopore and macropore for efficient organic adsorbents. *Chemical Engineering Journal*, 416, 129147. https://doi.org/10.1016/j.cej.2021.129147.

Ji, L., Wan, Y., Zheng, S., Zhu, D., 2011. Adsorption of tetracycline and sulfamethoxazole on crop residue-derived ashes: Implication for the relative importance of black carbon to soil sorption. *Environmental Science & Technology*, 45, 5580–5586. https://doi.org/10.1021/es200483b.

Kaikiti, K., Stylianou, M., Agapiou, A., 2021. Development of food-origin biochars for the adsorption of selected volatile organic compounds (VOCs) for environmental matrices. *Bioresource Technology*, 342, 125881. https://doi.org/10.1016/j.biortech.2021.125881.

Keck, M., Mager, K., Weber, K., Keller, M., Frei, M., Steiner, B., Schrade, S., 2018. Odor impact from farms with animal husbandry and biogas facilities. *Science of The Total Environment*, 645, 1432–1443. https://doi.org/10.1016/j.scitotenv.2018.07.182.

Lebrero, R., Bouchy, L., Stuetz, R., Muñoz, R., 2011. Odor assessment and management in wastewater treatment plants: A review. *Critical Reviews in Environmental Science and Technology*, 41, 915–950. https://doi.org/10.1080/10643380903300000.

Le-Minh, N., Sivret, E.C., Shammay, A., Stuetz, R.M., 2018. Factors affecting the adsorption of gaseous environmental odors by activated carbon: A critical review. *Critical Reviews in Environmental Science and Technology*, 48, 341–375. https://doi.org/10.1080/10643389.2018.1460984.

Li, L., Liu, S., Liu, J., 2011. Surface modification of coconut shell based activated carbon for the improvement of hydrophobic VOC removal. *Journal of Hazardous Materials*, 192, 683–690. https://doi.org/10.1016/j.jhazmat.2011.05.069.

Li, X., Zhang, L., Yang, Z., Wang, P., Yan, Y., Ran, J., 2020. Adsorption materials for volatile organic compounds (VOCs) and the key factors for VOCs adsorption process: A review. *Separation and Purification Technology*, 235, 116213. https://doi.org/10.1016/j.seppur.2019.116213.

Li, Y., Ma, J., Yong, X., Luo, L., Wong, J.W.C., Zhang, Y., Wu, H., Zhou, J., 2022. Effect of biochar combined with a biotrickling filter on deodorization, nitrogen retention, and microbial community succession during chicken manure composting.

Bioresource Technology, 343, 126137. https://doi.org/10.1016/j.biortech.2021.126137.

Lim, J.-H., Cha, J.-S., Kong, B.-J., Baek, S.-H., 2018. Characterization of odorous gases at landfill site and in surrounding areas. *Journal of Environmental Management*, 206, 291–303. https://doi.org/10.1016/j.jenvman.2017.10.045.

Liu, H., Guo, H., Guo, X., Wu, S., 2021. Probing changes in humus chemical characteristics in response to biochar addition and varying bulking agents during composting: A holistic multi-evidence-based approach. *Journal of Environmental Management*, 300, 113736. https://doi.org/10.1016/j.jenvman.2021.113736.

Liu, N., Zhou, J., Han, L., Ma, S., Sun, X., Huang, G., 2017. Role and multi-scale characterization of bamboo biochar during poultry manure aerobic composting. *Bioresource Technology*, 241, 190–199. https://doi.org/10.1016/j.biortech.2017.03.144.

Liu, Y., Lu, W., Wang, H., Gao, X., Huang, Q., 2019. Improved impact assessment of odorous compounds from landfills using Monte Carlo simulation. *Science of The Total Environment*, 648, 805–810. https://doi.org/10.1016/j.scitotenv.2018.08.213.

Ma, Q., Chen, W., Jin, Z., Chen, L., Zhou, Q., Jiang, X., 2021. One-step synthesis of microporous nitrogen-doped biochar for efficient removal of CO_2 and H_2S. *Fuel*, 289, 119932. https://doi.org/10.1016/j.fuel.2020.119932.

Mao, H., Zhang, H., Fu, Q., Zhong, M., Li, R., Zhai, B., Wang, Z., Zhou, L., 2019. Effects of four additives in pig manure composting on greenhouse gas emission reduction and bacterial community change. *Bioresource Technology*, 292, 121896. https://doi.org/10.1016/j.biortech.2019.121896.

Marrakchi, F., Ahmed, M.J., Khanday, W.A., Asif, M., Hameed, B.H., 2017. Mesoporous-activated carbon prepared from chitosan flakes via single-step sodium hydroxide activation for the adsorption of methylene blue. *International Journal of Biological Macromolecules*, 98, 233–239. https://doi.org/10.1016/j.ijbiomac.2017.01.119.

Maurer, D.L., Koziel, J.A., Kalus, K., Andersen, D.S., Opalinski, S., 2017. Pilot-scale testing of non-activated biochar for swine manure treatment and mitigation of ammonia, hydrogen sulfide, odorous volatile organic compounds (VOCs), and greenhouse gas emissions. *Sustainability*, 9, 929. https://doi.org/10.3390/su9060929.

Maziarka, P., Wurzer, C., Arauzo, P.J., Dieguez-Alonso, A., Mašek, O., Ronsse, F., 2021. Do you BET on routine? The reliability of N_2 physisorption for the quantitative assessment of biochar's surface area. *Chemical Engineering Journal*, 418, 129234. https://doi.org/10.1016/j.cej.2021.129234.

Ministry of Ecology and Environment of the People's Republic of China. 2018. Letter on soliciting opinions on the national environmental protection standard "Emission Standard of Odor Pollutants (Draft for Comment)". https://www.mee.gov.cn (accessed 11.19.22).

Ministry of the Environment, Government of Japan. 2006. *The offensive odor control law* | URL https://www.env.go.jp/en/laws/air/offensive_odor (accessed 11.17.22).

Miyazaki, M., Yamashita, T., Suzuki, Y., Saito, Y., Soeta, S., Taira, H., Suzuki, A., 2006. A major urinary protein of the domestic cat regulates the production of felinine, a putative pheromone precursor. *Chemistry & Biology*, 13, 1071–1079. https://doi.org/10.1016/j.chembiol.2006.08.013.

Nguyen, M.K., Lin, C., Hoang, H.G., Sanderson, P., Dang, B.T., Bui, X.T., Nguyen, N.S.H., Vo, D.-V.N., Tran, H.T., 2022. Evaluate the role of biochar during the organic waste composting process: A critical review. *Chemosphere*, 299, 134488. https://doi.org/10.1016/j.chemosphere.2022.134488.

Odor, 2022. *Wikipedia*. https://en.wikipedia.org/wiki/Odor.

Papurello, D., Boschetti, A., Silvestri, S., Khomenko, I., Biasioli, F., 2018. Real-time monitoring of removal of trace compounds with PTR-MS: Biochar experimental investigation. *Renewable Energy*, 125, 344–355. https://doi.org/10.1016/j.renene.2018.02.122.

Papurello, D., Lanzini, A., Bressan, M., Santarelli, M., 2020. H_2S removal with sorbent obtained from sewage sludges. *Processes*, 8, 130. https://doi.org/10.3390/pr8020130.

Piccardo, M.T., Geretto, M., Pulliero, A., Izzotti, A., 2022. Odor emissions: A public health concern for health risk perception. *Environmental Research*, 204, 112121. https://doi.org/10.1016/j.envres.2021.112121.

Pongkua, W., Dolphen, R., Thiravetyan, P., 2018. Effect of functional groups of biochars and their ash content on gaseous methyl tert-butyl ether removal. *Colloids and Surfaces A: Physicochemical and Engineering Aspects*, 558, 531–537. https://doi.org/10.1016/j.colsurfa.2018.09.018.

Premarathna, K.S.D., Rajapaksha, A.U., Sarkar, B., Kwon, E.E., Bhatnagar, A., Ok, Y.S., Vithanage, M., 2019. Biochar-based engineered composites for sorptive decontamination of water: A review. *Chemical Engineering Journal*, 372, 536–550. https://doi.org/10.1016/j.cej.2019.04.097.

Raymundo-Piñero, E., Cazorla-Amorós, D., Salinas-Martinez de Lecea, C., Linares-Solano, A., 2000. Factors controlling the SO_2 removal by porous carbons: relevance of the SO2 oxidation step. *Carbon*, 38, 335–344. https://doi.org/10.1016/S0008-6223(99)00109-8.

Ro, K.S., Lima, I.M., Reddy, G.B., Jackson, M.A., Gao, B., 2015. Removing Gaseous NH3 Using Biochar as an Adsorbent. *Agriculture*, 5, 991–1002. https://doi.org/10.3390/agriculture5040991.

Sadowska-Rociek, A., Kurdziel, M., Szczepaniec-Cięciak, E., Riesenmey, C., Vaillant, H., Batton-Hubert, M., Piejko, K., 2009. Analysis of odorous compounds at municipal landfill sites. *Waste Management and Research*, 27, 966–975. https://doi.org/10.1177/0734242X09334616.

Sánchez-García, M., Alburquerque, J.A., Sánchez-Monedero, M.A., Roig, A., Cayuela, M.L., 2015. Biochar accelerates organic matter degradation and enhances N mineralisation during composting of poultry manure without a relevant impact on gas emissions. *Bioresource Technology*, 192, 272–279. https://doi.org/10.1016/j.biortech.2015.05.003.

Sanchez-Monedero, M.A., Cayuela, M.L., Roig, A., Jindo, K., Mondini, C., Bolan, N., 2018. Role of biochar as an additive in organic waste composting. *Bioresource Technology*, 247, 1155–1164. https://doi.org/10.1016/j.biortech.2017.09.193.

Sánchez-Monedero, M.A., Sánchez-García, M., Alburquerque, J.A., Cayuela, M.L., 2019. Biochar reduces volatile organic compounds generated during chicken manure composting. *Bioresource Technology*, 288, 121584. https://doi.org/10.1016/j.biortech.2019.121584.

Sardar, M.F., Zhu, C., Geng, B., Ahmad, H.R., Song, T., Li, H., 2021. The fate of antibiotic resistance genes in cow manure composting: shaped by temperature-controlled composting stages. *Bioresource Technology*, 320, 124403. https://doi.org/10.1016/j.biortech.2020.124403.

Seoul Solution. 2010. https://seoulsolution.kr/en/envi (accessed 11.19.22).

Shen, S., Wang, Q., Chen, Y., Zuo, X., He, F., Fei, S., Xie, H., 2019. Effect of Landfill Odorous Gas on Surrounding Environment: A Field Investigation and Numerical Analysis in a Large-Scale Landfill, in: Hangzhou, China, Zhan, L., Chen, Y., Bouazza, A. (Eds.), *Proceedings of the 8th International Congress on Environmental Geotechnics Volume 2, Environmental Science and Engineering*. Springer, Singapore, pp. 51–59. https://doi.org/10.1007/978-981-13-2224-2_7.

Steiner, C., Melear, N., Harris, K., Das, K., 2011. Biochar as bulking agent for poultry litter composting. *Carbon Management*, 2, 227–230. https://doi.org/10.4155/cmt.11.15.

Su, L., Chen, M., Zhuo, G., Ji, R., Wang, S., Zhang, L., Zhang, M., Li, H., 2021. Comparison of biochar materials derived from coconut husks and various types of livestock manure, and their potential for use in removal of H_2S from biogas. *Sustainability*, 13, 6262. https://doi.org/10.3390/su13116262.

Sun, C., Wang, Z., Yang, Y., Wang, M., Jing, X., Li, G., Yan, J., Zhao, L., Nie, L., Wang, Y., Zhong, Y., Liu, Y., 2023. Characteristics, secondary transformation and odor activity evaluation of VOCs emitted from municipal solid waste incineration power plant. *Journal of Environmental Management*, 326, 116703. https://doi.org/10.1016/j.jenvman.2022.116703.

Sun, K., Kang, M., Ro, K.S., Libra, J.A., Zhao, Y., Xing, B., 2016. Variation in sorption of propiconazole with biochars: The effect of temperature, mineral, molecular structure, and nano-porosity. *Chemosphere*, 142, 56–63. https://doi.org/10.1016/j.chemosphere.2015.07.018.

Sun, K., Kang, M., Zhang, Z., Jin, J., Wang, Z., Pan, Z., Xu, D., Wu, F., Xing, B., 2013. Impact of deashing treatment on biochar structural properties and potential sorption mechanisms of phenanthrene. *Environmental Science & Technology*, 47, 11473–11481. https://doi.org/10.1021/es4026744.

Surra, E., Costa Nogueira, M., Bernardo, M., Lapa, N., Esteves, I., Fonseca, I., 2019. New adsorbents from maize cob wastes and anaerobic digestate for H_2S removal from biogas. *Waste Management*, 94, 136–145. https://doi.org/10.1016/j.wasman.2019.05.048.

Talaiekhozani, A., Bagheri, M., Goli, A., Talaei Khoozani, M.R., 2016. An overview of principles of odor production, emission, and control methods in wastewater collection and treatment systems. *Journal of Environmental Management*, 170, 186–206. https://doi.org/10.1016/j.jenvman.2016.01.021.

Toledo, M., Gutiérrez, M.C., Peña, A., Siles, J.A., Martín, M.A., 2020. Co-composting of chicken manure, alperujo, olive leaves/pruning and cereal straw at full-scale: Compost quality assessment and odor emission. *Process Safety and Environmental Protection*, 139, 362–370. https://doi.org/10.1016/j.psep.2020.04.048.

Toledo, M., Gutiérrez, M.C., Siles, J.A., Martín, M.A., 2019. Odor mapping of an urban waste management plant: Chemometric approach and correlation between physico-chemical, respirometric and olfactometric variables. *Journal of Cleaner Production*, 210, 1098–1108. https://doi.org/10.1016/j.jclepro.2018.11.109.

Vaughn, S.F., Winkler-Moser, J.K., Berhow, M.A., Byars, J.A., Liu, S.X., Jackson, M.A., Peterson, S.C., Eller, F.J., 2020. An odor-reducing, low dust-forming, clumping cat litter produced from Eastern red cedar (Juniperus virginiana L.) wood fibers and biochar1. *Industrial Crops and Products*, 147, 112224. https://doi.org/10.1016/j.indcrop.2020.112224.

Vikrant, K., Kim, K.-H., Peng, W., Ge, S., Sik Ok, Y., 2020. Adsorption performance of standard biochar materials against volatile organic compounds in air: A case study using benzene and methyl ethyl ketone. *Chemical Engineering Journal*, 387, 123943. https://doi.org/10.1016/j.cej.2019.123943.

Wang, N., Huang, D., Bai, X., Lin, Y., Miao, Q., Shao, M., Xu, Q., 2022. Mechanism of digestate-derived biochar on odorous gas emissions and humification in composting of digestate from food waste. *Journal of Hazardous Materials*, 434, 128878. https://doi.org/10.1016/j.jhazmat.2022.128878.

Wang, S., Xie, C., Wang, Shupei, Hang, F., Li, W., Li, K., Mann, A., Sarina, S., Doherty, W., Shi, C., 2022. Facile ultrasonic-assisted one-step preparation of sugarcane bagasse carbon sorbent for bio-based odor removal cat litter formulation. *Industrial Crops and Products*, 187, 115493. https://doi.org/10.1016/j.indcrop.2022.115493.

Wang, X., Liu, X., Wang, Z., Sun, G., Li, J., 2022. Greenhouse gas reduction and nitrogen conservation during manure composting by combining biochar with wood vinegar. *Journal of Environmental Management*, 324, 116349. https://doi.org/10.1016/j.jenvman.2022.116349.

Wang, Z., Han, L., Sun, K., Jin, J., Ro, K.S., Libra, J.A., Liu, X., Xing, B., 2016. Sorption of four hydrophobic organic contaminants by biochars derived from maize straw, wood dust and swine manure at different pyrolytic temperatures. *Chemosphere*, 144, 285–291. https://doi.org/10.1016/j.chemosphere.2015.08.042.

Xiao, J., Zhang, Y., Zhang, T.C., Yuan, S., 2023. Prussian blue-impregnated waste pomelo peels-derived biochar for enhanced adsorption of NH_3. *Journal of Cleaner Production*, 382, 135393. https://doi.org/10.1016/j.jclepro.2022.135393.

Xu, X., Cao, X., Zhao, L., Sun, T., 2014. Comparison of sewage sludge- and pig manure-derived biochars for hydrogen sulfide removal. *Chemosphere*, 111, 296–303. https://doi.org/10.1016/j.chemosphere.2014.04.014.

Xu, X., Huang, D., Zhao, L., Kan, Y., Cao, X., 2016. Role of inherent inorganic constituents in SO_2 sorption ability of biochars derived from three biomass wastes. *Environmental Science & Technology*, 50, 12957–12965. https://doi.org/10.1021/acs.est.6b03077.

Xu, X., Zhao, Y., Sima, J., Zhao, L., Mašek, O., Cao, X., 2017. Indispensable role of biochar-inherent mineral constituents in its environmental applications: A review. *Bioresource Technology*, 241, 887–899. https://doi.org/10.1016/j.biortech.2017.06.023.

Xu, Z., He, M., Xu, X., Cao, X., Tsang, D.C.W., 2021. Impacts of different activation processes on the carbon stability of biochar for oxidation resistance. *Bioresource Technology*, 338, 125555. https://doi.org/10.1016/j.biortech.2021.125555.

Yang, Y., Sun, C., Huang, Q., Yan, J., 2022. Hierarchical porous structure formation mechanism in food waste component derived N-doped biochar: Application in VOCs removal. *Chemosphere*, 291, 132702. https://doi.org/10.1016/j.chemosphere.2021.132702.

Ye, S., Zeng, G., Wu, H., Liang, J., Zhang, C., Dai, J., Xiong, W., Song, B., Wu, S., Yu, J., 2019. The effects of activated biochar addition on remediation efficiency of co-composting with contaminated wetland soil. *Resources, Conservation and Recycling*, 140, 278–285. https://doi.org/10.1016/j.resconrec.2018.10.004.

Ying, D., Chuanyu, C., Bin, H., Yueen, X., Xuejuan, Z., Yingxu, C., Weixiang, W., 2012. Characterization and control of odorous gases at a landfill site: A case study in Hangzhou, China. *Waste Management*, 32, 317–326. https://doi.org/10.1016/j.wasman.2011.07.016.

Zhang, L.-H., Calo, J.M., 2001. Thermal desorption methods for porosity characterization of carbons and chars. *Colloids and Surfaces A: Physicochemical and Engineering Aspects*, 187–188, 207–218. https://doi.org/10.1016/S0927-7757(01)00633-1.

Zhang, X., Gao, B., Creamer, A.E., Cao, C., Li, Y., 2017a. Adsorption of VOCs onto engineered carbon materials: A review. *Journal of Hazardous Materials*, 338, 102–123. https://doi.org/10.1016/j.jhazmat.2017.05.013.

Zhang, X., Gao, B., Zheng, Y., Hu, X., Creamer, A.E., Annable, M.D., Li, Y., 2017b. Biochar for volatile organic compound (VOC) removal: Sorption performance and governing mechanisms. *Bioresource Technology*, 245, 606–614. https://doi.org/10.1016/j.biortech.2017.09.025.

Zhang, X., Miao, X., Xiang, W., Zhang, J., Cao, C., Wang, H., Hu, X., Gao, B., 2021. Ball milling biochar with ammonia hydroxide or hydrogen peroxide enhances its adsorption of phenyl volatile organic compounds (VOCs). *Journal of Hazardous Materials*, 403, 123540. https://doi.org/10.1016/j.jhazmat.2020.123540.

Zhang, X., Zheng, H., Li, G., Gu, J., Shao, J., Zhang, S., Yang, H., Chen, H., 2021. Ammoniated and activated microporous biochar for enhancement of SO_2 adsorption. *Journal of Analytical and Applied Pyrolysis*, 156, 105119. https://doi.org/10.1016/j.jaap.2021.105119.

Zhang, Y., Ning, X., Li, Y., Wang, J., Cui, H., Meng, J., Teng, C., Wang, G., Shang, X., 2021. Impact assessment of odor nuisance, health risk and variation originating from the landfill surface. *Waste Management*, 126, 771–780. https://doi.org/10.1016/j.wasman.2021.03.055.

Zhang, Y., Xiao, J., Zhang, T.C., Ouyang, L., Yuan, S., 2022. Synthesis of $CuSiO_3$-loaded P-doped porous biochar derived from phytic acid-activated lemon peel for enhanced adsorption of NH_3. *Separation and Purification Technology*, 283, 120179. https://doi.org/10.1016/j.seppur.2021.120179.

Zhao, X., Zeng, X., Qin, Y., Li, X., Zhu, T., Tang, X., 2018. An experimental and theoretical study of the adsorption removal of toluene and chlorobenzene on coconut shell derived carbon. *Chemosphere*, 206, 285–292. https://doi.org/10.1016/j.chemosphere.2018.04.126.

Zhao, Y., Lu, W., Wang, H., 2015. Volatile trace compounds released from municipal solid waste at the transfer stage: Evaluation of environmental impacts and odor pollution. *Journal of Hazardous Materials*, 300, 695–701. https://doi.org/10.1016/j.jhazmat.2015.07.081.

Zhao, Z., Wang, B., Theng, B.K.G., Lee, X., Zhang, X., Chen, M., Xu, P., 2022. Removal performance, mechanisms, and influencing factors of biochar for air pollutants: a critical review. *Biochar*, 4, 30. https://doi.org/10.1007/s42773-022-00156-z.

Zhu, L., Shen, D., Luo, K.H., 2020. A critical review on VOCs adsorption by different porous materials: Species, mechanisms and modification methods. *Journal of Hazardous Materials*, 389, 122102. https://doi.org/10.1016/j.jhazmat.2020.122102.

Zhu, Y., Gao, Jihui, Li, Y., Sun, F., Gao, Jianmin, Wu, S., Qin, Y., 2012. Preparation of activated carbons for $SO2$ adsorption by $CO2$ and steam activation. *Journal of the Taiwan Institute of Chemical Engineers*, 43, 112–119. https://doi.org/10.1016/j.jtice.2011.06.009.

16

Fate, Transport, and Impact of Biochar in the Environment

Deng Pan[1], Yuqing Sun[2], and Daniel C.W. Tsang[3]

[1] EIT Institute for Advanced Study, Ningbo, China
[2] School of Agriculture, Sun Yat-Sen University, Guangzhou, Guangdong, China
[3] State Key Laboratory of Clean Energy Utilization, Zhejiang University, China

In the past decades, many reviews have systematically discussed the physical and chemical properties of different biochar and their environmental benefits. However, nowadays, when the application of biochar is driven to maturity, people increasingly need to characterize and classify materials more accurately to optimize the producing conditions and to understand their environmental behavior better. The stability of biochar in the environment, how it interacts with the environment, and what happens to nutrient elements and compounds will all be impacted by the change in biochar's properties. In addition, biochar's related environmental impact includes the release of pollutants, their toxicity to the organism, and their transport in the environment (Lian and Xing, 2017), which is an important concern before the widespread application of biochar in environmental practices. In this chapter, the transport mechanism of biochar in the environment is briefly introduced, and the stability and potential environmental impact of biochar are discussed.

16.1 Transport Mechanism of Biochar in the Environment

Biochar is conveyed through soil matrixes by decomposition, runoff, and infiltration in applications for soil supplements and wastewater treatment systems (Figure 16.1).

In the interfaces between soils/sediments and biochar or with aqueous phase interactions with biochar, biogeochemical cycle activities may take place (Odinga et al., 2020). Predominant processes include sorption/desorption, ecological element change, and interactions with plant roots, microbes, and soil organic matter and minerals. As a result, soluble biochar's organic molecules can participate in a variety of biological and chemical redox events as well as metal complexation reactions in both soils and the rhizosphere (Gomez-Eyles et al., 2013). These organic substances, therefore, have an impact on the vital biogeochemical processes along the soil-microbe-plant continuum that has been outlined above, including microbial ecology and function, nutrient cycling, nutrient absorption by the roots, and free radical scavenging (Husson, 2013). These can all have a major influence on the environment, whether in a positive or negative way.

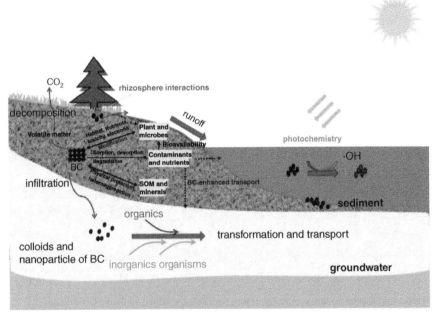

Figure 16.1 The transport mechanism of biochar in the environment (Adapted from Xia et al., 2022).

16.2 Stability of Biochar

16.2.1 Physical Degradation of Biochar

The physical decomposition of biochar is a significant but neglected process in the current biochar research. It is speculated that physical degradation will affect the life of biochar in the soil and its potential input to the river system (Spokas et al., 2014). The physical degradation of biochar occurs through a variety of mechanisms. Biochar is generally considered to be stronger than the original biomass mechanically, but structural fracture will occur under lower strain than the original biomass. In addition, this mechanical strength will decrease with aging (Théry-Parisot et al., 2010). When biochar is subjected to additional mechanical stress, fragments will eventually arise as a result of these structural defects (Gao and Wu, 2014). These smaller biochar fractions will have obvious environmental behavior and obstinacy relative to bulk biochar (Lian and Xing, 2017). Spokas et al. (2014) proposed that after exposure to drying and wetting cycles, the biochar with a high O/C ratio will have a rapid dissolution. The mechanism behind this is summarized as the adsorption of water, and steam will cause pressure on the physical structure of biochar. Due to the expansion of the exothermic graphite sheet, the adsorption of water and steam will cause pressure on the physical structure of biochar (Théry-Parisot et al., 2010).

However, though crucial for the stability, biochar's physical fragility in soil has not drawn as much scientific attention as the chemical and microbiological degradation. The dynamics and internal mechanism of biochar physical decomposition in soil are largely unknown. At the same time, the interaction between biochar particles and other soil fractions, such as NOM and other minerals, need to be further studied (Lian and Xing, 2017).

16.2.2 Chemical Decomposition of Biochar

Chemical decomposition is an important mechanism of biochar degradation in the environment, mostly consisting of chemical oxidation and carbonate dissolution. One of the major ways that biochar releases inorganic carbon is by carbonate dissolution (Jones et al., 2011). In addition, chemical oxidation is a significant mechanism to regulate the chemical decomposition (Zimmerman, 2010). According to Wang et al. (2016), the oxidation process can increase the oxygen and hydrogen concentration on the surface of biochar, encourage the creation of oxygen-containing groups (such as hydroxyl, carboxyl, and carbon oxygen double bond), and alter the amphoteric characteristics, all of which can hasten biochar's chemical decomposition. Oxidation is also conducive to microbial metabolism of highly aromatic biochar, for example, by making biochar less hydrophobic and encouraging it to break up into smaller pieces, which is better for microbial accessibility (Sand, 1997).

The molecular structure of the biochar and the surrounding environment has a significant impact on its chemical stability. It is thought that the aromaticity and level of aromatic carbon condensation in biochar are directly connected to its biochemical recalcitrance (Wiedemeier et al., 2015). According to Nguyen et al. (2010), the carbon loss and possible cation exchange capacity of biochar were highly associated with its O/C ratio. It is recommended that biochar with an O/C molar ratio below 0.2 can have a half-life of at least 1000 years in the environment (Spokas, 2010).

Other biochar properties, such as the size, configuration, and heteroatom of the polyaromatic ring, also have a great impact on its oxidation resistance. For instance, the presence of heteroatoms and the arrangement of polyaromatic rings will influence the production and activity of free radicals on the carbon surface, which will have a substantial effect on how quickly biochar decomposes. (Duan et al., 2015; Nia et al., 2017).

It should be emphasized that soil and environmental factors including temperature, natural organic matter (NOM), and minerals have an impact on the chemical stability of biochar. Under the alternating saturated and unsaturated situation, the degradation of biochar will be accelerated because carboxylic and OH functional groups are increased, and aliphatic groups are decreased (Nguyen and Lehmann, 2009). The relatively high temperature will also significantly accelerate the decomposition of biochar in the environment by accelerating the oxidation reaction (Fang et al., 2015). The effects of NOM and minerals on biochar stability are more complex because the ternary system involves multiple mechanisms. Biochar is an effective NOM adsorbent because of its porous nature, particularly for smaller aliphatic compounds. The lower adsorption of larger aromatic organic compounds mostly results from the size exclusion effect. (Smebye et al., 2016). In addition to hydrophobic adsorption and size exclusion, a few recently revealed processes, include EDA contact and negatively charged-assisted hydrogen bonding, may also contribute to the combination of NOM and biochar (Lian and Xing, 2017). After adsorption, NOM enhances the stability of biochar by preventing the enzyme from entering the surface and the oxidation and dissolution process (Wang et al., 2016). Soil minerals can also combine with biochar surface in a short period of time, and it is recommended to form a complex to increase biochar stability (Yang et al., 2016).

16.2.3 Microbial Decomposition of Biochar

Microbial degradation is also an important way of biochar decomposition. It mainly has two stages. In the first stage, volatile components in biochar are rapidly degraded, so this stage is called the rapid degrading stage. The volatile components of biochar can rapidly provide sufficient energy for the community of microbial in the environment, and increase microbial activity in the soil (Singh et al., 2012). Therefore, in short-term incubations, adding biochar to soil can induce a positive priming effect and significantly increase the mineralization rate of soil organic carbon (Luo et al., 2011). Next is the slow-degrading stage, in which the aromatic ring structure is slowly degraded (Ameloot et al., 2013). It is worth noting that during the long-term slow, degrading stage, the organic matter can be adsorbed on the surface of the biochar and the micro soil aggregate so as to protect the biochar physically (Maestrini et al., 2015). Therefore, the degradation results under short-term incubations (less than two years) cannot completely summarize the microbial degradation of biochar in the environment (Spokas, 2010).

Biochar and microbial communities interact with each other. On the one hand, microbial communities affect the decomposition mechanism and dynamics of biochar in the environment; On the other hand, biochar may potentially affect the structure and activity of microbial communities (Tian et al., 2016). However, in different studies, biochar has different effects on microbial communities. Mitchell et al. (2015) reported that after biochar application, the proportion of bacteria/fungi in forest soil increased, but the proportion of Gram-negative/Gram-positive bacteria decreased, indicating that biochar changed the microbial community. Similar experimental results have also been obtained by Liao et al. (2016). Nevertheless, a six-year field experiment showed that biochar only increased the metabolic activity of microorganisms but could not change their community structure (Tian et al., 2016). Different physical and chemical properties of biochar, such as the stability of biochar, pore volume, surface area, and different nutrients, are the reasons for this result. More efforts are needed to further explore the relationship between biochar and microbial communities and other potential microbial decomposition mechanisms for biochar.

16.3 Contaminants in Biochar and the Environmental Impact

During the production of biochar, contaminants can be generated or enriched and finally remain in the biochar, such as polycyclic aromatic hydrocarbons (PAHs) (Odinga et al., 2021), volatile organic compounds (VOCs), heavy metal particles (HMs), dioxins, persistent free radicals (PFRs) (Odinga et al., 2020), and metal cyanide (MCN). Potential risks to the environment and human health will result from these substances. In general, contaminants' level in biochar highly depends on the raw materials and the pyrolysis temperature, but the proper design of the pyrolysis unit can significantly reduce the contaminant contant, so as to achieve safe biochar production. Therefore, it is essential to systematically evaluate the potential impact of contaminants in biochar on the environment.

16.3.1 Polycyclic Aromatic Hydrocarbons (PAHs)

During pyrolysis, polycyclic aromatic hydrocarbons (PAHs) are produced spontaneously as a result of the decomposition of biomass followed by recombination mechanisms (Ledesma et al., 2002). Despite the fact that pyrolysis can produce different kinds of PAHs, only 16 types of PAHs are typically reported for use in determining the risk of the biochar in the environment (De la Rosa et al., 2016; Stefaniuk et al., 2018). The processes of PAH synthesis are linked to a series of ring-building reactions (Ledesma et al., 2002); when the value of activation energy climbed from 50 to 110 kcal/mol, and ring counts grew from one to five rings while PAH yields declined. At temperatures below 500 °C, PAHs with low molecular weight (2-ring or 3-ring) are produced (Wang et al., 2017). PAHs with high molecular weight (5-ring or 6 rings) are generated during condensation at temperatures over 500 °C

The amount of PAHs in biochar is influenced by variables such as the type of the raw materials, the temperature, and the type of the reactor (José et al., 2019; Nguyen et al., 2019). The majority of studies show that PAH production increased when the temperature was increased, particularly above 700 °C, but the contrary was also shown (Keiluweit et al., 2012). Other research has found that the greatest concentration of PAHs in biochar occurs between 400 and 500 °C (Devi and Saroha, 2015). It must be remembered that PAHs are produced during pyrolysis, but they may also be broken down and evaporated. For instance, the production of biochar from sewage sludge reduces the PAH concentration of the sludge by 8–25 times (Zielińska and Oleszczuk, 2015).

Oleszczuk et al. (2013) assessed the PAH concentration in commercial biochar and its ecotoxicological qualities and came to the conclusion that there was no direct relationship between the toxicity and the PAH level. Rombolà et al. (2015) hypothesized that instead of PAHs, water-soluble elements, such as organic acids, are to blame for the elevated toxicity of biochar made by poultry litter. The interaction between PAHs and other organic pollutants in biochar makes it challenging to establish a direct link between PAH levels and the toxicity of biochar, even though research suggests that PAHs are only considered to be of minimal significance in terms of their deleterious effects (Buss et al., 2015).

The fate of PAHs in biochar after the application is shown in Figure 16.2. Hale et al. (2012) mentioned that only 162 ng/L of water-extractable PAHs were found in biochar generated under different circumstances, suggesting a comparatively low level of bioavailable PAHs in the biochar. Biochar made from wood or wheat straw exhibited soil PAH values between 0.1 mg/kg and 1.5 mg/kg, which is under the limit of agricultural soil. (De la Rosa et al., 2016; Kuśmierz et al., 2016). The availability of PAHs was 42% lower in the commercial biochar-amended soil than in the control soil, according to long-term (851 days) field tests (Oleszczuk et al., 2016). In addition, microbial breakdown and leaching in biochar-amended soil

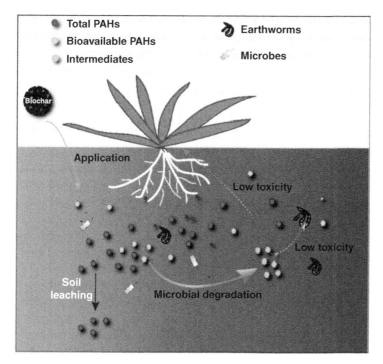

Figure 16.2 Fate of PAHs in biochar after application (Han et al., 2022 / with permission from Elsevier).

cause PAHs to be eliminated and reduced to background soil levels after 105 days (Kuśmierz et al., 2016). Therefore, the conclusion can be drawn that although biochar production will always result in the generation of PAHs, the risk to the environment is relatively low since the quantities of bioavailable PAHs are low.

16.3.2 Heavy Metals (HMs)

Another significant source of contamination in biochar is HMs, which include As, Cd, Cr, Co, Cu, Pb, Hg, Ni, Zn, and Mo (Shaheen et al., 2019). The majority of HMs produced during pyrolysis are contained in biochar, leading to higher HM levels than in the feedstock material (Shen et al., 2020). Most kinds of biochar made by plant biomass have relatively low HM levels. (Freddo et al., 2012). Nevertheless, some biochar made from marginal biomass, such as sewage sludge, included HM concentrations over the advised threshold values for soils and biochar (Buss et al., 2016b; Jin et al., 2016).

Many studies have shown that the level of HMs in biochar contributes to its toxicity. Buss et al. (2016a) reported that there was no correlation between the total or accessible HM levels in biochar and its phytotoxicity of it. This was supported by another study made by Stefaniuk et al. (2016), who found that biochar made from biogas residues had harmful effects on cress despite having lower amounts of Cr, Zn, and Cu than biochar made from pig dung, which had no effect on the germination index of cress-growing cabbage seeds.

After pyrolysis, available HM fractions are eventually converted into components that are stable and pose no threat to leaching or plant uptake (Buss et al., 2016a; Mendez et al., 2012; Zhang et al., 2020). When applied at usual rates, the produced biochar is less hazardous than the biomass that has not been pyrolyzed, even if threshold HM values are surpassed (Frišták et al., 2018; Roberts et al., 2017).

16.3.3 Persistent Free Radicals (PFRs)

Persistent free radicals (PFRs) refer to surface-stabilized metal-radical complexes that form incomplete combustion that is durable and stable in a variety of environmental matrices (Stephenson et al., 2016; Vejerano et al., 2018).

PFRs are numerous and pervasive in biochar, regardless of the type of primary components (Liao et al., 2014). The mechanism is shown in Figure 16.3. When cellulose and hemicellulose are pyrolyzed, their own monomers and monomeric radicals are created, which are then used to break down glucosidic linkages (Nzihou et al., 2013). The homolytic cleavage of the C–C and C–O links, as well as the production of radicals during lignin pyrolysis, leads to the formation of radical coupling products (Kibet et al., 2012). By transferring the electron to transition metals, abundant phenol or quinone moieties in biomass cause the synthesis of surface-bound PFRs in biochar (Fang et al., 2014). PFR content in biochar rises at higher pyrolysis temperatures (300–600 °C), and then sharply falls at 700 °C (Tao et al., 2020). Similar results were found in the manufacture of biochar made from rice husks; PFRs vanished as a result of high temperatures (700 °C), destroying the radical structures (Qin et al., 2017).

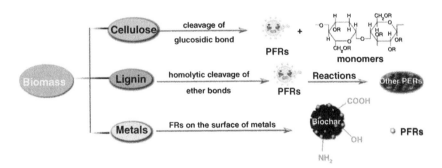

Figure 16.3 Formation mechanism of persistent free radicals (PFRs) in biochar (Han et al., 2022 / with permission from Elsevier).

Due to their prevalence throughout the pyrolysis processes, PFRs are considered a novel pollutant in biochar of considerable importance (Kiruri et al., 2014). Reactive oxygen species (ROS) are produced in water as a result of PFRs, and these ROS assault DNA to cause illnesses in the heart and lungs (Dugas et al., 2016). When tested on corn stalks, rice, and wheat straw biochar, modest concentrations of PFRs demonstrated significant cytotoxicity and phytotoxicity in contrast to conventional toxins such as metals and PAHs (Dugas et al., 2016). An aquatic species named *Scenedesmus obliquus* had its ROS and superoxide dismutase levels increased as a result of applying PFR-containing biochar (Zhang et al., 2019). Additionally, a positive association between *S. obliquus* toxicity and PFR concentration in pine needle biochar was observed (Zhang et al., 2019). Recent research on the model organism *Caenorhabditis elegans* has revealed the neurotoxic effects of straw biochar, including impairments in defecation and the ability to recognize barriers to chemical attractants, showing that the risk of biochar has been underappreciated (Lieke et al., 2018).

16.3.4 Dioxins

Dioxin-like compounds are typically produced when biomass is incompletely combusted. Polychlorinated dibenzo-p-dioxins (PCDD), polychlorinated dibenzofurans (PCDF), and dioxin-like polychlorinated biphenyls (DL-PCBs) are typically produced during incomplete pyrolysis of biomass (Stanmore, 2004). Diverse biochar produced at pyrolysis temperatures between 250 and 900 °C has been shown to contain these dioxins (Wu et al., 2019). According to a study by Hale et al. (2012), 14 samples of biochar contained hazardous dioxin levels ranging from 0.008 to 1.2 pg/g TEQ, which is less than the Swedish government's recommended limits of 250 pg/g TEQ (Hale et al., 2012). The TEQ of 7 ng/kg found in sawdust biochar was also lower than that of the IBI (9 ng/kg), and the EBC (20 ng/kg) (Lyu et al., 2016).

Dioxins and aryl hydrocarbon receptors (AhR) interact, which alters normal hormone signaling pathways and causes abnormalities in development and reproduction (Kulkarni et al., 2008). These environmental pollutants are thus categorized as probable carcinogen (Eichbaum et al., 2014). In a test using the hepatoma cell line H4IIE-luc, biochar generated at temperatures higher than 400 °C had decreased efficacy for AhR-mediated reactions (Lyu et al., 2016). Overall, dioxins in biochar provide little environmental danger (Weidemann et al., 2018).

16.3.5 Metal Cyanide (MCN)

Metal cyanides (MCN) and other novel N-containing metal compounds can be created by the interaction between organic nitrogen and metals (Luo et al., 2019).

According to recent research by Luo et al., MCN may be found in biochar made from food scraps, sludge, fungus, and algae. Phycocyanin-derived biochar has the highest amount of MCN at 85.87 g/kg (Luo et al., 2020). Information on the ecological danger of MCN from biochar is lacking as of now (Han et al., 2022).

16.3.6 Volatile Organic Compounds (VOCs)

Volatile organic compounds (VOCs) can be produced in biochar by the recondensation of pyrolysis gases during the pyrolysis of biomass (Buss and Mašek, 2014). These substances, which can surpass 100 μg/g if problems arise (Buss et al., 2015), mostly comprise 2,4-dimethylphenol, o-, m-, and p-cresol, organic acids, aldehydes, furans, ketones, alcohols, and organic acids (Cordella et al., 2012). With increasing pyrolysis temperature, VOC levels in biochar often decrease. Cross seed germination was completely hindered by the fumes and leachates from high-VOC biochar, showing the possible environmental risk of the high-VOC biochar (Bargmann et al., 2013). The pyrolysis device should be set more reasonably and controlled to prevent excessive VOC from generating during pyrolysis.

References

Ameloot, N., Graber, E.R., Verheijen, F.G., De Neve, S., 2013. Interactions between biochar stability and soil organisms: Review and research needs. *European Journal of Soil Science*, 64(4), 379–390.

Bargmann, I., Rillig, M.C., Buss, W., Kruse, A., Kuecke, M., 2013. Hydrochar and biochar effects on germination of spring barley. *Journal of Agronomy and Crop Science*, 199(5), 360–373.

Buss, W., Graham, M.C., Shepherd, J.G., Mašek, O., 2016a. Risks and benefits of marginal biomass-derived biochars for plant growth. *Science of the Total Environment*, 569, 496–506.

Buss, W., Graham, M.C., Shepherd, J.G., Mašek, O., 2016b. Suitability of marginal biomass-derived biochars for soil amendment. *Science of the Total Environment*, 547, 314–322.

Buss, W. and Mašek, O., 2014. Mobile organic compounds in biochar–a potential source of contamination–phytotoxic effects on cress seed (Lepidium sativum) germination. *Journal of Environmental Management*, 137, 111–119.

Buss, W., Mašek, O., Graham, M., Wüst, D., 2015. Inherent organic compounds in biochar–their content, composition and potential toxic effects. *Journal of Environmental Management*, 156, 150–157.

Cordella, M., Torri, C., Adamiano, A., Fabbri, D., Barontini, F., Cozzani, V., 2012. Bio-oils from biomass slow pyrolysis: A chemical and toxicological screening. *Journal of Hazardous Materials*, 231, 26–35.

De la Rosa, J.M., Paneque, M., Hilber, I., Blum, F., Knicker, H.E., Bucheli, T.D., 2016. Assessment of polycyclic aromatic hydrocarbons in biochar and biochar-amended agricultural soil from Southern Spain. *Journal of Soils and Sediments*, 16(2), 557–565.

Devi, P. and Saroha, A.K., 2015. Effect of pyrolysis temperature on polycyclic aromatic hydrocarbons toxicity and sorption behaviour of biochars prepared by pyrolysis of paper mill effluent treatment plant sludge. *Bioresource Technology*, 192, 312–320.

Duan, X., Ao, Z., Sun, H., Indrawirawan, S., Wang, Y., Kang, J., Liang, F., Zhu, Z.H., Wang, S., 2015. Nitrogen-doped graphene for generation and evolution of reactive radicals by metal-free catalysis. *ACS Applied Materials & Interfaces*, 7(7), 4169–4178.

Dugas, T.R., Lomnicki, S., Cormier, S.A., Dellinger, B., Reams, M., 2016. Addressing emerging risks: Scientific and regulatory challenges associated with environmentally persistent free radicals. *International Journal of Environmental Research and Public Health*, 13(6), 573.

Eichbaum, K., Brinkmann, M., Buchinger, S., Reifferscheid, G., Hecker, M., Giesy, J.P., Engwall, M., van Bavel, B., Hollert, H., 2014. In vitro bioassays for detecting dioxin-like activity—Application potentials and limits of detection, a review. *Science of the Total Environment*, 487, 37–48.

Fang, G., Gao, J., Liu, C., Dionysiou, D.D., Wang, Y., Zhou, D., 2014. Key role of persistent free radicals in hydrogen peroxide activation by biochar: Implications to organic contaminant degradation. *Environmental Science & Technology*, 48(3), 1902–1910.

Fang, Y., Singh, B., Singh, B.P., 2015. Effect of temperature on biochar priming effects and its stability in soils. *Soil Biology and Biochemistry*, 80, 136–145.

Freddo, A., Cai, C., Reid, B.J., 2012. Environmental contextualisation of potential toxic elements and polycyclic aromatic hydrocarbons in biochar. *Environmental Pollution*, 171, 18–24.

Frišták, V., Pipíška, M., Soja, G., 2018. Pyrolysis treatment of sewage sludge: A promising way to produce phosphorus fertilizer. *Journal of Cleaner Production*, 172, 1772–1778.

Gao, X. and Wu, H., 2014. Aerodynamic properties of biochar particles: Effect of grinding and implications. *Environmental Science & Technology Letters*, 1(1), 60–64.

Gomez-Eyles, J.L., Beesley, L., Moreno-Jimenez, E., Ghosh, U., Sizmur, T., 2013. The potential of biochar amendments to remediate contaminated soils. *Biochar and Soil Biota*, 4, 100–133.

Hale, S.E., Lehmann, J., Rutherford, D., Zimmerman, A.R., Bachmann, R.T., Shitumbanuma, V., O'Toole, A., Sundqvist, K.L., Arp, H.P.H., Cornelissen, G., 2012. Quantifying the total and bioavailable polycyclic aromatic hydrocarbons and dioxins in biochars. *Environmental Science & Technology*, 46(5), 2830–2838.

Han, H., Buss, W., Zheng, Y., Song, P., Khalid Rafiq, M., Liu, P., Mašek, O., Li, X., 2022. Contaminants in biochar and suggested mitigation measures – A review. *Chemical Engineering Journal*, 429, 132287. https://doi.org/10.1016/j.cej.2021.132287.

Husson, O., 2013. Redox potential (Eh) and pH as drivers of soil/plant/microorganism systems: A transdisciplinary overview pointing to integrative opportunities for agronomy. *Plant and Soil*, 362(1), 389–417.

Jin, J., Li, Y., Zhang, J., Wu, S., Cao, Y., Liang, P., Zhang, J., Wong, M.H., Wang, M., Shan, S., 2016. Influence of pyrolysis temperature on properties and environmental safety of heavy metals in biochars derived from municipal sewage sludge. *Journal of Hazardous Materials*, 320, 417–426.

Jones, D., Murphy, D., Khalid, M., Ahmad, W., Edwards-Jones, G., DeLuca, T., 2011. Short-term biochar-induced increase in soil CO_2 release is both biotically and abiotically mediated. *Soil Biology and Biochemistry*, 43(8), 1723–1731.

José, M., Sánchez-Martín, Á.M., Campos, P., Miller, A.Z., 2019. Effect of pyrolysis conditions on the total contents of polycyclic aromatic hydrocarbons in biochars produced from organic residues: Assessment of their hazard potential. *Science of the Total Environment*, 667, 578–585.

Keiluweit, M., Kleber, M., Sparrow, M.A., Simoneit, B.R., Prahl, F.G., 2012. Solvent-extractable polycyclic aromatic hydrocarbons in biochar: Influence of pyrolysis temperature and feedstock. *Environmental Science & Technology*, 46(17), 9333–9341.

Kibet, J., Khachatryan, L., Dellinger, B., 2012. Molecular products and radicals from pyrolysis of lignin. *Environmental Science & Technology*, 46(23), 12994–13001.

Kiruri, L.W., Khachatryan, L., Dellinger, B., Lomnicki, S., 2014. Effect of copper oxide concentration on the formation and persistency of environmentally persistent free radicals (EPFRs) in particulates. *Environmental Science & Technology*, 48(4), 2212–2217.

Kulkarni, P.S., Crespo, J.G., Afonso, C.A., 2008. Dioxins sources and current remediation technologies—a review. *Environment International*, 34(1), 139–153.

Kuśmierz, M., Oleszczuk, P., Kraska, P., Pałys, E., Andruszczak, S., 2016. Persistence of polycyclic aromatic hydrocarbons (PAHs) in biochar-amended soil. *Chemosphere*, 146, 272–279.

Ledesma, E.B., Marsh, N.D., Sandrowitz, A.K., Wornat, M.J., 2002. Global kinetic rate parameters for the formation of polycyclic aromatic hydrocarbons from the pyrolyis of catechol, a model compound representative of solid fuel moieties. *Energy & Fuels*, 16(6), 1331–1336.

Lian, F. and Xing, B., 2017. Black carbon (biochar) in water/soil environments: Molecular structure, sorption, stability, and potential risk. *Environmental Science & Technology*, 51(23), 13517–13532. https://doi.org/10.1021/acs.est.7b02528.

Liao, N., Li, Q., Zhang, W., Zhou, G., Ma, L., Min, W., Ye, J., Hou, Z., 2016. Effects of biochar on soil microbial community composition and activity in drip-irrigated desert soil. *European Journal of Soil Biology*, 72, 27–34.

Liao, S., Pan, B., Li, H., Zhang, D., Xing, B., 2014. Detecting free radicals in biochars and determining their ability to inhibit the germination and growth of corn, wheat and rice seedlings. *Environmental Science & Technology*, 48(15), 8581–8587.

Lieke, T., Zhang, X., Steinberg, C.E., Pan, B., 2018. Overlooked risks of biochars: Persistent free radicals trigger neurotoxicity in Caenorhabditis elegans. *Environmental Science & Technology*, 52(14), 7981–7987.

Luo, J., Jia, C., Shen, M., Zhang, S., Zhu, X., 2019. Enhancement of adsorption and energy storage capacity of biomass-based N-doped porous carbon via cyclic carbothermal reduction triggered by nitrogen dopants. *Carbon*, 155, 403–409.

Luo, J., Lin, L., Liu, C., Jia, C., Chen, T., Yang, Y., Shen, M., Shang, H., Zhou, S., Huang, M., 2020. Reveal a hidden highly toxic substance in biochar to support its effective elimination strategy. *Journal of Hazardous Materials*, 399, 123055.

Luo, Y., Durenkamp, M., De Nobili, M., Lin, Q., Brookes, P., 2011. Short term soil priming effects and the mineralisation of biochar following its incorporation to soils of different pH. *Soil Biology and Biochemistry*, 43(11), 2304–2314.

Lyu, H., He, Y., Tang, J., Hecker, M., Liu, Q., Jones, P.D., Codling, G., Giesy, J.P., 2016. Effect of pyrolysis temperature on potential toxicity of biochar if applied to the environment. *Environmental Pollution*, 218, 1–7.

Maestrini, B., Nannipieri, P., Abiven, S., 2015. A meta-analysis on pyrogenic organic matter induced priming effect. *Gcb Bioenergy*, 7(4), 577–590.

Mendez, A., Gomez, A., Paz-Ferreiro, J., Gasco, G., 2012. Effects of sewage sludge biochar on plant metal availability after application to a Mediterranean soil. *Chemosphere*, 89(11), 1354–1359.

Mitchell, P.J., Simpson, A.J., Soong, R., Simpson, M.J., 2015. Shifts in microbial community and water-extractable organic matter composition with biochar amendment in a temperate forest soil. *Soil Biology and Biochemistry*, 81, 244–254.

Nguyen, B.T. and Lehmann, J., 2009. Black carbon decomposition under varying water regimes. *Organic Geochemistry*, 40(8), 846–853. https://doi.org/10.1016/j.orggeochem.2009.05.004.

Nguyen, B.T., Lehmann, J., Hockaday, W.C., Joseph, S., Masiello, C.A., 2010. Temperature sensitivity of black carbon decomposition and oxidation. *Environmental Science & Technology*, 44(9), 3324–3331.

Nguyen, V.-T., Nguyen, T.-B., Chen, C.-W., Hung, C.-M., Chang, J.-H., Dong, C.-D., 2019. Influence of pyrolysis temperature on polycyclic aromatic hydrocarbons production and tetracycline adsorption behavior of biochar derived from spent coffee ground. *Bioresource Technology*, 284, 197–203.

Nia, Z.K., Chen, J.-Y., Tang, B., Yuan, B., Wang, X.-G., Li, J.-L., 2017. Optimizing the free radical content of graphene oxide by controlling its reduction. *Carbon*, 116, 703–712.

Nzihou, A., Stanmore, B., Sharrock, P., 2013. A review of catalysts for the gasification of biomass char, with some reference to coal. *Energy*, 58, 305–317.

Odinga, E.S., Gudda, F.O., Waigi, M.G., Wang, J., Gao, Y., 2021. Occurrence, formation and environmental fate of polycyclic aromatic hydrocarbons in biochars. *Fundamental Research*, 1(3), 296–305. https://doi.org/10.1016/j.fmre.2021.03.003.

Odinga, E.S., Waigi, M.G., Gudda, F.O., Wang, J., Yang, B., Hu, X., Li, S., Gao, Y., 2020. Occurrence, formation, environmental fate and risks of environmentally persistent free radicals in biochars. *Environment International*, 134, 105172. https://doi.org/10.1016/j.envint.2019.105172.

Oleszczuk, P., Jośko, I., Kuśmierz, M., 2013. Biochar properties regarding to contaminants content and ecotoxicological assessment. *Journal of Hazardous Materials*, 260, 375–382.

Oleszczuk, P., Kuśmierz, M., Godlewska, P., Kraska, P., Pałys, E., 2016. The concentration and changes in freely dissolved polycyclic aromatic hydrocarbons in biochar-amended soil. *Environmental Pollution*, 214, 748–755.

Qin, J., Chen, Q., Sun, M., Sun, P., Shen, G., 2017. Pyrolysis temperature-induced changes in the catalytic characteristics of rice husk-derived biochar during 1, 3-dichloropropene degradation. *Chemical Engineering Journal*, 330, 804–812.

Roberts, D.A., Cole, A.J., Whelan, A., de Nys, R., Paul, N.A., 2017. Slow pyrolysis enhances the recovery and reuse of phosphorus and reduces metal leaching from biosolids. *Waste Management*, 64, 133–139.

Rombolà, A.G., Marisi, G., Torri, C., Fabbri, D., Buscaroli, A., Ghidotti, M., Hornung, A., 2015. Relationships between chemical characteristics and phytotoxicity of biochar from poultry litter pyrolysis. *Journal of Agricultural and Food Chemistry*, 63(30), 6660–6667.

Sand, W., 1997. Microbial mechanisms of deterioration of inorganic substrates—a general mechanistic overview. *International Biodeterioration & Biodegradation*, 40(2–4), 183–190.

Shaheen, S.M., Niazi, N.K., Hassan, N.E., Bibi, I., Wang, H., Tsang, D.C., Ok, Y.S., Bolan, N., Rinklebe, J., 2019. Wood-based biochar for the removal of potentially toxic elements in water and wastewater: A critical review. *International Materials Reviews*, 64(4), 216–247.

Shen, X., Zeng, J., Zhang, D., Wang, F., Li, Y., Yi, W., 2020. Effect of pyrolysis temperature on characteristics, chemical speciation and environmental risk of Cr, Mn, Cu, and Zn in biochars derived from pig manure. *Science of the Total Environment*, 704, 135283.

Singh, B.P., Cowie, A.L., Smernik, R.J., 2012. Biochar carbon stability in a clayey soil as a function of feedstock and pyrolysis temperature. *Environmental Science & Technology*, 46(21), 11770–11778.

Smebye, A., Alling, V., Vogt, R.D., Gadmar, T.C., Mulder, J., Cornelissen, G., Hale, S.E., 2016. Biochar amendment to soil changes dissolved organic matter content and composition. *Chemosphere*, 142, 100–105.

Spokas, K.A., 2010. Review of the stability of biochar in soils: Predictability of O: C molar ratios. *Carbon Management*, 1(2), 289–303.

Spokas, K.A., Novak, J.M., Masiello, C.A., Johnson, M.G., Colosky, E.C., Ippolito, J.A., Trigo, C., 2014. Physical disintegration of Biochar: An overlooked process. *Environmental Science & Technology Letters*, 1(8), 326–332. https://doi.org/10.1021/ez500199t.

Stanmore, B., 2004. The formation of dioxins in combustion systems. *Combustion and Flame*, 136(3), 398–427.

Stefaniuk, M., Oleszczuk, P., Bartmiński, P., 2016. Chemical and ecotoxicological evaluation of biochar produced from residues of biogas production. *Journal of Hazardous Materials*, 318, 417–424.

Stefaniuk, M., Tsang, D.C., Ok, Y.S., Oleszczuk, P., 2018. A field study of bioavailable Polycyclic Aromatic Hydrocarbons (PAHs) in sewage sludge and biochar amended soils. *Journal of Hazardous Materials*, 349, 27–34.

Stephenson, E.J., Ragauskas, A., Jaligama, S., Redd, J.R., Parvathareddy, J., Peloquin, M.J., Saravia, J., Han, J.C., Cormier, S.A., Bridges, D., 2016. Exposure to environmentally persistent free radicals during gestation lowers energy expenditure and impairs skeletal muscle mitochondrial function in adult mice. *American Journal of Physiology-Endocrinology and Metabolism*, 310.

Tao, W., Duan, W., Liu, C., Zhu, D., Si, X., Zhu, R., Oleszczuk, P., Pan, B., 2020. Formation of persistent free radicals in biochar derived from rice straw based on a detailed analysis of pyrolysis kinetics. *Science of the Total Environment*, 715, 136575.

Théry-Parisot, I., Chabal, L., Chrzavzez, J., 2010. Anthracology and taphonomy, from wood gathering to charcoal analysis. A review of the taphonomic processes modifying charcoal assemblages, in archaeological contexts. *Palaeogeography, Palaeoclimatology, Palaeoecology*, 291(1–2), 142–153.

Tian, J., Wang, J., Dippold, M., Gao, Y., Blagodatskaya, E., Kuzyakov, Y., 2016. Biochar affects soil organic matter cycling and microbial functions but does not alter microbial community structure in a paddy soil. *Science of the Total Environment*, 556, 89–97.

Vejerano, E.P., Rao, G., Khachatryan, L., Cormier, S.A., Lomnicki, S., 2018. Environmentally persistent free radicals: Insights on a new class of pollutants. *Environmental Science & Technology*, 52(5), 2468–2481.

Wang, C., Wang, Y., Herath, H., 2017. Polycyclic Aromatic Hydrocarbons (PAHs) in biochar–Their formation, occurrence and analysis: A review. *Organic Geochemistry*, 114, 1–11.

Wang, J., Xiong, Z., Kuzyakov, Y., 2016. Biochar stability in soil: Meta-analysis of decomposition and priming effects. *Gcb Bioenergy*, 8(3), 512–523.

Weidemann, E., Buss, W., Edo, M., Mašek, O., Jansson, S., 2018. Influence of pyrolysis temperature and production unit on formation of selected PAHs, oxy-PAHs,

N-PACs, PCDDs, and PCDFs in biochar—a screening study. *Environmental Science and Pollution Research*, 25(4), 3933–3940.

Wiedemeier, D.B., Abiven, S., Hockaday, W.C., Keiluweit, M., Kleber, M., Masiello, C.A., McBeath, A.V., Nico, P.S., Pyle, L.A., Schneider, M.P., 2015. Aromaticity and degree of aromatic condensation of char. *Organic Geochemistry*, 78, 135–143.

Wu, J.-L., Ji, F., Zhang, H., Hu, C., Wong, M.H., Hu, D., Cai, Z., 2019. Formation of dioxins from triclosan with active chlorine: A potential risk assessment. *Journal of Hazardous Materials*, 367, 128–136.

Xia, X., Teng, Y., Zhai, Y., 2022. Biogeochemistry of iron enrichment in groundwater: An indicator of environmental pollution and its management. *Sustainability*, 14(12), 7059.

Yang, F., Zhao, L., Gao, B., Xu, X., Cao, X., 2016. The interfacial behavior between biochar and soil minerals and its effect on biochar stability. *Environmental Science & Technology*, 50(5), 2264–2271.

Zhang, P., Zhang, X., Li, Y., Han, L., 2020. Influence of pyrolysis temperature on chemical speciation, leaching ability, and environmental risk of heavy metals in biochar derived from cow manure. *Bioresource Technology*, 302, 122850.

Zhang, Y., Yang, R., Si, X., Duan, X., Quan, X., 2019. The adverse effect of biochar to aquatic algae-the role of free radicals. *Environmental Pollution*, 248, 429–437.

Zielińska, A. and Oleszczuk, P., 2015. The conversion of sewage sludge into biochar reduces polycyclic aromatic hydrocarbon content and ecotoxicity but increases trace metal content. *Biomass and Bioenergy*, 75, 235–244.

Zimmerman, A.R., 2010. Abiotic and microbial oxidation of laboratory-produced black carbon (biochar). *Environmental Science & Technology*, 44(4), 1295–1301.

17

Environmental and Economic Evaluation of Biochar Application in Wastewater and Sludge Treatment

Claudia Labianca[1], Sabino De Gisi[1], Michele Notarnicola[1], Xiaohong Zhu[3], and Daniel C.W. Tsang[2]

[1] *Department of Civil, Environmental, Land, Building Engineering and Chemistry (DICATECh), Polytechnic University of Bari, Bari, Italy*
[2] *State Key Laboratory of Clean Energy Utilization, Zhejiang University, China*
[3] *Department of Civil and Environmental Engineering, The Hong Kong Polytechnic University, Hong Kong, China*

17.1 Introduction

Wastewater and sludge contamination have recently become a predominantly severe environmental concern (Zhao et al., 2020). Researchers have discovered that wastewater treatment facilities (WWTFs) are an important source of microplastics, heavy metals, and organic pollutants (Chen et al., 2020), although they can purify wastewater in compliance with the law. These pollutants must be removed from wastewater and sludge before being discharged into water or used for soil applications (Jin et al., 2016).

Industries have increasingly used the term *carbon neutrality* to describe the absence of net carbon dioxide emissions in their production cycles. Several trials have been made to advance innovative technologies that convert waste into resources, materials, and energy (Bae and Kim, 2021).

As almost 4% of the non-CO_2 greenhouse gases (GHGs) are emitted by WWTFs, and as above-stated wastewater contains various organic/inorganic pollutants that need to be properly treated, the development of energy-efficient water

Biochar Applications for Wastewater Treatment, First Edition. Edited by Daniel C.W. Tsang and Yuqing Sun.
© 2023 John Wiley & Sons, Inc. Published 2023 by John Wiley & Sons, Inc.

treatment technologies and resource recovery from wastewater has become extremely important (Sun et al., 2022).

The recovery of chemicals, nutrients, energy, and even water itself from WWTFs has improved the energy self-sufficiency of wastewater treatment plants, making them more sustainable. Borzooei et al. (2019) demonstrated that adding a thickening stage upstream of anaerobic digestion would improve its energy efficiency and greenhouse gas emissions with a potential saving of up to 5000 MWh of the annual plant energy consumption. In the first place, the volume of sludge entering the digestion process was reduced, resulting in a lower level of thermal energy consumption. A second benefit was the increased specific biogas production as a result of the sludge pretreatment. The process demonstrated a favorable greenhouse gas balance, meaning that avoided emissions from replacing natural gas with electricity will be higher than emissions produced by process maintenance. By using life cycle assessment (LCA), Hao et al. (2019) investigated the environmental impact of resource recovery from WWTFs, concluding that (i) water reuse via effluent recycling is not enough for a carbon-neutral impact and (ii) thermal energy recovery can play a substantial role, which is different from phosphorus recovery. In another LCA study about the removal of pharmaceuticals and personal care products from effluents, it was shown that ozonation had the lowest life cycle costs, while pyrolysis had a better return on investment than anaerobic digestion, wet air oxidation, and composting (Tarpani and Azapagic, 2018). In this regard, biochar has also been proposed as a pyrolysis product produced from agricultural wastes and sludge from WWTFs (Shin et al., 2021b, 2021a). Organic and heavy-metal pollutants were efficiently removed by using biochar including a reduction of GHGs from agricultural and environmental industries and producing clean water at a cost-effective rate.

A wide range of functional groups exist in agricultural residue wastes, including cellulose, hemicellulose, lipids, sugars, and proteins, that may be physically activated by pyrolysis, steam or CO_2-based treatment to enhance their adsorption capacity (Qambrani et al., 2017). Therefore, in response to climate change concerns, biochar technologies are gaining much attention due to their benefits in carbon sequestration, energy production, and waste management.

Figure 17.1 shows the main covered subject areas by using two different lists of keywords. From an LCA point of view, it is evident that over the last decade, more attention was given to the application of biochar in wastewater and sludge management/treatment in the environmental science and energy fields.

In this chapter, we will describe the fundamentals of LCA related to biochar production and applications; after that, the main LCA studies of biochar applications in wastewater and sludge treatments will be reviewed.

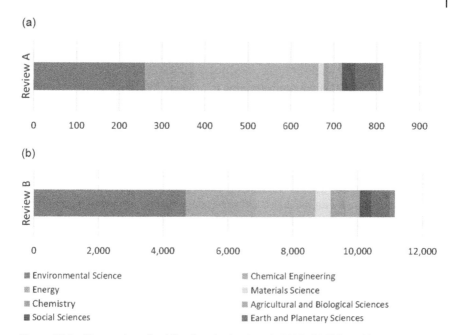

Figure 17.1 The number of publications in the decade 2013–2023 by subject areas using different research keywords (a) Review A: biochar, wastewater, sludge, life cycle assessment; (b) Review B: biochar, wastewater, sludge.

17.2 Environmental Evaluation

LCA is a standardized methodology (ISO, 2006a, 2006b) that evaluates the environmental impacts related to a material/process. It includes the following stages: (i) goal and scope definition, (ii) life cycle inventory analysis, (iii) life cycle impact assessment, and (iv) interpretation of results.

This section examines some aspects of biochar production and its application to wastewater and sludge treatment.

17.2.1 LCA Insights into Biochar Production and Applications

Biochar has lately attracted attention due to its promising role in many environmental applications. It can be produced via slow and fast pyrolysis, gasification, or combustion processes (Qambrani et al., 2017) and from a huge variety of organic materials such as plant tissue, raw pine chips, forage plant biomass, woody debris, peanut hulls and pecan shells, corn stalks, pine chips, poultry litter, paper mill, citrus wood, cottonseed hulls, empty fruit bunches, rubber wood sawdust, rice husks, sewage biosolids, poultry manure, goat manure, human manure, swine manure, and agro-industrial biomass (Qambrani et al., 2017).

The different types of biomass, together with the thermochemical conditions used to pyrolyze it, greatly influence the final quality of biochar, its possible applications, and the total energy consumption to produce it (Chen et al., 2011). Some scientists have published various studies describing the environmental benefits associated with biochar used in different applications, including wastewater treatment. However, the environmental benefits of biochar strictly depend on its production method and utilization.

Commonly, the performance of biochar can be compared with conventional materials, depending on the application target. For example, the conventional material compared with biochar is chlorine when biochar is used in water filtration (Uddin and Wright, 2022). It is also common to compare its performance with activated carbon, which is energy intensive and expensive (Thompson et al., 2016).

Furthermore, based on the targeted use and goal of the study, a variety of functional units can be defined for biochar LCA analyses. If the aim of the analysis is waste management, upstream functional units, such as 1 tonne of wood waste are usually used (Miranda et al., 2021; Sun et al., 2022). If the aim is to produce energy from bio-oil or syngas, the functional unit could be 1 MJ of energy produced (Uddin and Wright, 2022). If the sole purpose of a study is biochar production, it seems appropriate to consider 1 tonne of biochar as the final product to be the most suitable functional unit (Muñoz et al., 2017).

Several system boundaries (e.g., feedstock source and availability, existing scopes with the feedstocks, utilization of co-products, and geographical location) could be included in the LCA study. Usually, feedstock production is not part of the system boundaries if the feedstock is a waste resource such as wastewater sludge, corn stover, or wood waste (Oldfield et al., 2018). Nevertheless, if the main scope of biochar production is energy production, then agricultural processes are recommended to be included in the system boundaries (Matuštík et al., 2020). It is common to also include the end-use of biochar to fully calculate the environmental advantages, such as soil amendment, combustion, or water and wastewater treatment like the application in this chapter. Starting from the "cradle," a complete LCA quantifies ideally emissions from each life cycle stage such as (i) feedstock production and collection, (ii) feedstock handling and pre-processing, (iii) thermochemical treatment, (iv) biochar distribution, and (v) end-use (Figure 17.2).

Another important aspect to highlight is that in most of the published LCA studies, biochar is not the main product of the process but it is considered as a by-product. In fact, pyrolysis represents a multi-output process in producing bio-oil, syngas, and biochar (Figure 17.2, Table 17.1). Therefore, some assumptions and system boundary conditions are usually adopted when biochar is used. Mass and economic allocations are the most adopted ones (Brassard et al., 2019; Yang et al.,

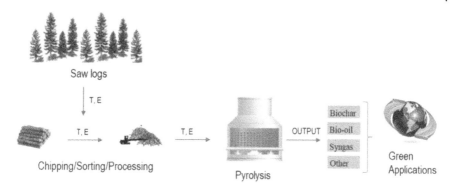

Figure 17.2 System boundaries (T = transport, E = emissions) of an LCA study.

Table 17.1 Operating parameters and product yields of pyrolysis (Adapted from (Brassard et al., 2017)).

	Pyrolysis details				Yields		
References	Feedstock type	Feedstock size (mm)	Residence time (s)	T (°C)	Bio-oil (%)	Biochar (%)	Syngas (%)
(Brassard et al., 2021)	Wood	10–100	61	559	58.3	26.5	15.2
(Yang et al., 2021)	Crop residues (grain, bean, tuber, oil crop, cotton, sugarcane and hemp)	–	–	300–800	30.1	35.2	34.8
(Solar et al., 2016)	Pine	0.5–2	1920	900	9.7	21	69.3
(Ferreira et al., 2015)	Medium density fiberboard	< 0.21	2040	450	25	39.7	–
			900	450	27	30.5	–
			540	450	26	24.9	–
			2040	600	40	17.3	–
			900	600	26	19	–
			540	600	23.9	25.5	–
(Joubert et al., 2015)	Eucalyptus	< 5	13	500	67.5	16.5	15.8

(Continued)

Table 17.1 (Continued)

	Pyrolysis details				Yields		
References	Feedstock type	Feedstock size (mm)	Residence time (s)	T (°C)	Bio-oil (%)	Biochar (%)	Syngas (%)
(le Roux et al., 2015)	Aspen wood	< 2	210	450	56.1	19.8	24.1
(Liang et al., 2015)	Potato peel	< 1	8	450	22.7	30.5	-
(Tomasi Morgano et al., 2015)	Beechwood	2–4.5	600	350	42.9	31.5	12.5
				400	37.2	24.4	15.7
				450	48.8	20.6	23.4
				500	39.8	18	31.6
(Veses et al., 2015)	Pine	< 15	420	400	41	36	23
				450	49	26	25
				500	50	21	26
(Li et al., 2014)	Pine wood	< 1	480	400	30.2	45.1	24.7
				500	34.	30.3	35.7
				600	33.1	25.2	41.7
(Veses et al., 2014)	Pine	< 15	420	450	49	26	26.5
(Agirre et al., 2013)	Fruits cutting	< 8	7200	900	-	22	-
		10–20			-	22.8	-
		< 5			-	21.3	-
(Martínez et al., 2013)	Waste tyres	2–4	180	550	42.6	40.5	16.9
(Sirijanusorn et al., 2013)	Cassava rhizome	0.25–0.425	-	500	40	22	38
			-	550	50	23	27
			-	650	44	13	43
		< 0.25	-	650	27	20	53
(Tröger et al., 2013)	Corn stover	< 5	13	500	43.5	20.8	28.8
	Rape stalks				37.2	28.8	32.4
	Sunflower stalks				31.1	31.4	30.9
	Wheat straw				43.9	23.7	24.8
	Softwood				64	13.9	20.3

Table 17.1 (Continued)

	Pyrolysis details				Yields		
References	Feedstock type	Feedstock size (mm)	Residence time (s)	T (°C)	Bio-oil (%)	Biochar (%)	Syngas (%)
(Brown and Brown, 2012)	Red oak	< 0.75	11	625	73.6	11	12.9
			11	425	42.2	35.7	22.1
(Liaw et al., 2012)	Douglas fir wood	< 2	60	320	59	18	23
(Pittman et al., 2012)	Corn stalks	0.5–5	55	400	35	29	13.5
			55	450	35	23.5	32

2021) but other types of allocations can also be found in the literature. However, ISO 14044 guidelines (ISO, 2006b) recommend adopting the system expansion method instead of allocation whenever feasible. In this way, the co-products are supposed to substitute other products with analogous functions outside the system boundary.

17.2.2 Main LCA Literature Studies of Biochar Applications in Wastewater and Sludge Treatments

The use of biochar in water and wastewater treatment is relatively new and is promising for the removal of organic and heavy metal contaminants (Table 17.2). Biochar can behave as a super-sorbent for the removal of both organic and inorganic contaminants in water and soil (Sharma et al., 2018).

As stated by Gongora et al. (2021), it is not always clear if the environmental benefits gained through the improved wastewater treatment system overcome the burdens associated with the manufacturing and use of biochar. A comparative analysis between aerobic treatment units (ATUs) and biochar-based constructed wetlands (CWs) was performed accordingly, and three biochar concentrations were tested in the substrate as 0%; 10%, and 20% v/v. They found that ATU technology delivered lower environmental performance than the CWs. As demonstrated by other authors (De Feo and Ferrara, 2017), the highest contribution came from the wastes generated at the end of CWs' life, after 20 years of operation, due to the disposal and treatment of spent substrate materials, such as sand and gravel. However, the inclusion of biochar in the substrate brought environmental benefits of global warming potential (GWP) owing to the carbon stored in the soil (Azzi et al., 2019). Also in other categories such as eutrophication potential, human

Table 17.2 Main literature studies of biochar LCA in wastewater/sludge treatments.

References	Scope of the study	FU	Methodology and database	Main results
(Choudhary and Philip, 2022)	To investigate biochar in removing four pharmaceutical and personal care products from secondary wastewater.	2000 m^3 of secondary treated wastewater	ReCiPe. Ecoinvent	The biochar-based system performed considerably better than GAC in terms of impacts. This can be ascribed to the use of waste as feedstock, which reduces energy consumption during the production and regeneration stages. Also, the energy demand for biochar production is 15 times lower than activated carbon. The estimated carbon sequestration for moderate-capacity wood biochar is very high.
(Alatrista Gongora et al., 2021)	To compare aerobic treatment unit with reinforced concrete and HDPE tank and constructed wetland with three biochar concentrations in the substrate (0%; 10%, and 20% v/v)	1 m^3 of treated wastewater, considering a lifetime of the treatment systems of 20 years	CML 2001; database not defined	The biochar-based scenarios in any of the configurations had better overall environmental performance than the aerobic treatment units. For the biochar-based systems, the construction stage had the highest impact, and the PVC pipes, coir peat, geomembrane, and electronic devices were the input materials with the highest contribution.
(Miranda et al., 2021)	To compare the effects of the addition of sulphuric acid, biochar or both to liquid cattle-slurry on global GHGs compared to the untreated system	1 tonne of fresh dairy cattle slurry	CML 2001	Sulphuric acid showed the highest efficiency in the mitigation of GHGs compared to biochar, reducing NH_3, CH_4, and CO_2 emissions by 61%, 98%, and 15%, respectively, when compared to the system with biochar.

(Tian et al., 2020)	To convert the residual organic wastes from treated wastewater into biochar and hydrochar for land application	1 million gallons of domestic WW	IPCC 2013 and ReCiPe. Ecoinvent v. 3.6 and CapdetWorks v. 4.0	Only the retrofit scenario with hydrothermal liquefaction outperformed the conventional design in terms of net present value, and the profit of the proposed retrofit scenarios would benefit from the government incentives for the use of hydrochar or biochar and encourage biogas prices. High impacts were related to nitrogen production, which was used for inert gas purging.
(Thompson et al., 2016)	To compare the environmental impacts of using biochar sorbents to coal-derived PAC for tertiary wastewater treatment	75% removal of sulfamethoxazole (SMX) from 47,300 m^3/day (12.5 mgd) of secondary wastewater effluent over 40 years	TRACI. US-EI v.2.2 and agri-footprint databases	Compared to PAC, low-capacity wood biochar had more benefits for global warming, respiratory effects, and noncancerous effects, but was less effective in five categories, due to higher biochar dose requirements.

toxicity, acidification potential, ozone depletion potential and others, the biochar addition positively influenced the results at different magnitudes. It is important to highlight that the authors considered a subsequent release of 50% of the carbon in the biochar to the atmosphere in the form of CO_2 within 100 years. A few authors have also demonstrated possible applications of biochar to sludge treatment. Miranda et al. (2021) compared the untreated scenario with the treated systems to evaluate the advantages of cattle-slurry treatment. After mechanical separation with a screw press separator, the liquid fraction was treated with sulphuric acid, biochar, or a combination of both. At the farm scale, the slurry is usually separated into a liquid fraction and a nutrient-rich solid fraction. As a result, nutrients can be managed more efficiently, in addition to reducing the volume and cost of slurry to be transported to nearby fields. The functional unit considered was 1 tonne of fresh dairy cattle slurry. It is interesting to note that the untreated scenario showed a lower efficiency in the mitigation of total GHGs and NH_3 emissions. However, sulphuric acid showed better performance in the mitigation of gas emissions compared to biochar. The highest impacts were in general related to the storage phase due to the added treatments; instead, the mechanical separation and field application had a negligible influence on the environmental profile.

In the study by Choudhary and Philip (2022), the biochar was prepared via slow pyrolysis of empty palm bunch at three temperatures between 250 and 750 °C. In order to further improve the adsorption capacity, H_2SO_4 was added and the adsorption behavior of target pollutants such as methylparaben, carbamazepine, ibuprofen, and triclosan was tested through batch experiments on the biochar-based sorbents. They included the life cycle impact of feedstock acquisition, production, use, hauling, transformation products, and sludge disposal phases. The environmental impacts were calculated for 2000 m^3 of secondary treated wastewater effluent and compared with commercial granulated activated carbon (GAC). The GWP of the biochar-based treatment column was nearly 200 times less than GAC. Similarly, its production and use reduced fossil depletion and human toxicity by 35.2% and 12.8%, respectively. Compared to GAC, the impact on climate change, fossils or ozone depletion for both the production and regeneration phases of the biochar-based system was still lower. This can be ascribed to the 15 times lower energy demand for biochar production than activated carbon (Kwon et al., 2020). As a result, the biochar-based treatment system contributed 20 times less to the total impacts than GAC. The developed sorbent was also more effective at removing the target pollutants than GAC, making it more efficient. This was in line with the findings by Lu et al. (2012), who reported that biochar was more effective than commercial GAC for lead adsorption. The main differences between the two sorbents stand in their feedstocks, preparation methods, and final physiochemical properties.

A similar comparison was carried out in the study by Thompson et al. (2016) where the performance of PAC (powdered activated carbon), wood biochar, and

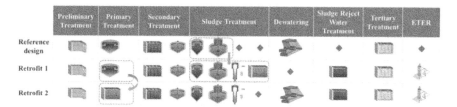

Figure 17.3 Overview of the WWTF reference and retrofitting designs investigated by Tian et al. (2020) / American Chemical Society.

biosolids biochar was compared and tested for tertiary wastewater treatment. They showed that wood biochar can have significant environmental advantages for climate change mitigation. The capacity wood biochar scenario sequestered about 6.5 gigagrams CO_2 eq/yr to offset all of the carbon emissions from energy and chemical use.

In the study by Tian et al. (2020), the scope was to compare some retrofitting designs of WWTFs such as advanced bioreactors, thermochemical conversion technologies, and effluent thermal energy recovery and to then unmask the full spectrum of life cycle environmental impacts of the resource recovery system from municipal sewage in WWTFs. In the retrofit with pyrolysis and anaerobic primary bioreactors, the residual biosolids were pyrolyzed in an inert atmosphere and converted into biocrude oil, biochar, and gas (Tian et al., 2020) (Figure 17.3). The biogas was then burnt to offset a fraction of electricity and heat demand. For the designs with pyrolysis, very high environmental impacts derived from nitrogen production, which was used for inert gas purging.

From a circular economy perspective, biochar can be produced from sewage sludge, making the entire effluent treatment cycle more sustainable. In the study by Menezes et al. (2022), different sludges were collected after primary, secondary, and tertiary sedimentation (overpassing anaerobic digestion or dewatering), and pyrolyzed at 450 °C in N_2 atmosphere, lowering the total costs. The final product yield ranges were 51.88–57.90%, 5.02–9.81%, and 32.51–42.90% for biochar, bio-oil, and syngas, respectively.

17.3 Technical, Economic, and Sustainability Considerations

A study by Banik et al. (2018) examined the effects of technical parameters like feedstock type, pyrolysis temperature, residence time, ash content, elemental composition and operating pH, along with the effects of treating the feedstock with aluminum chloride prior to pyrolysis on the biochar zero salt effect (PZSE), anion exchange capacity (AEC), cation exchange capacity (CEC), and zero net charge (PZNC). The authors highlighted that the biochar prepared at 500 °C showed high CEC, as well as low AEC, PZSE, and PZNC. The study by Moradi-Choghamarani et al. (2019) demonstrated that any increase in pyrolysis temperature could lead to biochar with a low CEC and high AEC, PZSE, and PZNC, due to the positive surface charge of the produced biochar. Fidel et al. (2018) proved that the ammonium and nitrate adsorption capability of biochar prepared from red oak and corn stover at the temperatures of 400 °C, 500 °C and 600 °C is extremely electrostatic and pH dependent. Other authors demonstrated that impregnation with magnetic particles can improve biochar magnetic properties and anion adsorption capacity (Choi et al., 2019; Wang et al., 2021). However, every additional treatment would influence the final impacts and only a detailed assessment would reveal if the overall balance would still be convenient in terms of environmental impacts.

Conversion of organic materials to biochar through pyrolysis supports an alternative way to manage a range of wastes (Hospido et al., 2005, Bridle and Pritchard, 2004). Compared with landfilling and direct agricultural utilisation, the pyrolysis of wastewater sludge could be an ideal method of managing waste, since it reduces the volume of solid residue and eliminates pathogens and organic compounds present in the sludge.

When used as a sorbent, biochar is capable of removing various pollutants simultaneously, so it can be used as the sole treatment method to achieve the desired result and improve the overall economic sustainability of the treatment. However, it is often difficult to combine all the information together; practitioners should strive to integrate risk assessment, environmental sustainability, costs, and social acceptance holistically.

Biochar-based systems need to consider emissions from biomass growth, collection, pyrolysis, transport, and application to consider it truly carbon negative. As a result of its ability to actively reduce the overall greenhouse gases in the atmosphere, biochar technology was included as a promising negative emission technology (NET) in the last IPCC special report published on October 8, 2018 (Hansson et al., 2021). Carbon stability in biochar through pyrolysis prevents its decomposition, ensuring that carbon remains away from the atmosphere for hundreds to thousands of years.

During the growth of biomass (e.g. trees), the carbon dioxide is removed from the atmosphere through photosynthesis; after harvesting and pyrolyzing the biomass, a portion of the carbon dioxide bound in the biomass is captured and stored (pyrogenic carbon capture and storage (PyCCS)) for a period of time that goes beyond the biogenic cycle duration (Gupta, 2021). Different authors have calculated the amount of carbon captured and stored in biochar, which varies with the organic carbon fraction of the biochar and the fraction of organic carbon remaining after a period of time of at least 100 years (Woolf et al., 2021). A range of

2–2.6 kg CO_2eq can be associated with PyCCs for every kg of biochar used (Azzi et al., 2019; Gupta, 2021; Woolf et al., 2021).

As well as carbon offsets or carbon certificates, carbon credits represent the avoidance, reduction, or sequestration of CO_2-eq. Credits derived from biochar represent the permanent sequestration of carbon since they capture carbon in a stable form for a prolonged period of time. Government initiatives may allow farmers, land managers and anybody using biochar in full-scale applications to earn carbon credits by reducing GHGs and storing carbon. The average American market price for biochar varied from $600 to $1300/ton in 2021, much lower than the price of $2850/ton in 2013 (Thengane et al., 2021).

Sorensen and Lamb (2018) estimated the purchasing and transporting costs for red oak biochar for full-scale application at $290/ton. Maroušek (2014) suggested an estimation of $500/ton for pristine BC. However,, Rosales et al. (2017) proposed a cost of around $0.076/kg for biochar production, which is about 3–6% of other commercial carbonaceous sorbents' price. Tian et al. (2020) highlighted that there are several uncertainties embedded in the techno-economic analysis because the price of biochar presents a wide range variation from $50 to $1900/tonne.

This highlights the potential of biochar production technologies making them attractive candidates for several applications in the treatment of polluted soils, wastewater, sludge, etc. For instance, Zhou et al. (2018) developed a low-cost in-situ treatment setup called "burning and soil covering" to be proposed to farmers on crop production residues. As previously mentioned, the impregnation with metallic compounds has also been used recently to improve the properties of biochar. In this regard, Chen et al. (2018) proposed a cost of $1.2/tonne for Mg-impregnated biochar with a high efficiency of 89.25% used for phosphorus removal. Kamali et al. (2019) underlined that under the current situation, the production costs are $0.03–1.21/g for TiO_2 nanoparticles and $0.05–0.10/g for nano zero-valent iron particles, which are quite high compared to those reported for biochar in the application of engineered nanomaterials for contaminated wastewater treatment.

17.4 Future Trends

Future research should better demonstrate all possible applications of biochar in an integrated holistic framework for water and wastewater treatment. From a circular economy perspective, converting sewage sludge into biochar via pyrolysis should be part of the treatment system as a prospective strategy for recycling nutrients and reducing environmental risks. Tarelho et al. (2020) analyzed the pyrolysis of biological sludge from pulp and paper industry wastewater treatment to produce biochar. Sun et al. (2022) have demonstrated that based on a 20-tonne industrial pyrolysis furnace at least 0.252 tonnes of CO_2 eq could be sequestered as stable C in sewage sludge-derived biochar, depending on the final application. Yet, the feasibility of pyrolyzing sewage sludge still remains uncertain from the perspective of carbon balance. In fact, the net C emissions from traditional sewage sludge treatments amounted to at least 2.432 tonnes of CO_2 eq per tonne of dry sludge under similar scenarios. Furthermore, they highlighted that pyrolysis could decrease the risk of contamination of the sewage sludge through the immobilization of heavy metals and the removal of organic contaminants and odorants. Shifting environmental impacts from one area to another is one of the risks associated when developing solutions for environmental problems. Also, biochar production can vary according to feedstock options, conversion technologies, and end uses (Anderson et al., 2016). The main technologies to convert organic waste to biochar include slow and fast pyrolysis, but also hydrothermal carbonization. Pyrolysis temperature can strongly influence the chemical properties of biochar with important effects on its performance as a sorbent or soil amendment. Geographically, these variations become even more pronounced with different supply chain contexts.

17.5 Conclusions

This chapter demonstrated that biochar can be considered as an economic and environmentally friendly solution to deal with polluted wastewater and sludge when compared to other conventional technologies. Biochar can be also adopted as an alternative sorbent to activated carbon in WWTFs. The production of biochar and other biofuels from sewage sludge through pyrolysis follows the circular economy principles, providing an innovative and sustainable way to create combustibles and sorbents. However, there may be some variations in biochar performance based on the pyrolysis conditions and feedstock materials. Future LCA should be tailored to specific applications. Lastly, a holistic approach is required to account for the total environmental impacts of any application or service at every stage of the biochar life cycle.

References

Agirre, I., Griessacher, T., Rösler, G., Antrekowitsch, J., 2013. Production of charcoal as an alternative reducing agent from agricultural residues using a semi-continuous semi-pilot scale pyrolysis screw reactor. *Fuel Processing Technology*, 106, 114–121. https://doi.org/10.1016/J.FUPROC.2012.07.010.

Alatrista Gongora, G.R., Lu, R.H., El Hanandeh, A., 2021. Comparative life cycle assessment of aerobic treatment units and constructed wetlands as onsite wastewater treatment systems in Australia. *Water Science and Technology*, 84, 1527–1540. https://doi.org/10.2166/wst.2021.316.

Anderson, N.M., Bergman, R.D., Page-Dumroese, D.S., 2016. A supply chain approach to biochar systems. London, U.C.P. (Ed.), No. Chapter 2. pp. 25–26.

Azzi, E.S., Karltun, E., Sundberg, C., 2019. Prospective life cycle assessment of large-scale biochar production and use for negative emissions in stockholm. *Environmental Science & Technology*, 53, 8466–8476. https://doi.org/10.1021/acs.est.9b01615.

Bae, S., Kim, Y. M., 2021. Carbon-neutrality in wastewater treatment plants: Advanced technologies for efficient operation and energy/resource recovery. *Energies*, 14(24), 8514.

Banik, C., Lawrinenko, M., Bakshi, S., Laird, D.A., 2018. Impact of pyrolysis temperature and feedstock on surface charge and functional group chemistry of biochars. *Journal of Environmental Quality*, 47, 452–461. https://doi.org/10.2134/jeq2017.11.0432.

Borzooei, S., Campo, G., Cerutti, A., Meucci, L., Panepinto, D., Ravina, M., ... Zanetti, M., 2019. Optimization of the wastewater treatment plant: From energy saving to environmental impact mitigation. *Science of the Total Environment*, 691, 1182–1189.

Brassard, P., Godbout, S., Hamelin, L., 2021. Framework for consequential life cycle assessment of pyrolysis biorefineries: A case study for the conversion of primary forestry residues. *Renewable and Sustainable Energy Reviews*, 138. https://doi.org/10.1016/j.rser.2020.110549.

Brassard, P., Godbout, S., Palacios, J.H., Le Roux, É., Alvarez-Chavez, B.J., Raghavan, V., Hamelin, L., 2019. Bio-based products from woody biomass pyrolysis for a sustainable bioeconomy, in: *27th European Biomass Conference (EUBCE)-Setting the Course for a Biobased Economy*. pp. 1234–1240.

Brassard, P., Godbout, S., Raghavan, V., 2017. Pyrolysis in auger reactors for biochar and bio-oil production: A review. *Biosystems Engineering*, 161, 80–92. https://doi.org/10.1016/j.biosystemseng.2017.06.020.

Bridle, T. R., Pritchard, D., 2004. Energy and nutrient recovery from sewage sludge via pyrolysis. *Water Sci Tech.*, 50, 69–175.

Brown, J.N., Brown, R.C., 2012. Process optimization of an auger pyrolyzer with heat carrier using response surface methodology. *Bioresource Technology*, 103, 405–414. https://doi.org/10.1016/J.BIORTECH.2011.09.117.

Chen, Q., Qin, J., Cheng, Z., Huang, L., Sun, P., Chen, L., Shen, G., 2018. Synthesis of a stable magnesium-impregnated biochar and its reduction of phosphorus leaching from soil. *Chemosphere*, 199, 402–408. https://doi.org/10.1016/J.CHEMOSPHERE.2018.02.058.

Chen, X., Chen, G., Chen, L., Chen, Y., Lehmann, J., McBride, M.B., Hay, A.G., 2011. Adsorption of copper and zinc by biochars produced from pyrolysis of hardwood and corn straw in aqueous solution. *Bioresource Technology*, 102, 8877–8884. https://doi.org/10.1016/J.BIORTECH.2011.06.078.

Chen, W.T., Haque, M.A., Lu, T., Aierzhati, A., Reimonn, G., 2020. A perspective on hydrothermal processing of sewage sludge. *Current Opinion in Environmental Science & Health*, 14, 63–73.

Choi, Y.K., Jang, H.M., Kan, E., Wallace, A.R., Sun, W., 2019. Adsorption of phosphate in water on a novel calcium hydroxide-coated dairy manure-derived biochar. *Environmental Engineering Research*, 24, 434–442. https://doi.org/10.4491/EER.2018.296.

Choudhary, V., Philip, L., 2022. Sustainability assessment of acid-modified biochar as adsorbent for the removal of pharmaceuticals and personal care products from secondary treated wastewater. *Journal of Environmental Chemical Engineering*, 10, https://doi.org/10.1016/j.jece.2022.107592.

De Feo, G., Ferrara, C., 2017. A procedure for evaluating the most environmentally sound alternative between two on-site small-scale wastewater treatment systems. *Journal of Cleaner Production*, 164, 124–136. https://doi.org/10.1016/j.jclepro.2017.06.205.

Ferreira, S.D., Altafini, C.R., Perondi, D., Godinho, M., 2015. Pyrolysis of medium density fiberboard (MDF) wastes in a screw reactor. *Energy Conversion and Management*, 92, 223–233. https://doi.org/10.1016/J.ENCONMAN.2014.12.032.

Fidel, R.B., Laird, D.A., Spokas, K.A., 2018. Sorption of ammonium and nitrate to biochars is electrostatic and pH-dependent. *Scientific Reports* 2018, 8(1), 1–10. https://doi.org/10.1038/s41598-018-35534-w.

Gupta, S., 2021. Carbon sequestration in cementitious matrix containing pyrogenic carbon from waste biomass: A comparison of external and internal carbonation approach. *Journal of Building Engineering*, 43, 102910. https://doi.org/10.1016/j.jobe.2021.102910.

Hansson, A., Haikola, S., Fridahl, M., Yanda, P., Mabhuye, E., Pauline, N., 2021. Biochar as multi-purpose sustainable technology: Experiences from projects in Tanzania. *Environment, Development and Sustainability*, 23, 5182–5214. https://doi.org/10.1007/s10668-020-00809-8.

Hao, X., Wang, X., Liu, R., Li, S., van Loosdrecht, M.C., Pauline, N., Jiang, H., 2019. Environmental impacts of resource recovery from wastewater treatment plants. *Water Research*, 160, 268–277.

Hospido, A., Moreira, T., Martín, M., Rigola, M., Feijoo, G., 2005. Environmental evaluation of different treatment processes for sludge from urban wastewater treatments: Anaerobic digestion versus thermal processes. *The International Journal of Life Cycle Assessment*, 10, 336–345.

ISO, 2006a. *ISO 14040: Environmental management: Life cycle assessment, principles and guidelines*. International Organization for Standardization, Geneva.

ISO, 2006b. *ISO 14044: International Standard. Environmental Management–Life Cycle Assessment- Requirements and Guidelines*. International Organisation for Standardization, Geneva, Switzerland.

Jin, H., Hanif, M.U., Capareda, S., Chang, Z., Huang, H., Ai, Y., 2016. Copper (II) removal potential from aqueous solution by pyrolysis biochar derived from anaerobically digested algae-dairy-manure and effect of KOH activation. *Environmental Chemical Engineering*, 4 (1), 365–372.

Joubert, J.E., Carrier, M., Dahmen, N., Stahl, R., Knoetze, J.H., 2015. Inherent process variations between fast pyrolysis technologies: A case study on Eucalyptus grandis. *Fuel Processing Technology*, 131, 389–395. https://doi.org/10.1016/J.FUPROC.2014.12.012.

Kamali, M., Persson, K.M., Costa, M.E., Capela, I., 2019. Sustainability criteria for assessing nanotechnology applicability in industrial wastewater treatment: Current status and future outlook. *Environment international*, 125, 261–276. https://doi.org/10.1016/J.ENVINT.2019.01.055.

Kwon, G., Bhatnagar, A., Wang, H., Kwon, E.E., Song, H., 2020. A review of recent advancements in utilization of biomass and industrial wastes into engineered biochar. *Journal of Hazardous Materials*, 400, 123242. https://doi.org/10.1016/J.JHAZMAT.2020.123242.

Le Roux, É., Chaouch, M., Diouf, P.N., Stevanovic, T., 2015. Impact of a pressurized hot water treatment on the quality of bio-oil produced from aspen. *Biomass and Bioenergy*, 81, 202–209.

Li, B., Lv, W., Zhang, Q., Wang, T., Ma, L., 2014. Pyrolysis and catalytic upgrading of pine wood in a combination of auger reactor and fixed bed. *Fuel*, 129, 61–67. https://doi.org/10.1016/J.FUEL.2014.03.043.

Liang, S., Han, Y., Wei, L., McDonald, A.G., 2015. Production and characterization of bio-oil and bio-char from pyrolysis of potato peel wastes. *Biomass Conversion and Biorefinery*, 5, 237–246. https://doi.org/10.1007/s13399-014-0130-x.

Liaw, S.S., Wang, Z., Ndegwa, P., Frear, C., Ha, S., Li, C.Z., Garcia-Perez, M., 2012. Effect of pyrolysis temperature on the yield and properties of bio-oils obtained from the auger pyrolysis of Douglas Fir wood. *Journal of Analytical and Applied Pyrolysis*, 93, 52–62. https://doi.org/10.1016/J.JAAP.2011.09.011.

Lu, H., Zhang, W., Yang, Y., Huang, X., Wang, S., Qiu, R., 2012. Relative distribution of Pb^{2+} sorption mechanisms by sludge-derived biochar. *Water Research*, 46, 854–862. https://doi.org/10.1016/J.WATRES.2011.11.058.

Maroušek, J., 2014. Significant breakthrough in biochar cost reduction. *Clean Technologies and Environmental Policy*, 16, 1821–1825. https://doi.org/10.1007/s10098-014-0730-y.

Martínez, J.D., Murillo, R., García, T., Veses, A., 2013. Demonstration of the waste tire pyrolysis process on pilot scale in a continuous auger reactor. *Journal of Hazardous Materials*, 261, 637–645. https://doi.org/10.1016/J.JHAZMAT.2013.07.077.

Matuštík, J., Hnátková, T., Kočí, V., 2020. Life cycle assessment of biochar-to-soil systems: A review. *Journal of Cleaner Production*, 259, 120998. https://doi.org/10.1016/J.JCLEPRO.2020.120998.

Menezes, L.N.B., Silveira, E.A., Mazzoni, J.V.S., Evaristo, R.B., Rodrigues, J.S., Lamas, G.C., Suarez, P.A.Z., Ghesti, G.F., 2022. Alternative valuation pathways for primary, secondary, and tertiary sewage sludge: Biochar and bio-oil production for sustainable energy. *Biomass Conversion and Biorefinery*, 1–14.

Miranda, C., Soares, A.S., Coelho, A.C., Trindade, H., Teixeira, C.A., 2021. Environmental implications of stored cattle slurry treatment with sulphuric acid and biochar: A life cycle assessment approach. *Environmental Research*, 194. https://doi.org/10.1016/j.envres.2020.110640.

Moradi-Choghamarani, F., Moosavi, A.A., Baghernejad, M., 2019. Determining organo-chemical composition of sugarcane bagasse-derived biochar as a function of pyrolysis temperature using proximate and Fourier transform infrared analyses. *Journal of Thermal Analysis and Calorimetry*, 138, 331–342. https://doi.org/10.1007/s10973-019-08186-9.

Muñoz, E., Curaqueo, G., Cea, M., Vera, L., Navia, R., 2017. Environmental hotspots in the life cycle of a biochar-soil system. *Journal of Cleaner Production*, 158, 1–7. https://doi.org/10.1016/J.JCLEPRO.2017.04.163.

Oldfield, T.L., Sikirica, N., Mondini, C., López, G., Kuikman, P.J., Holden, N.M., 2018. Biochar, compost and biochar-compost blend as options to recover nutrients and sequester carbon. *Journal of Environmental Management*, 218, 465–476. https://doi.org/10.1016/J.JENVMAN.2018.04.061.

Pittman, C.U., Mohan, D., Eseyin, A., Li, Q., Ingram, L., Hassan, E.B.M., Mitchell, B., Guo, H., Steele, P.H., 2012. Characterization of bio-oils produced from fast pyrolysis of corn stalks in an auger reactor. *Energy and Fuels*, 26, 3816–3825. https://doi.org/10.1021/EF3003922.

Qambrani, N.A., Rahman, M.M., Won, S., Shim, S., Ra, C., 2017. Biochar properties and eco-friendly applications for climate change mitigation, waste management, and wastewater treatment: A review. *Renewable and Sustainable Energy Reviews*, https://doi.org/10.1016/j.rser.2017.05.057.

Rosales, E., Meijide, J., Pazos, M., Sanromán, M.A., 2017. Challenges and recent advances in biochar as low-cost biosorbent: From batch assays to continuous-flow systems. *Bioresource Technology*, 246, 176–192. https://doi.org/10.1016/J.BIORTECH.2017.06.084.

Sharma, M., Singh, J., Baskar, C., Kumar, A., 2018. A comprehensive review on biochar formation and its utilization for wastewater treatment. *Pollution Research*, 37, S1–S18.

Shin, J., Kwak, J., Lee, Y.G., Kim, S., Choi, M., Bae, S., Choi, M.,2021. Competitive adsorption of pharmaceuticals in lake water and wastewater effluent by pristine and NaOH-activated biochars from spent coffee wastes: Contribution of hydrophobic and π-π interactions. *Environmental Pollution*, 270, 116244.

Shin, J., Bae, S., Chon, K., 2021. Fenton oxidation of synthetic food dyes by Fe-embedded coffee biochar catalysts prepared at different pyrolysis temperatures: A mechanism study. *Chemical Engineering Journal*, 421, 129943.

Sirijanusorn, S., Sriprateep, K., Pattiya, A., 2013. Pyrolysis of cassava rhizome in a counter-rotating twin screw reactor unit. *Bioresource Technology*, 139, 343–348. https://doi.org/10.1016/J.BIORTECH.2013.04.024.

Solar, J., de Marco, I., Caballero, B.M., Lopez-Urionabarrenechea, A., Rodriguez, N., Agirre, I., Adrados, A., 2016. Influence of temperature and residence time in the pyrolysis of woody biomass waste in a continuous screw reactor. *Biomass Bioenergy*, 95, 416–423. https://doi.org/10.1016/J.BIOMBIOE.2016.07.004.

Sorensen, R.B., Lamb, M.C., 2018. Return on investment from biochar application. *Crop, Forage & Turfgrass Management*, 4, 1–6. https://doi.org/10.2134/CFTM2018.02.0008.

Sun, H., Luo, L., Wang, D., Liu, W., Lan, Y., Yang, T., Gai, C., Liu, Z., 2022. Carbon balance analysis of sewage sludge biochar-to-soil system. *Journal of Cleaner Production*, 358, 132057. https://doi.org/10.1016/J.JCLEPRO.2022.132057.

Tarelho, L.A.C., Hauschild, T., Vilas-Boas, A.C.M., Silva, D.F.R., Matos, M.A.A., 2020. Biochar from pyrolysis of biological sludge from wastewater treatment. in: *Energy Reports*, Elsevier Ltd, 757–763, Sá Caetano, N., Borrego, C., Nunes, M. I., Felgueiras, M. C. https://doi.org/10.1016/j.egyr.2019.09.063.

Tarpani, R.R.Z., Azapagic, A., 2018. Life cycle environmental impacts of advanced wastewater treatment techniques for removal of pharmaceuticals and personal care products (PPCPs). *Journal of Environmental Management*, 215, 258–272.

Thengane, S.K., Kung, K., Hunt, J., Gilani, H.R., Lim, C.J., Sokhansanj, S., Sanchez, D.L., 2021. Market prospects for biochar production and application in California. *Biofuels, Bioproducts and Biorefining*, 15, 1802–1819. https://doi.org/10.1002/BBB.2280.

Thompson, K.A., Shimabuku, K.K., Kearns, J.P., Knappe, D.R.U., Summers, R.S., Cook, S.M., 2016. Environmental comparison of biochar and activated carbon for tertiary wastewater treatment. *Environmental Science & Technology*, 50, 11253–11262. https://doi.org/10.1021/acs.est.6b03239.

Tian, X., Richardson, R.E., Tester, J.W., Lozano, J.L., You, F., 2020. retrofitting municipal wastewater treatment facilities toward a greener and circular economy by virtue of resource recovery: Techno-economic analysis and life cycle assessment. *ACS Sustainable Chemistry & Engineering*, 8, 13823–13837. https://doi.org/10.1021/acssuschemeng.0c05189.

Tomasi Morgano, M., Leibold, H., Richter, F., Seifert, H., 2015. Screw pyrolysis with integrated sequential hot gas filtration. *Journal of Analytical and Applied Pyrolysis*, 113, 216–224. https://doi.org/10.1016/J.JAAP.2014.12.019.

Tröger, N., Richter, D., Stahl, R., 2013. Effect of feedstock composition on product yields and energy recovery rates of fast pyrolysis products from different straw types. *Journal of Analytical and Applied Pyrolysis*, 100, 158–165. https://doi.org/10.1016/J.JAAP.2012.12.012.

Uddin, M.M., Wright, M.M., 2022. Chapter 21, Life Cycle Analysis of Biochar Use in Water Treatment Plants. in: *Sustainable Biochar for Water and Wastewater*

Treatment, Elsevier, Mohan, D., Pittman, C.U. Mlsna, T.E. 705–735. https://doi.org/10.1016/B978-0-12-822225-6.00012-9.

Veses, A., Aznar, M., López, J.M., Callén, M.S., Murillo, R., García, T., 2015. Production of upgraded bio-oils by biomass catalytic pyrolysis in an auger reactor using low cost materials. *Fuel*, 141, 17–22. https://doi.org/10.1016/J.FUEL.2014.10.044.

Veses, A., Aznar, M., Martínez, I., Martínez, J.D., López, J.M., Navarro, M.V., Callén, M.S., Murillo, R., García, T., 2014. Catalytic pyrolysis of wood biomass in an auger reactor using calcium-based catalysts. *Bioresource Technology*, 162, 250–258. https://doi.org/10.1016/J.BIORTECH.2014.03.146

Wang, W.D., Cui, Y.X., Zhang, L.K., Li, Y.M., Sun, P., Han, J.H., 2021. Synthesis of a novel $ZnFe_2O_4$/porous biochar magnetic composite for Th(IV) adsorption in aqueous solutions. *International Journal of Environmental Science and Technology*, 18, 2733–2746. https://doi.org/10.1007/s13762-020-03023-1.

Woolf, D., Lehmann, J., Ogle, S., Kishimoto-Mo, A.W., McConkey, B., Baldock, J., 2021. Greenhouse gas inventory model for biochar additions to soil. *Environmental Science & Technology*, 55, 14795–14805. https://doi.org/10.1021/acs.est.1c02425.

Yang, Q., Mašek, O., Zhao, L., Nan, H., Yu, S., Yin, J., Li, Z., Cao, X., 2021. Country-level potential of carbon sequestration and environmental benefits by utilizing crop residues for biochar implementation. *Applied Energy* 282, https://doi.org/10.1016/j.apenergy.2020.116275.

Zhao, S., Ta, N., Wang, X., 2020. Absorption of Cu (II) and Zn (II) from aqueous solutions onto biochars derived from apple tree branches. *Energies*, 13(13), 3498. https://doi.org/10.1021/acs.est.1c02425.

Zhou, Q., Houge, B.A., Tong, Z., Gao, B., Liu, G., 2018. An in-situ technique for producing low-cost agricultural biochar. *Pedosphere*, 28, 690–695. https://doi.org/10.1016/S1002-0160(17)60482-X.

Index

Note: page numbers in *italics* refer to figures; those in **bold** to tables.

a

acenaphthene   72
acid activation   14–15
acidification potential   298
acid modification   10, 14, 24, 56, 57, 217
activated carbon   42–43, 62, 72, 77, 245, 252, 292, 298
activation,
  acid   14–15
  alkaline   15–16
  chemical   178, 179
  gas   11–12
  physical   178–179
  steam   10–11
adsorption,
  and functional group   252–253
  of heavy metals   41–51, 68
  inhibitors   91–92
  methods   23–24, 41
  of nutrients   29–38
  and odor control   254–256, **257**
  of pharmaceutical and personal care products (PPCPs)   53–59
  physical   43–44, 57
advanced oxidation processes (AOPs),
  future perspectives   204–205
  peroxide-based   193–205
  persulfate-based   213–223
  sulfate radical-based (SR-AOP)   213–215, 219, 222, 223
aerobic composting   244
aerobic/oxic (A/O) process   30
aerobic treatment units (ATUs)   295
agricultural waste   2, 7
AhR *see* aryl hydrocarbon receptors
aldehydes   245, **247**, 282
algal blooms   30, 64
alkali modification   10, 14, 50, 57, 68, 217
alkaline activation   15–16
alkylphenols   72
"aluminum bridge"   33
aluminum chloride   299
aluminum-doped biochar   33
aluminum oxides   31
ammonia (NH_3)   11–12, 20–21, 298
  free   107, 111–112
  inhibition   91
  odor control   243, 244, 251, 255, 260
  removal from wastewater   30, 36
  sludge composting   144–146
ammonium (NH_4)   21
  desorption   38
  removal from wastewater   31–32, 36–37
  see also ammonium oxidizing bacteria (AOB); anaerobic ammonium oxidation (anammox)

Biochar Applications for Wastewater Treatment, First Edition. Edited by Daniel C.W. Tsang and Yuqing Sun.
© 2023 John Wiley & Sons, Inc. Published 2023 by John Wiley & Sons, Inc.

Index

ammonium oxidizing bacteria (AOB) *106*, 108
ammonium persulfate [(NH$_4$)$_2$S$_2$O$_8$] 14
amphoteric characteristics 58, 204, 275
anaerobic ammonium oxidation (anammox) 30, 105–115
   constraints 107–108
   electron transfer 112–113
   future perspectives 114–115
   microbial immobilization 113–114
   overview 105–108
   pH buffering 111–112
   role of biochar in promoting 108–114
anaerobic digestion 89–100, 290
   adsorption of inhibitors 91–92
   biochar as an additive 90–99
   digestate quality 99–100
   microbial growth and activities 92–99
   odor control 244, 255
   pH buffering 90–91
Anaerolineaceae 98
anammox bacteria (AnAOB) 106–114
   difficulties in enrichment 107–108
   interactions with other microbial groups 108
AnAOB *see* anammox bacteria
aniline 20, 22
anion exchange capacity (AEC) 299–300
anions 37–38
annamox *see* anaerobic ammonium oxidation
antibiotic-resistant bacteria (ARB) 54
antibiotic-resistant genes (ARG) 54
antibiotics 54
   see also tetracycline
antifreeze 72
AOB *see* ammonium oxidizing bacteria
A/O process *see* aerobic/oxic (A/O) process
AOPs *see* advanced oxidation processes
aquatic life 64
ARB *see* antibiotic-resistant bacteria
archaea 97–98
ARG *see* antibiotic-resistant genes
arginine 160
arsenic (As) 47–50, 174
aryl hydrocarbon receptors (AhR) 281
atenolol 56
ATR *see* attenuated total reflection

atrazine (ATZ) 74, 220, 232–235
attenuated total reflection (ATR) 8
ATUs *see* aerobic treatment units
ATZ *see* atrazine
Australia 61
azo dyes 181

b

ball milling 13–14, 198, 252
bentonites 18, 31, 32
best management practice (BMP) 61
BET *see* Brunauer-Emmett-Teller (BET) surface area
BIM *see* biochar–immobilizing microorganisms
bimetallic catalysts 197–198, 220
biochar-immobilizing microorganisms (BIM) 23
biochar/layered double hydroxide composites 46–47
biochar modification 9–25
   acid 10, 14, 24, 56, 57, 217
   alkali 10, 14, 24, 50, 57, 68, 217
   chemical 14
   and heavy metal adsorption 46–51
   magnetic 24, 56, 57
   mineral 24
   physical 10–14, 57
   and PPCPs adsorption 57–58
   sustainability 24–25
   thermal 216–217
biochar production 1, 66–67, 100, 159–160
   life cycle assessment (LCA) analysis 291–295
   overview 2–4
biochar properties 4–9
   magnetic 47–48, 58, 300
   and odor control 247–254
   physicochemical 4, 66, 67, 78, 196–197
biochar recycling 51
biochar stability 275–277
biofertilizers 244
biofiltration 62, 245
   removal of metal/metalloids 67–69
   removal of nutrients 70–72
   removal of organic pollutants 72–75
   stormwater 61–78

Index

biogas   89, 159, 255, 256, 280, 290
biomass,
   heating rate   159
   microbial   162
   performance   292
   pyrolysis   3–5
   thermochemical decomposition of 1, 3
bioreactors   245, 299
bioretention *see* biofiltration
bioscrubbers   245
biosolids   4, 5, 6, 67
bioswales   62
biotechnologies   245
bisphenol A (BPA)   75, 196, 203, 222, 233
Boehm titration method   7, 236
boron   20, 198
Boudouart reaction   11
BPA *see* bisphenol A
Brunauer-Emmett-Teller (BET) surface area   13
bulk conductivity   173
"burning and soil covering" technology   301
2-butanone   255
butyrate   98–99

C

cadmium (Cd)   46, 47, 48, 50
Caenorhabditis elegans   281
calcium phosphate   35
Canada   61
Candidatus Anammoximicrobium   106
Candidatus Anammoxoglobus   106
Candidatus Brocadia   106
Candidatus Jettenia   106
Candidatus Kuenenia   106
Candidatus Scalindua   106
capacitive deionization   173–176, 179, 182
capillary suction time (CST)   122, 126
carbamazepine   298
carbon,
   activated   42–43, 62, 72, 77, 245, 252, 292, 298
   capture and storage   1, 300–301
   credits   301
   electrodes   172

   nanofibers   12
   neutrality   289
   porous   179–180, 250, 252
   sequestration   290, 302
   see also carbon nanotubes (CNTs); dissolved organic carbon (DOC)
carbonaceous adsorbents   252
carbonaceous materials   14, 22–23, 62, 194, 221, 237
carbonaceous nanocomposites   22–23
carbon dioxide (CO_2)   11, 50, 144, 216, 289, 300
carbon disulphide   247
carbonization   126
   hydrothermal   2, 20, 127, 139, 177–178, 183, 253
   in preparation of electrode materials   177–178
carbonized organic matter (COM)   253
carbon nanotubes (CNTs) 22–24, 172
carcinogens   281
catalysts   195–199, 220
cation exchange capacity (CEC)   6, 18, 31, 32, 34, 45, 67, 162, 299–300
cations   37
cat litter   245, 255
CEC *see* cation exchange capacity
cellulose   3–4, 67, 76, 231, 280
cerussite ($PbCO_3$)   46
chemical decomposition   275–276
chemical functional groups   252–253
chemical modification   14
chemical pollutants   64, 72–75
chemical scrubbers   245
chemisorption   38, 43
China   61, 138, 247, **248**
chitosan   22
chlorine   292
chloroform   235
4-chlorophenol   203
chromium (Cr)   45–47, 50, 174
ciprofloxacin   180
citric acid   15
classical capacitive deionization (CDI)   175, 176, 179, 180
clay–biochar composites   18–19, 58
clays   18–19, 58
climate change   1

clofibric acid   202
Clostridiaceae   92
^{13}C NMR spectra   8
C/N ratios   5, 105, 138, 141–142, 256
CNTs *see* carbon nanotubes
cobalt   231
cobalt-based biochar   219–220
COM *see* carbonized organic matter
composites   16–24, 180
   biochar/layered double
      hydroxide   46–47
   heteroatom-doped   20–22
   layered double
      hydroxide–biochar   19–20
   magnetic-biochar   47–48, 58
   metal–biochar   16–18, 217–220
   microorganism–biochar   23–24
   mineral–biochar   18–19
   nZVI/biochar   48–49
   see also nanocomposites
composting,
   aerobic   244
   biochar as additives   256–260
   microplastics   138
   see also sludge composting
constructed wetlands (CWs)   295
copper-based biochar   219
copper oxide (CuO)   198, 219–220
co-pyrolysis   24–25, 221
corn-core powder   125
corrosion inhibitors   72
costs   77, 301
covalent bonding   23, 24, 43
CPMAS ^{13}C NMR (solid-state ^{13}C cross-polarization magic-angle nuclear magnetic resonance)   7
cresol   3, 282
Cr(OH)$_3$, 46
CST *see* capillary suction time
Cu(II)   45, 46, 47, 48, 50
CWs *see* constructed wetlands
cyclodextrin   55

d

decomposition   275–277
   chemical   275–276
   microbial   276–277
   physical   275
   thermochemical   1, 3

denitrification   30, 70, 71, 105, 144
density functional theory (DFT)   222
desorption,
   ammonium   38
   heavy metals   51
   nitrate   38–39
   nutrients   38–39
   phosphorous   39
2,4-dichlorophenoxy acetic acid   197
(2,4-D) dichlorophenoxyacetic acid 74, 232–234
DIET *see* direct interspecies electron transfer
diffuse reflection *see* DRIFT
dimethyl disulfide (DMDS)   22, 255
2,4-dimethylphenol   282
dimethyl sulfide   247
dimethyl trisulfide (DMTS)   255
dioxin-like polychlorinated biphenyls (DL-PCBs)   281
dioxins   277, 281
direct interspecies electron transfer (DIET)   90, 92, 97–99
dissolved organic carbon (DOC)   75, 143
Diuron (DIU)   74
DL-PCBs *see* dioxin-like polychlorinated biphenyls
DOC *see* dissolved organic carbon
DRIFT (diffuse reflectance)   8
dry absorption   245
dyestuffs   230
   see also azo dyes; methyl orange; Prussian blue

e

EBRT *see* empty bed residence time
EC *see* electrical conductivity
economic evaluation   299–301
   see also costs
EDS *see* energy dispersive spectroscopy
electrical conductivity (EC)   98–99, 217
electric field-assisted pyrolysis   20
electro-adsorption   182–183
   see also capacitive deionization
electrochemical behaviours   173–177
electrochemical deposition   182, 184
electrochemical disinfection   183, 184
electrochemical oxidation   181–182, 184

electrochemical wastewater treatment   171–184
  future perspectives   183–184
  preparation of electrode materials   177–180
  technologies   181–183
electrode materials   177–180
electron acceptors   174
electron donors   174
electron exchange   173–174
electron paramagnetic resonance (EPR)   215, 220
electron transfer (ET)   49, 89, 98–99, 112–113, 202, 215, 237–238
  see also direct interspecies electron transfer (DIET); indirect interspecies electron transfer (IIET)
electrosorption capacity   174–177, 180
electrostatic interaction,
  adsorption of heavy metals   44–45
  adsorption of PPCPs   55, 56
Emission Standards for Odor Pollutants (China)   247
empty bed residence time (EBRT)   245
endocrine disruptors   230
energy dispersive spectroscopy (EDS)   6
energy production   290
Enhanced Biological Phosphorus (EBPR)   30
environmental evaluation   289–299
environmental impact   273–282
  biochar stability   275–277
  contaminants in biochar   277–282
  transport mechanism of biochar   274
environmentally persistent free radicals (EPFRs)   114
  see also persistent free radicals (PFRs)
EPR *see* electron paramagnetic resonance
EPS *see* extracellular polymeric substances
ET *see* electron transfer
ethanol (EtOH)   13, 215
ethyl acetate   252
eutrophication   30, 147, 295
extracellular polymeric substances (EPS)   112, 121, 126, 130

f

Fe-loaded biochar   131, 233, 238–239
Fenton-like reactions   122, 193–194, 197–201, *204*, 205
Fenton process   180, 181
fertilizer,
  biochar as   29, 39–40, 162–163, 256
  organic   156, 256
  pollution from   64
  sludge compost as   137–138, 144–145, 156
  see also biofertilizers
Fick's law   97
fipronil (FPR)   74
fish   64
flow electrode-capacitive deionization (FCDI)   175–176
folding point chlorination   30
food waste (FW)   93, 98, 99, 244
forestry waste   2
Fourier-transform infrared spectroscopy (FTIR)   8, 9
FPR *see* fipronil
free radical reaction   3
free radicals   3, 12, 114, 215, 220
  see also environmentally persistent free radicals (EPFRs); persistent free radicals (PFRs)
FTIR *see* Fourier-transform infrared spectroscopy
fulvic acid (FA)   145–146
furfuryl alcohol   215

g

GAC *see* granulated activated carbon
gas activation   11–12
gasification   2
Geobacter   98
global warming potential (GWP)   295
D-glucopyranose   3
granulated activated carbon (GAC)   98–100, 298
graphene   23, 181
graphite   7
graphitic biochar   20, 173–174, 197–198, 202
graphitization   20, 173–174, 216
greenhouse gases (GHG) emissions   1, 105, 289–290, 298, 301
  sludge composting   138, 143–145

wastewater treatment facilities (WWTFs)   289–290
green infrastructure (GI)   61
GWP *see* global warming potential

h

heating rates   159
heavy metals (HMs)   15
   adsorption of   41–51, 68
   and biofiltration systems   67–69
   desorption from biochar   51
   environmental impact   277, 279–280
   and sludge composting   142, 147
   and sludge dewatering   114, 130–131
   in stormwater   64
hemicellulose   3–4, 231, 280
heptane   13
herbicides   64, 72, 230
heteroatom doping   20–22, 195, 198–199, 220–222
heterotrophic bacteria   107–108
hexamethylcyclotrisiloxane   255
hexane   13
HMs *see* heavy metals
HOCs *see* hydrophobic organic compounds
HTT   4, 6, 7, 9
human toxicity   295, 298
humic acid (HA)   58, 145–146, 234
humic substances (HS)   145
humification process   140, 145, 258
humins (HU)   145
hydrocarbons   11, 72
   see also polycyclic aromatic hydrocarbons (PAHs)
hydrocerrusite [$Pb_3(CO_3)_2(OH)_2$]   46
hydrochar   2
hydrochloric acid (HCl)   14, 217
hydrogen bonding   56, 58, 180, 276
hydrogen peroxide (H_2O_2)   14, 16, 193, 213
hydrogen sulfide (H_2S)   138, 143
   odor control   244, 247, 249, 251, 253–255, 258, 260
hydrophilicity   9, 10, 57, 58, 178, 231
hydrophobicity   25, 34, 35, 57, 58, 93, 128, 159
hydrophobic organic compounds (HOCs)   72

hydrophobic partitioning   55
hydroxyl radicals 181, 193, 199, 213, 233–235

i

IBI *see* International Biochar Initiative
ibuprofen   59, 298
IET *see* interspecies electron transfer
IFAS-EBPR   30
IIET *see* indirect interspecies electron transfer
indirect interspecies electron transfer (IIET)   97
industrial production   244–245
inhibition   91
inhibitors   91–92, 107, 111–112
insecticides   64
Integrated Fixed-Film Activated Sludge Systems (IFAS)   30
International Biochar Initiative (IBI)   2
interspecies electron transfer (IET)   89–90
intraparticle diffusion   55
ion exchange,
   adsorption of heavy metals   45, 46
   adsorption of nutrients   34
   adsorption of PPCPs   55
ion storage   175–176, 182
iron-biochar   218
   see also Fe-loaded biochar; iron-coated biochar (FeBC); nanoscale zero-valent iron (nZVI)
iron carbide (Fe_3C)   218–219
iron-coated biochar (FeBC)   68, 72
iron oxide (Fe_2O_3)   50, 74, 125, 126, 218, 230

j

Japan   247, **248**

k

kaolinite   18
ketones   245, **247**, 282
ketoprofen   236, *237*

l

Lachnospiraceae   92
landfilling   244, 300

Index | **315**

layered double hydroxide (LDH) 19–20, 46–47
layered double hydroxide–biochar composites 19–20
LCA *see* life cycle assessment
LDH *see* layered double hydroxide
lead (Pb) 44, 46–50, 67, 68, 298
life cycle assessment (LCA) 290–299
   system boundaries 292, *293*
light absorption wavelengths 9
lignin 3, 67, 76, 231, 280
lignocellulosic materials 2, 5, 15, 178
lime 122
limonene 255
livestock industry 24
livestock manure 2, 4, 6, 7, 49, 64, 67, 71
   biological treatment 244
   composting 256–260
   odor control 255
living cells 23
low impact development (LID) 61
low impact urban design and development (LIUDD) 61

m

macropores 11, 43, 93, 176, 250–251
magnesium 33
magnesium ammonium phosphate (MAP) 30
magnesium phosphate 35
magnetic-biochar composites 47–48, 58
magnetic modification 24, 56, 57
magnetite (Fe_3O_4) 17, 71, 76, 98, 122, 218, 220
manganese 131, 220, 238–239
manganese dioxide (MnO_2) 180
manganese ferrite ($MnFe_2O_{4)}$ 131, 234–236
manganese oxide 50, 230
MAP *see* magnesium ammonium phosphate
MCCP *see* modified corn-core powder
MCN *see* metal cyanides 281–282
mechanochemical technique 13
melamine 20, 22
membrane-capacitive deionization (MCDI) 175

3-mercapto-3-methyl-1-butanol (MMB) 255–256
3-mercapto-3-methylbutyl-formate 255
mercury (Hg) 48, 50, 68
mesopores 43, 123, 160, 250–251
metal–biochar composites 16–18, 217–220
metal cyanides (MCN) 281–282
metalloids,
   adsorption of 41–51
   removal in biofiltration systems 67–69
metal oxides 17, 33, 36, 49–50, 197, 218, 229
metal salts 16–17
methane (CH_4),
   anaerobic digestion 89, 91, 97–99
   sludge composting 143–144
methanogens 89–91, 93, 97–98, 173
methanol 24
Methanosaeta 97, 98
Methanosarcina 93, 97
methyldiethanolamine 253
methylene blue (MB) 204, 233
methyl mercaptan 243, 244, 247
3-methyl-3-(2-methyldisul-fanyl)-1-butanol 255
3-methyl-3-methylthio-1-butanol 255
methyl orange 181
methylparaben 298
methyl sulfide 247
methyl tert-butyl ether (MTBE) 256
microbial colonization 92–97
microbial communities 23, 91, 98, 101, 114, 145, 277
microbial decomposition 276–277
microbial immobilization 23, 92–93, 113–114
microorganism–biochar composites 23–24
microorganisms,
   adsorption of PPCPs 55
   anaerobic digestion 92–99
   composting process 260
   sludge composting 145
   in stormwater 64

see also ammonium oxidizing bacteria (AOB); anammox bacteria (AnAOB); antibiotic-resistant bacteria (ARB); heterotrophic bacteria; microorganism–biochar composites; nitrite oxidizing bacteria (NOB); nitrogen-fixing bacteria
microplastics (MPs)   25, 75–76, 138, 289
 composting   138
 in urban runoff   75–76
micropores   7, 9, 11, 15–16, 44, 75, 159, 217, 250–251
microwave-assisted pyrolysis   12–13, 178, 199
microwave digestion   127
microwave pretreatment   231
mineral–biochar composites   18–19
mineral components   45, 174, 231, 253–254
MLE see modified Ludzak-Ettinger (MLE) process
MMB see 3-mercapto-3-methyl-1-butanol
modified corn-core powder (MCCP)   125–126
modified Ludzak-Ettinger (MLE) process   30
montmorillonite   19, 56–57
MPs see microplastics
MSW see municipal solid waste
MTBE see methyl tert-butyl ether
municipal solid waste (MSW)   2, 244
municipal solid waste incineration power plants (MSWIPP)   244

n

nanocarbons   20
nanocomposites   22–23
nanomaterials   20
nanoscale zero-valent iron (nZVI)   24, 46–49, 68, 197, 218
naphthalene   72
National Stormwater Quality Database (NSQD)   64, **65–66**
natural organic matter (NOM)   75, 214, 233–234, 275, 276
N-doping see nitrogen doping

negative emission technology (NET)   300
NET see negative emission technology
Netherlands   106
New Zealand   61
nickel (Ni)   48, 220
nitrate,
 desorption   38–39
 removal from wastewater   32–33
nitric acid (HNO_3)   14, 16, 21, 51, 217
nitrification   30, 70, 71, 105, 144
nitrite oxidizing bacteria (NOB)   *106*, 108
nitrogen,
 circulation   *106*
 removal from stormwater   64, 70–72
 removal from wastewater   29–31, 105
 in sludge-derived biochar (SDBC)   160–161
 in the soil   162–163
 see also nitrogen doping (N-doping); nitrogen-fixing bacteria; nitrogen removal rate (NRR)
nitrogen doping (N-doping)   20–22, 198–199, 221
nitrogen-fixing bacteria   145, 146
nitrogen removal rate (NRR)   105, 114
nitrous oxide (N_2O)   144, 146, 162
NMR analysis   7–8
NOB see nitrite oxidizing bacteria
NOM see natural organic matter; noncarbonized organic matter
noncarbonized organic matter (NOM)   253
nonsteroidal anti-inflammatory drugs (NSAIDs)   54
NRR see nitrogen removal rate
NSAIDs see nonsteroidal anti-inflammatory drugs
nutrients,
 adsorption   29–38
 biochars as suppliers of   39–40
 desorption   38–39
 removal from wastewater   31–38, 70–72
 in SDBC   160–163
 in sludge   155–164
 in stormwater   64, 70–72
nZVI see nanoscale zero-valent iron

Index | 317

O

odor,
    abatement technologies   245
    causes and treatment   244–244
    evaluation of emission   245
    man-made sources   244
    natural sources   244
    policies   247
    profiles   243, **246–247**
    thresholds   **246–247**
    see also odor control; odor pollutants
odor control   243–261
odor pollutants   245–247
Odor Prevention Act (Japan)   247
ofloxacin   204
organic pollutants,
    ozonation sewage treatment   233–234
    sludge composting   138, 140, 142–143
    in stormwater   72–75
organic waste   2–3, 31, 256, 258, 302
ORR *see* oxygen reduction reaction
overflow   63
oxalic acid   15
oxidation   16
    in chemical decomposition of biochar   275–276
    electrochemical   181–182, 184
see also advanced oxidation processes (AOPs); anaerobic ammonium oxidation (anammox)
oxygen reduction reaction (ORR)   181
ozonation   229–239
    effects of process conditions   233–235
    efficacy on sewage treatment   232
    implementation prospects   235–239
    life cycle costs   290
    preparation of catalyst   230–232
    technical mechanism   235–239
ozone (O_3)   14, 229, 232–238
    decomposition rate   235
    depletion potential   298

P

PAC *see* powdered activated carbon (PAC)
PAHs *see* polycyclic aromatic hydrocarbons
PAOs *see* phosphate-accumulating organisms
paper industry   301

PCBs *see* polychlorinated biphenyls
PCDDs *see* polychlorinated dibenzo-p-dioxins
PCDFs *see* polychlorinated dibenzofurans
PDS *see* peroxydisulfate
peak flow   61, 63
percarbonate   193
perfluorooctane sulfonic acid (PFOS)   74
perfluorooctanoic acid (PFOA)   74
peroxide-based AOP   193–205
peroxydisulfate (PDS)   193, 196, 198, 201–203, 213–214, 220–222
peroxymonosulfate (PMS)   125, 127, 131, 193, 196, 198, 201–202, 213–214, 217–222
persistent free radicals (PFRs)   130, 195, 203, 214–216
    environmental impact   277, 280–281
    see also environmentally persistent free radicals (EPFRs)
persulfate (PS)   126, 128
    see also persulfate-based AOP
persulfate-based AOP   193, 201–203, 213–223
pesticides   64, 72
PFOA *see* perfluorooctanoic acid
PFOS *see* perfluorooctane sulfonic acid
pH   5–6, 14
    and adsorption of nutrients   36
    and adsorption of PPCPs   59
    in anaerobic ammonium oxidation (anammox)   111–112
    in anaerobic digestion   90–91
    and removal of phosphorus in biofiltration systems   70
    sludge composting   141
    of the soil   162
pharmaceutical and personal care products (PPCPs),
    adsorption of   53–59
    properties of   58–59
    removal from effluents   290
phenanthrene   72, 74, 254
phenol   4, 252, 280
phosphate   33, 36, 37, 59
phosphate-accumulating organisms (PAOs)   30
phosphoric acid (H_3PO_4)   14, 15, 217

phosphorus,
  desorption 39
  removal 29–31, 35, 301
  in sludge-derived biochar (SDBC) 161
  in stormwater 64, 70
photocatalysis 203–204
phthalates 72
phthalic acid esters (PAEs) 142
phycocyanin 282
phyllosilicates 18
physical adsorption 43–44, 57
  see also physisorption
physical decomposition 275
physical modification 10–14, 57
physisorption 38–39, 249
  see also physical adsorption
phytic acid 252
π-π interaction 55–56, 112
Planctomycetes 106, 113
plastics 25, 138
  see also microplastics (MPs)
PMS *see* peroxymonosulfate
point of zero charge (pH$_{zpc}$) 44, 232
polychlorinated biphenyls (PCBs) 11, 64, 72, 142
polychlorinated dibenzofurans (PCDFs) 142, 281
polychlorinated dibenzo-p-dioxins (PCDDs) 142, 281
polycyclic aromatic hydrocarbons (PAHs) 11, 64, 72, 74, 142–143
  environmental impact 277–279
polyethyleneimine 22
pore filling 55–57, 74
pore size distribution 75, *124*, 125, 176
  and odor control 250–251
pore volume (PV),
  and odor control 249, 251, 254
potassium 161
potassium hydroxide (KOH) 15–16, 57, 179, 183, 197, 198, 217
potassium permanganate (KMnO$_4$) 14, 16, 125, 130
powdered activated carbon (PAC) 298–299
PPCPs *see* pharmaceutical and personal care products

precipitation 30, 35, 45–46
PRFs *see* persistent free radicals
propiconazole 254
propionate 98–99
Prussian blue 251, 252
Pseudomonadaceae 92
pyrene 72
pyridine 253
pyrogenic carbon capture and storage (PyCCS) 300
pyrolysis 2, 66
  biomass 3–5
  cellulose 3, 67, 280
  electric field-assisted 20
  fast 159, 177
  lignin 3, 67, 280
  microwave-assisted 11–12, 178, 199
  operating parameters **293–294**
  product yields **293–294**
  secondary 216–217
  slow 3, 159, 177, 298
  in synthesis of composites 17, 18
  temperature 4–7, 9, 11, 32, 35, 57, 67, 126, 139, 159, 215, 216
  see also co-pyrolysis
pyrrole 253
PZNC *see* zero net charge
PZSE *see* zero salt effect

r

rain gardens 62
raw materials 1–7, 159
reactive oxygen species (ROS) 114, 181, 193–194, 197–198, 203, 215, 230, 281
redox metal-loaded biochar 197–198
redox reactions 49
reduction 46
retention time 160
reverse osmosis 55
reverse voltage desorption (RVD) 175
ROS *see* reactive oxygen species
Rotterdam-Dokhaven Sewage Treatment Plant 106
runoff 29, 61–64, **65**, 67–68, 70, 72, 74–77

s

sand filtration systems 76
SBB *see* sludge-based biochar

Index | 319

scanning electron microscopy (SEM)   6–7, 35, *239*
Scenedesmus obliquus   281
SCMs *see* stormwater control measures
SDBC *see* sludge-derived biochar
seed germination   282
SEM *see* scanning electron microscopy
septic tank systems   64
sewage sludge management   155–156
   see also sludge
SFG *see* surface functional groups
skeleton builders   122, 127
sludge   2, 4, 7, 11, 138
   activated   155
   dewaterability   122, 126–127, 131, *132*
   filterability   126
   moisture content   121, 141–142
   nutrient recovery   155–164
   odor control   244
   pH   141
   primary   155
   secondary   155
   see also sludge-based biochar (SBB); sludge composting; sludge conditioning; sludge dewatering
sludge-based biochar (SBB)123–129, 131
   see also sludge-derived biochar
sludge composting   137–147
   future perspectives   146–147
   gaseous emissions   143–145
   heavy metals   142, 147
   microbial growth and activities   145
   organic pollutants   138, 140, 142–143
   parameters of   138–142
sludge conditioning   122–128
   and physicochemical characteristics   127–128
   preparation of conditioner   123–126
   and sludge dewaterability   126–127
sludge-derived biochar (SDBC)   156
   as a catalyst   203
   challenges   163–164
   nutrients in   160–161
   properties   156–160
   see also sludge-based biochar (SBB)
sludge dewatering   121–133
   closed-loop recycling system   131, *132*
   technical mechanism and implementation   128–133
Smithella   98

sodium borohydride ($NaBH_4$)   24
sodium hydroxide (NaOH)   15–16, 51, 217
soil,
   amendment   161–163
   biochar decomposition in   275–277
   improvement   63
   transport mechanism of biochar   274
South Korea   247, **248**
specific resistance of filtration (SRF)   122–123, 126–127
specific surface area (SSA)   34, 55, 74, 93, 100, 196–197, 216, 217
   and removal of odor pollutants   249
Sponge City   61
SR-AOP *see* sulfate radical-based advanced oxidation technology
SRF *see* specific resistance of filtration
SSA *see* specific surface area
steam activation   10–11
stormwater   64, **65–66**, 67–75
   biofiltration systems   61–78
   management   62
   organic pollutants in   72–75
   removal of chemical pollutants   64, 72–75
   removal of heavy metals   64
   removal of microorganisms   63
   removal of nutrients   64
   runoff   29, 61–62, 64, **65**, 72
stormwater control measures (SCMs)   62
styrene   247
sulfamerazine   181
sulfamethoxazole (SMX)   59, 197, 203
sulfate radical-based advanced oxidation technology (SR-AOP)   213–215, 219, 222, 223
sulfathiazole   181
sulfur   20, 198, 222
sulfur dioxide (SO_2)   250–251, 253–254
sulfur-doping   222
sulfuric acid (H_2SO_4)   14–16, 51, 68, 217, 222, 251, 298
sulfurization   22
superoxide dismutase   281
surface complexation,
   adsorption of heavy metals   6, 43, 45, 46
   adsorption of PPCPs   55
surface functional groups (SFG)   1, 34–35
sustainability   24–25
sustainable drainage systems (SuDS)   61

swales 62
Syntrophomonas 98

t

tannic acid 131
tars 11
TBA *see* tert-butyl alcohol
TCEP *see* tris(2-chloroethyl) phosphate
TCH *see* tetracycline hydrochloride
TEM *see* transmission electron microscopy
temperature,
  adsorption process 37, 57
  graphitization 173
  pyrolysis 4–7, 9, 11, 32, 35, 57, 67, 126, 139, 159, 215, 216
  sludge composting 141
template method 179–180
tert-butyl alcohol (TBA) 215, 235
tetracycline 56, 254
tetracycline hydrochloride (TCH) 199, 204
thermal energy recovery 294, 299
thermochemical conversion 2, 299
thermochemical decomposition 1, 3
titanium 220
titanium oxide 230
titration 7, 236
TOC 232–235
toluene 250, 252, 255
TOrCs 77
torrefaction 1, 2
total suspended solids (TSS) 64
trace organic pollutants (TROCs) 72, 74–75
transmission electron microscopy (TEM) 7
trichloroethylene (TCE) 218
triclosan 59, 298
trimethylamine 247
tris(2-chloroethyl) phosphate (TCEP) 74
TROCs *see* trace organic pollutants
TSS *see* total suspended solids

u

United Kingdom 61
United States 61
urban runoff 64, 67–68, 70, 72, 74, 77
  microplastics in 75–76

urea 20, 22, 217
urea phosphate 253

v

vacancy defects 198, *199*, 203
vanillin 3
vehicle emissions 256
VFAs *see* volatile fatty acids
VOCs *see* volatile organic compounds
volatile fatty acids (VFAs) 89, 90–91, 93, 99, 258
volatile organic compounds (VOCs) 11, 42, 64
  environmental impact 277, 282
  odor control 243–245, 249–250, 253–260
volatile sulfur compounds (VSCs) 244
VSCs *see* volatile sulfur compounds

w

waste activated sludge (WAS) 93, 98
wastewater treatment facilities (WWTFs) 289–290, 299
  see also wastewater treatment plants (WWTPs)
wastewater treatment plants (WWTPs) 30, 244
  see also wastewater treatment facilities (WWTFs)
water filtration 282
water-sensitive urban design (WSUD) 61
WWTFs *see* wastewater treatment facilities
WWTPs *see* wastewater treatment plants

x

X-ray photoelectron spectroscopy (XPS) 35, 231

z

zeolites 31, 55
zero net charge (PZNC) 299–300
zero salt effect (PZSE) 299–300
zero-valent iron (ZVI) 218
  see also nanoscale zero-valent iron (nZVI)
zero voltage desorption (ZVD) 175
zinc chloride ($ZnCl_2$) 14, 123–126
ZVI *see* zero-valent iron

Printed and bound by CPI Group (UK) Ltd, Croydon, CR0 4YY
21/08/2023

08102759-0004